GENETICS

By
MANJU YADAV
Lecturer
Department of Zoology
M.M.H. College
Ghaziabad (U.P.)
(India)

DISCOVERY PUBLISHING HOUSE
NEW DELHI-110002

Published by:
Namit Wasan
DISCOVERY PUBLISHING HOUSE PVT. LTD.
4383/4B, Ansari Road, Darya Ganj
New Delhi-110 002 (India)
Phone : +91-11-23279245; 23253475; 43596065
E-mail : discoverybooksindia@gmail.com
discoverypublishinghouse@gmail.com
namitwasan9@gmail.com
web : www.discoverypublishinggroup.com

Edition: **2020**

ISBN: 978-81-7141-711-7

Genetics

Printed at:
Infinity Imaging Systems
Delhi

Preface

Genetic is one of the most fascinating scientific disciplines. It has been advancing very rapidly and has got wide practical applications in increasing the usefulness of plants, animals and micro-organisms for the mankind. The present title Genetics is intended for beginners who wish to understand the fundamentals of Genetics. The whole gamut of subject has been described in a simple and systematic manner suited for readers who have not been exposed sufficiently to the subject. It is an adequate text for all requirement in this area for the students of all Indian Universities. The primary aim throughout has been clarity, simplicity and the high standard. It will be a valuable study aid to teachers, students and research scholar in the related field.

The author has freely consulted various articles, discussion notes, reviews and extracts from the various scientific journals in the preparation of the present book, in able to make it comprehensive and upto date.

Though every care has been taken by the printer, publisher and me, it is quite likely that some errors might have found their way into the book but I hope these are of very insignificant nature. However, suggestions to improve book and pointing out of errors and mistakes, if any, will gratefully be appreciated.

The author expresses grateful to her friends and colleagues whose constant inspiration have initiated her in bringing out this book.

Special thanks are given to Mr. Wasan and staff of M/s Discovery Publishing House for their whole hearted co-operation in the publication of this book.

Author

CONTENTS

1

MENDEL AND HIS LAWS

You have seen how the nuclei of eukaryotic cells divide mitotically and meiotically. You know that by these processes cells pass copies of their genetic information to their descendants. By thinking about meiosis and sexual reproduction, you can account for the fact that off-spring of the same parents may differ. But what if no one understood cell division–especially the complexities of meiosis? How could there be any understanding of how traits pass from parents to offspring without such background information? The answer is the subject of this chapter.

The term *Mendelian genetics* refers to certain basic inheritance patterns. The term honors the Austrian monk GregorJohann Mendel (1822-1844), the person who first made rigorous, quantitative observations of the patterns of inheritance and proposed plausible mechanisms to explain them. In organisms that reproduce sexually and have more than one chromosome (and orderly meiosis), many traits pass from parent to offspring in accord with these patterns. When Mendel began his work with the garden pea in his monastery garden, little was known about the sex lives of plants or the consequences of sexual reproduction for the inheritance of traits. Chromosomes, mitosis, and meiosis were unknown.

THE ROAD TO MENDEL

Like many great puzzles, the riddle of heredity seems simple now that it has been solved. The solution was not an easy one to find, however. Our present understanding is the result of a long history of thought, surmise, and investigation. At every stage we have learned more, and as we have done so, the models used to describe the mechanisms of heredity have been changed to encompass new facts.

Two concepts provided the basis for most of the thinking about heredity before the twentieth century. The first is that *heredity occurs within species.* For a very long time people believed that it was possible to obtain bizarre composite animals by breeding (crossing) widely different species. The minotaur of Creatn mythology, a creature with the body of a bull and the torso and head of a man, is one example. The giraffe was thought to be another; its scientific name, *Giraffa camelopardalis,* suggests that it was believed to be the result of a cross between a camel and a leopard. From the Middle Ages onward, however, people discovered that such extreme crosses were not possible and that variation and heredity occur mainly within the boundaries of a particular species. Species were thought to have been maintained without significant change from the time to time of their creation.

The second early concept related to heredity is that *traits are transmitted directly.* When variation is inherited by off-spring from their parents, *what* is transmitted? The ancient Greeks suggested that parts of the bodies of parents were transmitted directly to their offspring. Hippocrates called this reproductive material *gonos,* meaning "seed". Hence, a characteristic such as a misshapen limb was the result of material that came from the misshapen limb of a parent. Information from each part of the body was thought to be passed along independently of the information from the other parts, and the child was formed after hereditary material from all parts of the parents' bodies had come together.

This idea was predominant until fairly recently. For example, in 1868 Charles Darwin proposed that all cells and tissues excrete microscopic granules, or "gemmules," that are passed along to offspring, guiding the growth of the corresponding part in the developing embryo. Most similar theories of the direct transmission of hereditary material assumed that the male and female contributions *blended* in the offspring. Thus, parents with red and brown hair would be expected to produce children with reddish brown hair, and tall and short parents would produce children of intermediate height.

Taken together, however, these two concepts lead to a paradox. If no variation enters a species from outside, and if the variation within each species is blended in every generation, then all members of a species should soon resemble one another exactly. Obviously, this does not happen. Individuals within most species differ widely from each other, and they differ in characteristics that are transmitted from generation to generation.

How could this paradox be resolved? Actually, the resolution has been provided long before Darwin, in the work of the German botanist Josef Koelreuter. In 1760 Koelreuter carried out the first successful *hybridizations* of plant species. He was able to cross different strains of tobacco and obtain fertile offspring. The hybrids differed in appearance from both of their parent strains. When crosses were made within the hybrid generation, the offspring were highly variable. Some of these offspring resembled plants of the hybrid generation (their parents), but a few resembled the original strains (their grandparents).

Koelreuter's work provided an important clue about how heredity works: the traits that he was studying were capable of being masked in one generation, only to reappear in the next. This pattern is not predicted by the theory of direct transmission. How could a characteristic that is transmitted directly be latent and then reappear? Nor were the traits of Koelreuter's plants blended. A contemporary account stated that the traits reappeared in the next generation "fully restored to all their original powers and properties."

It is worth repeating that the offspring of Koelreuter's crosses were not identical to one another. Some resembled the hybrid generation, while others did not. The alternative forms of the traits Koelreuter was studying were distributing themselves among the offspring. A modern geneticist would say the alternative forms of each trait were *segregating* among the progeny of a single mating, meaning that some offspring exhibited one alternative form of a trait (e.g., hairy leaves), while other offspring from the same mating exhibited a different alternative (e.g., smooth leaves). This segregation of alternative forms of a trait provided the clue that led Mendel to his understanding of the nature of heredity.

Over the next hundred years, Koelreuter's work was elaborated on by other investigators. Prominent among them were English gentleman farmers who were trying to improve varieties of agricultural plants. In one such series of experiments, carried out in the 1790s, T.A. Knight crossed two *true-breeding* varieties (varieties that are uniform from one generation to the next) of the garden pea, *Pisum sativum*. One of these varieties had purple flowers, and the other had white flowers. All of the progeny of the cross had purple flowers. Among the offspring of these hybrids, however, were some plants with purple flowers and others, less common, with white flowers. Just as in Koelreuter's earlier studies, a trait from one of the parents was hidden in one generation, only to reappear in the next.

In these deceptively simple results were the makings of a scientific revolution. Nevertheless, another century passed before the process of gene segregation was appreciated properly. Why did it take so long? One reason was that early workers did not quantify their results. A numerical record of results proved to be crucial to understanding the process. Knight and later experimenters who carried out other crosses with pea plants noted that some traits had a "stronger tendency" to appear than others, but they did not record the numbers of the different classes of progeny. Science was young then, and it was not obvious that the numbers were important.

Mendel and the Garden Pea

The first quantitative studies of inheritance were carried out by Gregor Mendel an Austrian monk. Born in 1822 to peasant parents, Mendel was educated in a monastery and went on to study science and mathematics at the University of Vienna, where he failed his examinations for a teaching certificate. He returned to the monastery and spent the rest of his life there, eventually becoming abbot. In the garden of the monastery, Mendel initiated a series of experiments on plant hybridization. The results of these experiments would ultimately change our views of heredity irrevocably.

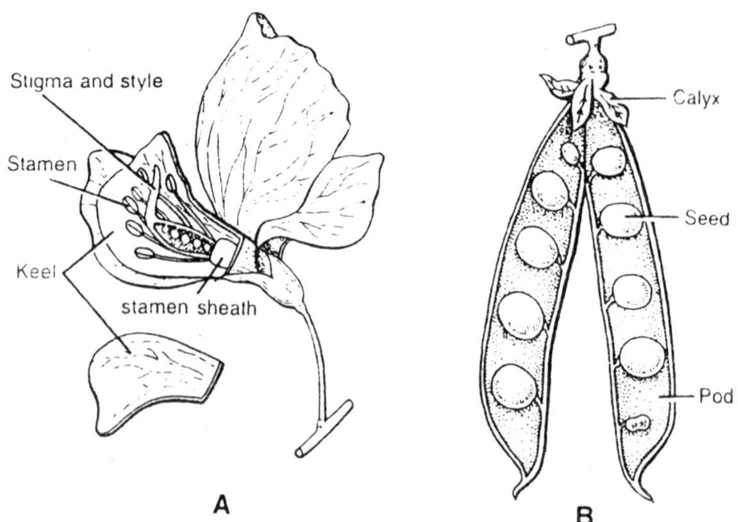

Fig. 1.1. A self-fertilizing flower of Pisum sativum. A–Portion of the keel which encloses the reproductive organs has been cut open to show the stamen and stigma; B–The mature fruit (pod) containing seeds that develop from the fertilized pea flower.

For his experiments, Mendel choose the garden pea, the same plant that Knight and many others had studied earlier. The choice was a good one for several reason. First, many earlier investigators had produced hybrid peas by crossing different varieties. Mendel knew that he could expect to observe segregation among the offspring. Second, a large number of true-breeding varieties of peas were available. Mendel initially examined 32. Then, for further study, he selected lines that differed with respect to seven easily distinguishable traits, such as smooth versus shriveled seeds and purple versus white flowers, a characteristic Knight had studied. Third, pea generation time. Thus, one can conduct experiments involving numerous plants, grow several generations in a single year, and obtain results relatively quickly.

A fourth advantage of studying peas is that the sexual organs of the pea are enclosed within the flower. The flowers of peas, like those of most flowering plants, contain both male and female sex organs. Furthermore, the gametes produced by the male and female parts of the same flower, unlike those of many flowering plants, can fuse to form viable offspring. Fertilization takes place automatically within an individual flower if it is not disturbed, resulting in offspring that are the progeny of a single individual. Therefore, one can either let *self-fertilization* take place within an individual flower, or remove the flower's male parts before fertilization and introduce pollen from a strain with alternative characteristics, thus performing *cross-fertilization.*

How Mendel Interpreted His Results

From these experiments Mendel was able to understand four things about the nature of heredity. First, plants exhibiting the traits he studied did not produce progeny of intermediate appearance when crossed, as a theory of blending inheritance would have predicted. Instead, alternatives were inherited intact, as discrete characteristics that either were or were not seen in a particular generation. Second, for each pair of alternative forms of a trait that Mendel examined, one alternative was not expressed in the F_1 hybrids, although it reappeared in some F_2 individuals. *The "invisible" trait must therefore have been latent (present but not expressed) in the F_1 individuals.* Third, the pairs of alternative forms of the traits that Mendel examined segregated among the progeny of a particular cross, some individuals exhibiting one form of a trait, some the other. Fourth, pairs of alternatives were expressed in the F_2 generation in the ratio of $^3/_4$ dominant to $^1/_4$ recessive. This characteristics 3:1 segregation is often referred to as the *Mendelian ratio.*

To explain these results, Mendel proposed a simple model. It has become one of the most famous models in the history of science, containing simple assumptions and making clear predictions. The model has five elements. For each, we will first state Mendel's assumption and then rephrase it in modern terms.

1. Parents do not transmit their physiological traits directly to their offspring. Rather, they transmit discrete information about the traits, that Mendel called "factors." These factors later act in the offspring to produce the trait. In modern terms, we would say that information about the alternative forms of traits that an individual expresses is *encoded* by the factors that it receives from its parents.
2. Each individual, with respect to each trait, contains two factors that may code for the same form of the trait or for two alternative forms of the trait. We now know that there are two factors for each trait present in each individual because these factors are carried on chromosomes, and each adult individual is *diploid.* When the individual forms gametes (eggs or sperms), only one of each kind of chromosome is included in each gamete: the gametes are *haploid.* Therefore, only one factor for each trait of the adult organism is included in the gamete. Which of the two factors for each trait is included in a particular gamete is randomly determined.
3. Not all copies of a factor are identical. The alternative forms of a factor, leading to alternative forms of a trait, are called *alleles.* When two haploid gametes containing exactly the same allele of a factor fuse during fertilization to form a zygote, the offspring that develops from that zygote is said to be *homozygous*; when the two haploid gametes contained different alleles, the individual offspring is *heterozygous.*

   In modern terminology, Mendel's factors are called *genes.* We now know that each gene is composed of a particular DNA nucleotide sequence. The position on a chromosome where a gene is located is often referred to as the gene's *locus* (plural, loci). Most genes exist in alternative versions, or alleles, resulting from differences at one or more nucleotide positions in the DNA. Different alleles of a gene are usually recognized by the change in the organism's appearance or function that results from the nucleotide differences.
4. The two alleles, one each contributed by the male and female gametes, do not influence each other in any way, In the cells that develop within the new individual, these alleles remain discrete. They neither blend with each other nor become altered by the

other. (Mendel referred to them as "uncontaminated.") Thus, when the individual matures and produces its own gametes, the alleles for each gene are segregated randomly into these gametes.

5. The presence of a particular allele does not ensure that the form of the trait encoded by it will actually be expressed in the individual carrying that allele. In heterozygous individuals, only one allele (the dominant one) achieves expression, while the other (recessive) allele is present but unexpressed. To distinguish between the presence of an allele and its expression, modern geneticists refer to the totality of alleles that an individual contains as the individuals *genotype* and to the physical appearance of an individual as its *phenotype.* The phenotype of an individual is the observable outward manifestation of its genotype, the result of the functioning of the enzymes and proteins encoded by the genes it carries. In other words, the genotype is the blueprint, and the phenotype is the realized outcome.

Mendel's results were clear because he was studying alternatives that exhibited complete dominance. Many traits in humans also exhibit dominant or recessive inheritance, in a manner similar to the traits Mendel studied in peas. Table 1.1 lists a few of the many human traits known to be caused by recessive or dominant alleles.

These five elements, taken together, constitute Mendel's model of the hereditary process. Does Mendel's model predict the results he actually obtained?

The F_1 Generation

Consider again Mendel's cross of purple-flowered with white-flowered plants. We will assign the symbol w to the recessive allele, associated with the production of white flowers, and the symbol *W* to the dominant allele, associated with the production of purple flowers. By convention, genetic traits are usually assigned a letter symbol referring to their less common state, in this case the letter "*W*" for white flower colour. The recessive allele (white flower colour) is written in lower case as w; the dominant allele (purple flower colour) is assigned the same symbol in upper case, *W.*

In this system, the genotype of an individual that is true-breeding for the recessive white-flowered trait would be designated *ww.* In such an individual, both copies of the allele specify the white flower phenotype. Similarly, the genotype of a true-breeding purple-flowered individual would be designated *WW*, and a heterozygote would be designated *Ww* (the dominant allele is usually written first). Using

Table 1.1. Some Dominant And Recessive Traits in Humans

Recessive Traits	*Phenotypes*	*Dominant Traits*	*Phenotypes*
Common baldness	M-shaped hairline reading with age	Mid-digital hair	Presence of hair on middle segment of fingers
Albinism	Lack of melanin pigmentation	Brachydactyly	Short fingers
Alkaptonuria	Inability to metabolize homogeneistic acid	Huntington's disease	Degeneration of nervous system, starting in middle age
Red-green colour blindness	Inability to distinguish red or green wavelengths of light	Phenylthiocarbamide (PTC) sensitivity	Ability to taste PTC as bitter
Cystic fibrosis	Abnormal gland secretion, leading to liver degeneration and lung failure	Camptodactyly	Inability to straighten the little finger
Duchenne muscular dystrophy	Wasting away of muscles during childhood	Hypercholesterolemia (the most common Mendelian disorder –1 : 500)	Elevated levels of blood cholesterol and risk of heart attack
Hemophilia	Inability of blood to clot	Polydactyly	Extra fingers and toes
Sickle-cell anemia	Defective hemoglobin that collapses red blood cells		

these conventions, and denoting a cross between two strains with ×, we can symbolize Mendel's original cross as *ww* × *WW*. Since a white-flowered parent can produce only *w* gametes and a pure purple-flowered (homozygous dominant) parent can produce only *W* gametes, the union of an egg and a sperm from the parents can produce only heterozygous *Ww* offspring in the F_1 generation. Because the *W* allele is dominant, all of these F_1 individuals are expected to have purple flowers. The *w* allele is present in these heterozygous individuals, but it is not phenotypically expressed.

The F_2 Generation

When F_1 individuals are allowed to self-fertilize, the *W* and *w* alleles segregate at random during gamete formation. Their subsequent union at fertilization to form F_2 individuals is also random, not being influenced by which alternative alleles the individual gametes carry. What will the F_2 individuals look like? The possibilities may be visualized in a simple diagram called a *Punnett square*, named after its originator, the English geneticist Reginald Crundall Punnett. Mendel's model, analyzed in terms of a Punnett square, clearly predicts that the F2 generation should consist of $^3/_4$ purple-flowerd plants and $^1/_4$ white-flowered plants, a phenotypic ratio of 3:1.

TERMINOLOGY

The following terms are commonly used in genetics and should be understood:

1. Gene

In modern sense an inherited factor that determines a biological character of an organism is called a gene. This is a functional unit of hereditary material. In the past, gene was believed to be a unit of structure also, but recent knowledge has shown that it is no longer a unit of structure.

2. Allelomorphs of Alleles, Homozygous and Heterozygous

Alleles, the abbreviated form of the term "allelomorphs" (meaning one form or the other) indicates alternative form of the same gene. For instance, in the above example "D" and "d" are two allelomorphs of the gene for plant height. In pure tall or pure dwarf plants, same allele is duplicated (DD or dd), while in hybrid tall both the alleles will be present (Dd). An individual, having only one allele or in other words two identical alleles, is known as homozygous (DD or dd). Similarly, an individual, having two different alleles will be called heterozygous or hybrid (Dd). In the present usage the term gene and the allele are

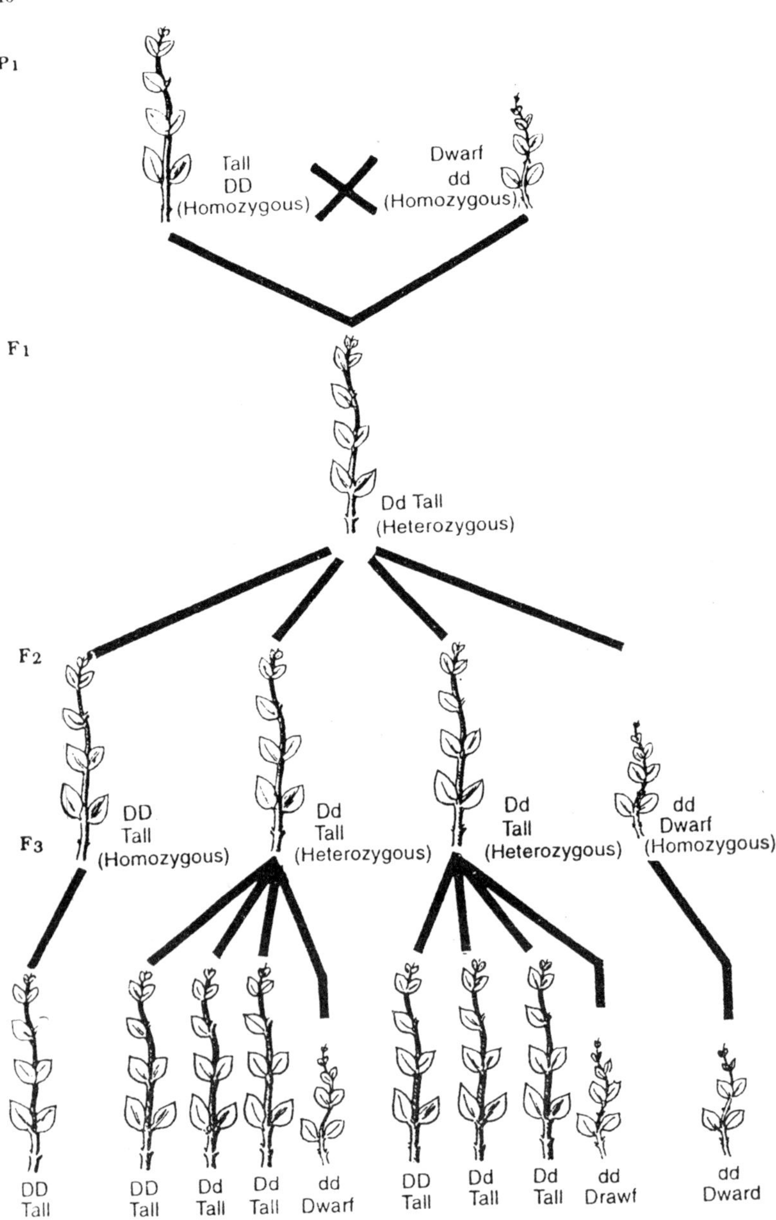

Fig. 1.2. A cross between a tall (TT) and a dwarf (tt) pea plant and their offsprings of F_1 and F_2 generations.

interchangeable, but while gene can be used for any factor, allele is used with reference to another allele. For instance, while "D" and "d" are alleles to each other, they can not be allelic to any other gene.

3. Monohybrid, Dihybrid and Trihybrid

In the examples discussed earlier in this chapter, single characters each controlled by a single genes or a pair of alleles were considered. Such crosses are known as monohybrid crosses and the F_2 ratio of 3 : 1 is known as the monohybrid ratio. Similarly crosses can be considered when two or three genes or pairs of alleles are involved. Such crosses will be called dihybrid and trihybrid crosses and the respective ratio as dihybrid and trihybrid ratios.

4. Reciprocal Crosses

A set of two reciprocal crosses means that the same two parents are used in two experiments in such a way that if in one experiment "A" is used as the female parent and "B" is used as the male parent, in the other experiment "A" will be used as the male parent and "B" as the female parent. This was earlier discussed in this chapter.

5. Backcross and Testcross

The F_1 individuals obtained in a cross are usually selfed to get the F_2 progeny. They can also be crossed with one of the two parents from which they were derived. Such a cross of F_1 individual with either of the two parents is known as a "backcross". In such backcrosses, when F_1 is backcrossed to the parent with dominant phenotype, no recessive individuals are obtained in the progeny. On the other hand, when it is crossed with recessive parent, both phenotypes appear in the progeny. While both these crosses are "backcrosses" only the cross with the recessive parent is knows as a "testcross". It is called a testcross, because it is used to test whether an individual is homozygous (pure) or heterozygous (hybrid).

This can be illustrated with the help of the example of tall (DD) and dwarf (dd) plants. In the F_1 generation only tall plants (Dd) appear. These plants can be backcrossed with either of the two parents as shown in Fig. 1.1.

The Principle of Dominance

To begin with, Mendel based his information on a carefully planned series of experiments and, more important, on a statistical analysis of the results. The use of mathematics to describe biological phenomena was a new concept. Clearly, Mendel's two years at the university had not been wasted.

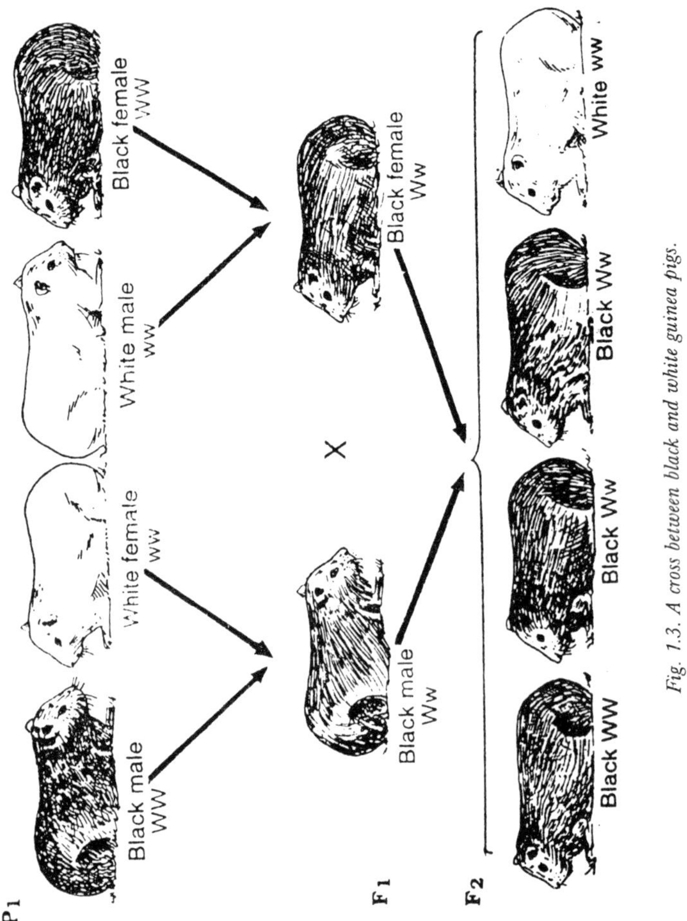

Fig. 1.3. A cross between black and white guinea pigs.

The care with which Mendel planned his projects is reflected in his selection of the common garden pea as his experimental subject. There were several advantages in this choice. Pea plants were readily available, fairly easy to grow, and Mendel had already developed some thirty-four pure strains. These strains differed from each other in very obvious ways, so there would be little difficulty in classifying the results of a given experiment. Mendel chose to study seven different pairs of characteristics:

1. Seed form–round or wrinkled
2. Colour of seed contents–yellow or green
3. Colour of seed coat–white or gray

4. Colour of unripe seed pods–green or yellow
5. Shape of ripe seed pods–inflated or constricted between seeds
6. Length of stem–short (9 to 18 inches) or long (6 to 7 feet)
7. Position of flowers–axial (along the stem) or terminal (at the end of the stem).

Mendel's approach, a novel one at that time, was to cross two true-breeding strains that differed in only one characteristic, such as seed colour. Peas ordinarily self-fertilize, so for this cross it was necessary to transfer the pollen by hand. Mendel called this original parent generation P_1 and designated their first-generation offspring the F_1 (first filial) generation. When the F_1 plants were allowed to self-pollinate, so that they crossed with each other at random, the offspring resulting from this cross were called the F_2 generation, and so on.

Now, when Mendel crossed his original P_1 plants, he found that the characteristics of the two parents didn't blend, as prevailing theory said they should. When plants with yellow seeds were crossed with plants that had green seeds, their F_1 offspring did not have yellow-green seeds. Instead, all of them had yellow seeds. Mendel termed the trait that appeared in the F_1 generation the *dominant trait*, but he was now left with a vexing question. What had happened to the trait that had disappeared in this cross, the *recessive trait*? After all, it had been passed along through countless generations so it couldn't have just disappeared.

The Principle of Segregation

In the next stage of his experiments, Mendel allowed the F_1 plants to randomly self-pollinate, and, lo and behold, the missing recessive trait reappeared in some of their F_2 offspring! Also, the ratio of F_1 plants with recessive traits to those with dominant traits was fairly constant, regardless of the particular characteristics involved.

Finally, in the third year of the experiments. Mendel allowed the F_2 plants to self-pollinate. He found that all those with recessive traits produced only recessive F_3 offspring. However the F_2 plants that showed a dominant trait produced both type of F_3 offspring. One-third of them produced *only* dominant offspring. The other two-thirds produced both dominant and recessive offspring, but they produced three times as many offspring with the dominant trait as they did with the recessive trait. In other words, the F_2 recessives and one-third of the F_2 dominants breed true (that is, passed their visible traits on to all offspring). The other two-thirds of these dominants produced mixed offspring–but in the same 3 : 1 ratio of dominant to recessive as in the plants their F_1

parents had produced. Let's consider a specific example of a cross between two pure lines–plants that breed true when they self-pollinate and are identical in all characteristics except one. In this case, suppose one strain always produces round seeds and the other always produces wrinkled seeds. We have already seen that the round form is dominant, so we'll designate that form as *R* and the recessive wrinkled form as *r*. Since both plants are true-breeding, we can assume that in each plant, the two genetic components (one derived from each of the two parents) are identical for their respective traits. That is, in one plant, the components are *RR*, and in the other, they are *rr*. Such plants are said to be *homozygous* for that particular characteristic; that is, having identical alleles for that trait. As a result, all the gametes they produce will be the same.

Remember that the plants are identical in all characteristics except one, so we only need to diagram the differing characteristics. Since half the genes in the F_1 generation come from each parent, the *genotype* (the assortment of genes that make up the gene component of an organism) of any F_1 plant would have to be *Rr*. Hence, these plants would be *heterozygous* for the characteristics in question: that is, having different alleles for that trait. Now, when these plants produce gametes ("eggs and sperm"), each gamete will carry either an *R* allele or an *r* allele, and the combination that occur when two gametes come together at fertilization will determine the genetic characteristics of the resulting F_2 plant.

Table 1.2. Mendel's Pea Plant Experiment

Dominant Form	*No. in F_2 Generation*	*Recessive Form*	*No. in F_2 Generation*	*Total Examined*	*Ratio*
Round seeds	5.474	Wrinkled seeds	1,850	7,324	2.96 : 1
Yellow seeds	6,022	Green seed coats	2,001	8,023	3.01 : 1
Gray seed coats	705	White seed coats	224	929	3.15 : 1
Green pods	428	Yellow pods	152	580	2.82 : 1
Inflated pods	882	Constricted pods	299	1,181	2.95 : 1
Long stems	787	Short stems	277	1,064	2.84 : 1
Axial flowers	651	Terminal flowers	207	858	3.14 : 1

The results we should expect the F_2 generation can be diagrammed in a form called a *Punnett square.* As you can see in figure 8.3, the expressed, observable characteristics, or *phenotypes,* of the F_2 seeds occur in a 3 : 1 ratio. However, a moment (or two) of reflection will show that the F_3 ratio Mendel obtained in the third year of his experiments could have been gotten only if every heritable characteristic is determined by two components, one from each parent. Thus, if two homozygous parents were represented by *RR* and *rr* (as for seed form), their offspring would have to be heterozygous *Rr,* since one component would have come from each parent; and when this *Rr* individual produced gametes, half would contain an *R* component and the other half would contain the *r* component. A cross between two *Rr* individuals could then be expected to yield offspring with the genotypes *RR, rR, Rr,* and *rr,* or a genotype ratio of 1 : 2 : 1. What would be the genotypes and the phenotypes of the offspring if an *Rr* crossed with a homozygous recessive *rr* instead?

Mendel's experiment with monohybrid crosses enabled him to show that whatever it is that is actually passed along (Mendel called it a factor, but it is now generally referred to as a gene. An allele is one form of a gene.) and that one allele is contributed by each parent. However, he had discovered that only one of these two alleles may actually be expressed. Hence, an individual can only show the effects of an allele passed along by one parent, the other allele being completely dormant. The principles of dominance and segregation implied by these results ruled out (in Mendel's mind, at least) the earlier notion that heritable characteristics were always blended in the offspring.

The Principle of Independent Assortment

In his next set of experiments, Mendel crossed *dihybrids,* plants differing in two characteristics. In one such experiment, plants with round (*R*), yellow (*Y*) seeds were crossed with a strain that had wrinkled (*r*), green (*y*) seeds. We saw in Table, that round and yellow are both dominant, so we shouldn't be astounded to learn that all the F_1 plants had round, yellow seeds. Since one P_1 parent was *RR YY* and the other was *rr yy,* all the F_1 plants would have to be *Rr Yy* or heterozygous for both characteristics. When these plants were allowed to self-pollinate, we would expect a random and independent assortment of the various characteristics to produce the F_2 generation.

It was apparent to Mendel from the results of this dihybrid cross that the segregation of genes for one characteristic was not affected by

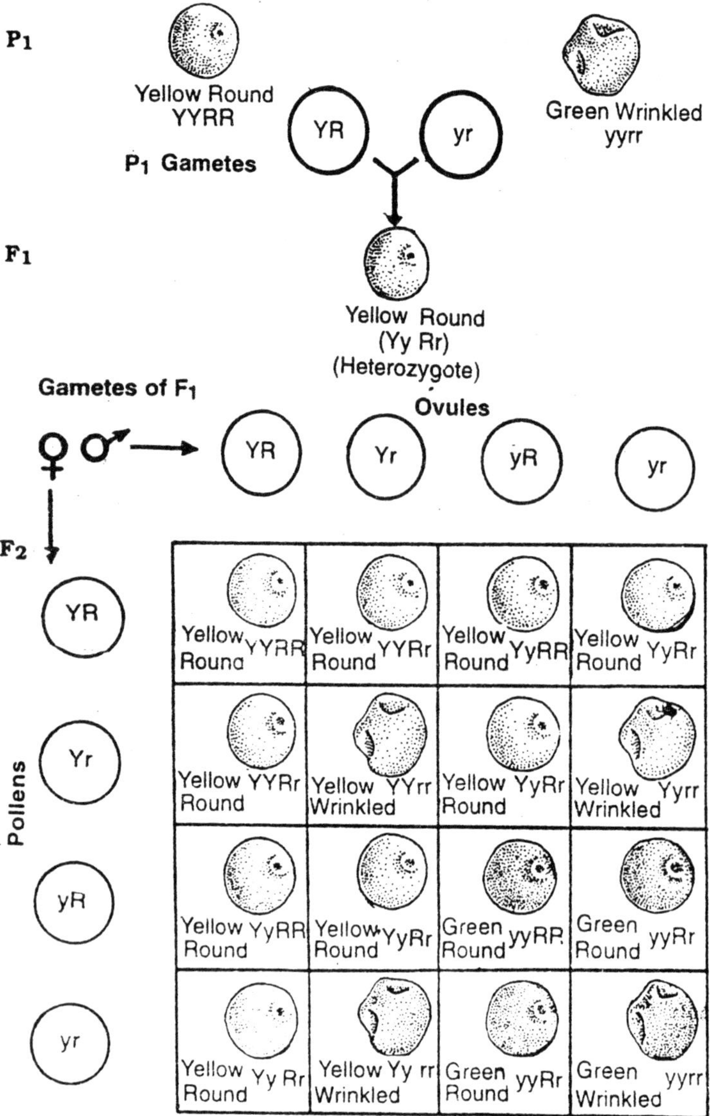

Fig. 1.4. A dihybrid cross between the vestigial-grey and long-black Drosophila.

the segregation of genes for the other characteristic. Thus, he was able to deduce that genetic combinations follow the principle of independent assortment. It should be pointed out here that if you were to run a single experiment to test Mendel's results, you might end up with all wrinkled, green seeds in the F_2 generation–just as when you toss a coin ten times. It might come up tails each time.

The Test Cross

Although Mendel had carried out numerous progeny tests for determining whether a dominant individual was homozygous or heterozygous, he soon devised a much simpler procedure, the *test cross.* The subject was simply crossed with a recessive individual. Recessives are always homozygous, so the predictions are straightforward. Let's use the traits round and wrinkled as an example:

1. If the dominant round individual in question is homozygous, then the test cross becomes *RR* × *rr*, and all of the progeny will be round (*Rr*).
2. If the dominant round individual is heterozygous, then the test cross becomes *Rr* × *rr*, and, statistically, half the offspring will be heterozygous round (*Rr*) and half will be wrinkled (*rr*).

The test cross is often applied today to test the pedigrees of plants and animals in agriculture.

Simple Mendelian Genetics in Humans

Other systems present some special problems in the application of Mendelian methodology. One of the most difficult, yet most interesting, is the human species. Obviously, controlled crosses cannot be made, so human geneticists must resort to a scrutiny of established matings in the hope that informative matings have been made by chance. The scrutiny of established matings is called *pedigree analysis.* A member of a family who first comes to the attention of a geneticist

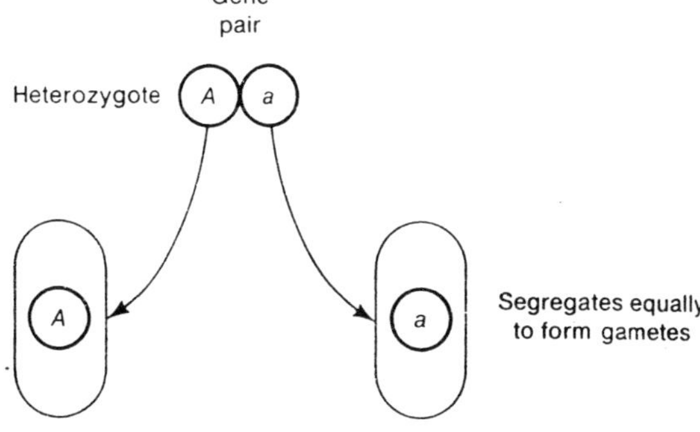

Fig. 1.5. Diagrammatic visualization of the equal segregation of one gene pair into gametes.

is called the *propositus.* Usually the phenotype of the propositus is exceptional in some way – for example, a dwarf. The investigator then traces the history of the character shown to be interesting in the propositus back through the history of the family, and a family tree of pedigree is drawn up using certain standard symbols. (The terms autosomal and sex-linked in the figure will be explained later; they are included to make the table complete).

Table 1.3. Rise in number of genotypic classes as the power of the number of segregating gene pairs

Number of segregating gene pairs	*Number of phenotypic classes*	*Number of genotypic classes*
1	2	3
2	4	9
3	8	27
4	16	81
.	.	.
.	.	.
.	.	.
n	2^n	3^n

Many human diseases and other exceptional conditions are determined by simple Mendelian recessive alleles. There are certain clues in the pedigree that must be sought. Characteristically the condition appears in progeny of unaffected parents. Furthermore, two affected individuals cannot have an unaffected child. Quite often such recessive alleles are revealed by consanguineous matings–for example, cousin marriages. This is particularly true of rare conditions where chance matings of heterozygotes are expected to be extremely rare. It has been estimated, for example, that first-cousin marriages account for about 18 to 24 percent of albino children and 27 to 53 percent of children with Tay-Sachs disease; both are rare recessive conditions. Some other examples of disease causing recessive alleles in humans are those for cystic fibrosis and phenylketonuria (PKU). Of course, variants that are not regarded as diseases also may be caused by recessive alleles. These may be rare as in albinism or common as in light eye colour (blue or green) in North American populations.

There are also examples of exceptional conditions caused by dominant alleles. (This is in contrast to the situation for such conditions

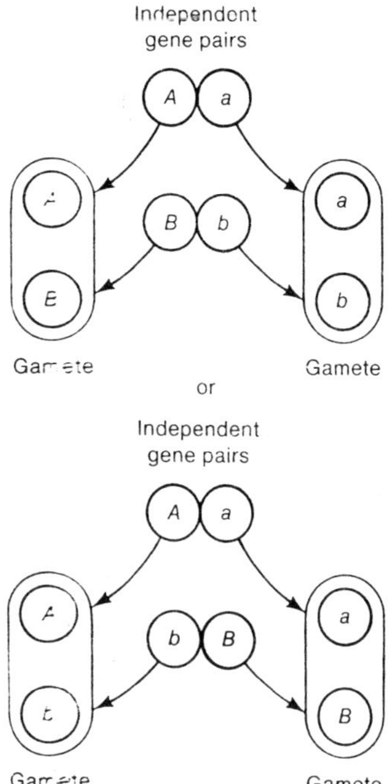

Fig. 1.6. Diagrammatic visualization of the segregation of two independent gene pairs into gametes.

as PKU, where the normal condition is attributable to the dominant allele, and the recessive allele causes PKU.) Once again, there are some simple rules to follow to discern from pedigrees a condition caused by a dominant allele: the condition typically occurs in every generation; unaffected individuals never transmit the condition to their offspring; two affected parents may have unaffected children; and the condition is passed, on average, to one-half of the children of an affected individuals. As with recessive alleles acting in a Mendelian manner, both sexes may be equally affected. Achondroplasia (a kind of dwarfism), Huntington's chorea, and brachydactyly (very short fingers) are examples of exceptional conditions in humans caused by dominant alleles.

Notice that, in this kind of Mendelian analysis, there is again the notion of identification of genes affecting major biological function, this time in humans. Pedigrees for PKU, for example, demonstrate that there is a kind of gene controlling the character we might call

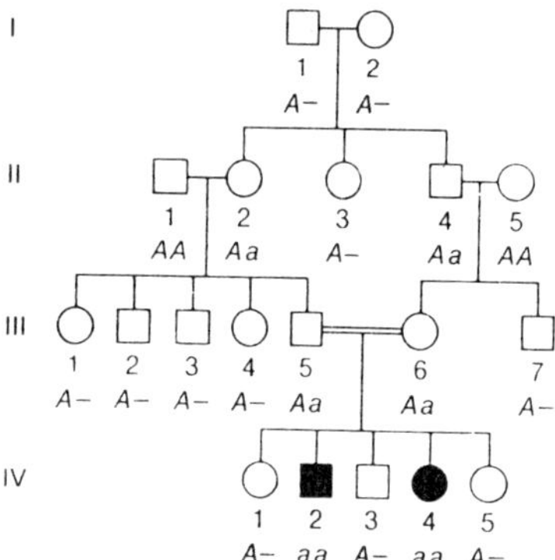

Fig. 1.7. Illustrative pedigree, involving an exceptional recessive phenotype determined by the recessive Mendelian allele a.

"normal PKU function." The two alleles stand for presence and absence. This identification is an important step toward discovery of the precise way in which the abnormal allele is failing and its possible correction. The medical applications of such genetic analysis obviously are far-reaching. In fact, medical genetics is today a key part of medical training. Simple pedigree analyses have extensive use, no only in such medical research but also in the day-to-day counselling of prospective parents who fear genetic disease in their children.

Simple Mendelian Genetics in Agriculture

As mentioned in Chapter 1, there has been an interest in the plant breeding since prehistoric times. The methods used by Neolithic farmers were probably the same as those used until the discovery of Mendelian genetics. Basically the approach was to select superior phenotypes from seeds or plants derived from natural populations. Particularly desirable were pure lines of favourable phenotype, because these lines produced constant results over generations of planting. Without the knowledge of Mendelian genetics, how is it possible to develop pure lines? It so happens that self-pollinating plants, such as many crop plants, naturally tend to be homozygous, and any heterozygous gene pairs become homozygous over generations of selfing. Thus pure lines have developed automatically over the years.

However, these less sophisticated breeding practices suffered from a major problem: the breeder was forced to rely on favourable combinations of genes that occurred in nature. With the advent of Mendelian genetics, it became evident that favourable qualities in

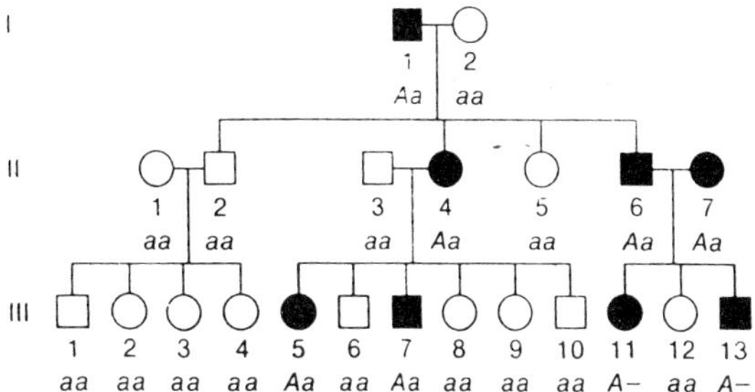

Fig. 1.8. Illustrative pedigree involving an exceptional dominant phenotype determined by the dominant Mendelian allele A. In this pedigree, most of the genotype can be deduced.

different lines could be combined through hybridization and subsequent gene reassortment. This procedure forms the basis of modern plant breeding.

For naturally self-pollinating plants, such as rice or wheat, two pure lines (each of different favourable genotype) are hybridized by manual cross-pollination, and hence an F_1 is developed. The F_1 is then allowed to self, and its heterozygous gene pairs assort to produce many different genotypes, some of which represent desirable new combinations of the parental genes. A small proportion of these new genotypes will be pure-breeding already, but if not, several generations of selfing will produce homozygosity of the relevant genes.

An example of genetic improvement in a species more familiar to most is the tomato. Anyone who has read a recent seed catalogue will be familiar with the abbreviations V, F, and N next to listed tomato variety. These represent, respectively, resistance to the pathogens *Verticillium*, *Fusarium*, and nematodes–resistance that has been crossed into the tomatoes, typically from wild forms of tomato. Another familiar phenomenon in tomatoes is determinate as opposed to indeterminate growth pattern. Determinate plants are bushier and more compact, and they do not need as much staking. Determinate growth is caused by a recessive allele *sp* (self-prunning), which has been crossed into modern varieties. Another useful allele is *u* (uniform repening); this allele eliminates the green patch or shoulder around the stem on the ripe fruit.

Such example could be listed for many pages. The point is that simple Mendelian genetics (as in this chapter) has provided agricultural plant breeding with its rationale and its modern methods. Entire complex genotypes may be constructed from an array of ancestral

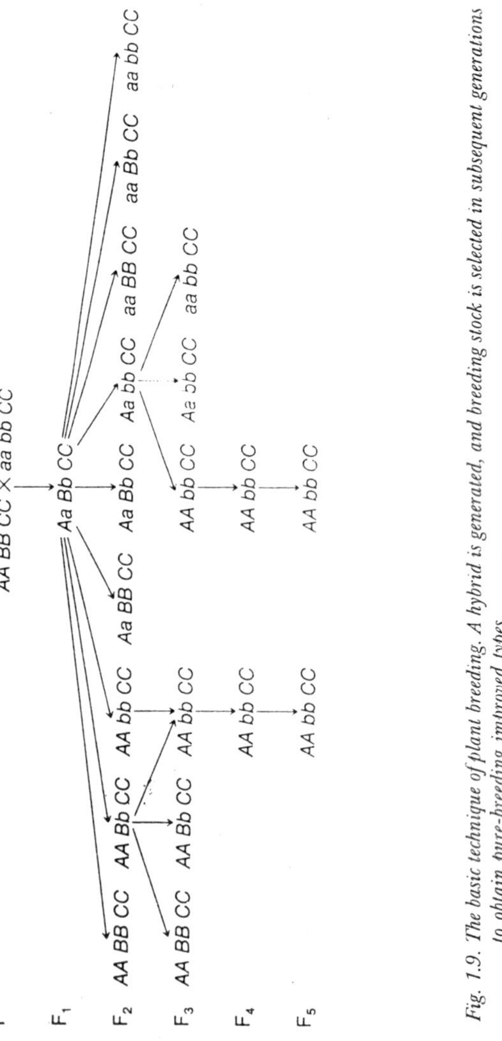

Fig. 1.9. The basic technique of plant breeding. A hybrid is generated, and breeding stock is selected in subsequent generations to obtain pure-breeding improved types.

lines, each showing some desirable feature. When we think of genetic engineering, we think of the genetic techniques of the 1970s and 1980s, but genetic engineering for plant improvement began long ago.

Mendelian genetics also has provided a formal theoretical basis for animal breeding, enabling a greater efficiency than under traditional practices. More recent techniques such as the use of frozen semen, artificial insemination, frozen embryos, and surrogate mothers in livestock species have enabled breeds to amplify the number of offspring of a specific genotype, a number normally limited by the lifespan of the maternal animal.

A Punnett square Predicts the Ratios of Genotypes and Phenotypes of the Offspring of a Cross

During meiosis in heterozygous (*Bb*) black guinea pigs, the chromosome containing the *B* allele becomes separated from its homologue (the chromosome containing the *b* allele), so each sperm or egg contains *B* or *b* but never both. Gametes containing *B* alleles and those containing *b* alleles are formed in equal numbers by heterozygous *Bb* individuals. Because no special attraction or repulsion occurs between an egg and a sperm containing the same allele, fertilization is a random process.

The possible combinations of eggs and sperm at fertilization may be represented in the form of a "checker-board" devised by an early geneticist, Sir Reginald Punnett, and known as a *Punnett square.* The types of gametes from one parent are represented across the top, and those from the other parent are indicated along the left side; the squares are then filled in with the resulting F_2 zygote combinations. Three fourths of all the F_2 offspring are genotypically *BB* or *Bb* and phenotypically black; one fourth are genotypically bb and phenotypically brown. The genetic mechanism responsible for the approximate 3:1 F_2 ratios (called *monohybrid F_2 phenotypic ratios*) obtained by Mendel in his pea-breeding experiments is again evident. The corresponding genotypic ratio is 1*BB* : 2 *Bb* : 1*bb.*

A Testcross can Detect Heterozygosity

One third of the black guinea pigs in the F_2 generation derived from the mating of F_1 hybrids are themselves homozygous, *BB*; the other two thirds are heterozygous, *Bb*. Guinea pigs with the genotypes *BB* and *Bb* are alike phenotypically; they both have black coats. Geneticists distinguish the homozygous (*BB*) and heterozygous (*Bb*) black-coated guinea pigs by a test cross in which each black guinea pig is mated with a homozygous brown (*bb*) guinea. In a test cross, the two types of gametes produced by the heterozygous parent are not "hidden" in the offspring by dominant alleles coming from the other parent. Therefore, through a test cross one can deduce the genotypes of all the classes of offspring directly from their phenotypes. If all the offspring were black, what inference would you make about the genotype of the black parent? If any of the offspring were brown, what conclusion would you draw regarding the genotype of the black parent?

Would you be more certain about one of these inferences than the other?

Mendel did just these sorts of experiments, breeding heterozygous tall (*Tt*) pea plants with homozygous recessive (*tt*) short ones. He predicted that the heterozygous parent would produce equal numbers of *T* and *t* gametes, whereas the homozygous short parent would produce only *t* gametes, and that this should lead to equal numbers of tall (*Tt*) and short (*tt*) individuals among the progeny. This was essentially a test of the hypothesis that there is 1 : 1 segregation of the alleles of the heterozygous parent. Thus, Mendel's principles of dominance and segregation not only explain the known facts, such as the monohybrid F_2 3:1 phenotypic ratio, but also enabled him to predict the results of other experiments, in this case the 1:1 test cross phenotypic ratio.

The Laws of Probability are used to Predict the Likelihood of Genetic Events

All genetic ratio are properly expressed in terms of probabilities. In the examples just discussed, among the offspring of two individuals heterozygous for the same gene pair, the ratio of the phenotypes of the dominant and recessive alleles is 3:1. A better way to express our expectations is to say that there are 3 chances in 4 (3/4) that any particular individual offspring of two heterozygous individuals will express the phenotype of the dominant allele and 1 chance in 4 (1/4)

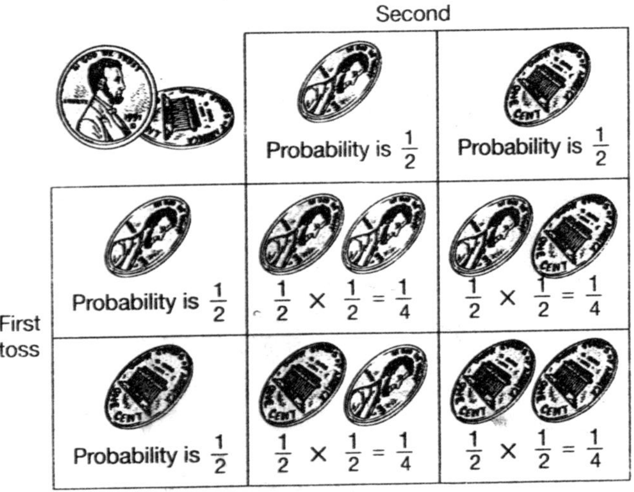

Fig. 1.10. The laws of probability can be illustrated by two successive coin tosses.

that it will express the phenotype of the recessive allele. Although we sometimes speak in term of percentages, probabilities must always be calculated as fractions (e.g., 3/4) or decimal fractions (e.g., 0.75). If an event is certain to occur, its probability is 1; if it is certain not to occur, its probability is 0. A probability can be 0, 1, of some number between 0 and 1.

Often we wish to combine two or more probabilities. The Punnett square, which we use to predict the results of genetic crosses, is a device that allows us to combine probabilities. When we use a Punnett square we are intuitively following two important rules, known as the *product law* and the *sum law.*

The product law predicts the combined probabilities of independent events

Events are independent if the occurrence of one does not affect he probability that the other will occur. For example, the probability of obtaining heads on the first toss of a con is 1/2; the probability of obtaining heads on the second toss (an independent event) is also 1/2. If two or more events are *independent* of each other, the probability of their both occurring is the *product* of their individual probabilities. If this seems strange to you, keep in mind that when we multiply two number that are less than 1, the product is a smaller number. The probability of obtaining heads first and also second on successive tosses of the coin is the product of their individual probabilities ($1/2 \times 1/2 = 1/4$), or 1 chance in 4).

Similarly, we can apply the product law to genetic events. If both parents are *Bb*, what is the probability that they will produce a child who is *bb*? For the child to be *bb*, he or she must receive a *b* gamete from each parent. The probability of a *b* egg is 1/2 and the probability of a *b* sperm is also 1/2. These probabilities are independent, so we combine them by the product rule ($1/2 \times 1/2 = 1/4$). You may wish to check this result using a Punnett square.

The sum law predicts the combined probabilities of mutually exclusive events

Events are mutually exclusive if the occurrence of one precludes the occurrence of the other Mutually exclusive events can be thought of as different ways of obtaining some specified result. Naturally, if there is more than one way to obtain a result, the chance of its being obtained are improved; we therefore combine the probabilities of mutually exclusive events by summing (adding) their individual probabilities.

For example, if we flip a coin twice, what is the probability that it will come up heads one time and tails the other time if we do not specify the order in which these events are to occur? There are two mutually exclusive ways to obtain this outcome. We could get heads the first time (probability 1/2) and tails the second (probability (1/2); we use the product law to calculate the combined probability of these independent events, which is 1/2 × 1/2 = 1/4.

Alternatively, we could also get tails the first time and heads the second; the probability of this occurring is also 1/4. We combine the probabilities of these mutually exclusive outcomes using the sum law: 1/4 + 1/4 = 1/2. That is, the probability of getting heads once (and only once) and tails once (and only once) on two successive tosses of the coin is 1/2.

We can also apply the sum law to genetic events. For example, if both parents are *Bb*, what is the probability that they will produce a child like themselves (*Bb*)? There are two mutually exclusive ways of obtaining a *Bb* child. A *B* egg can combine with a *b* sperm the probability of this outcome is 1/4 (calculated by the product rule). A *b* egg can combine with a *B* sperm; this probability is also 1/4. Because these two ways of obtaining a *Bb* child are mutually exclusive, we combine their probabilities using the sum law (1/4 + 1/4 = 1/2). Again, a Punnett square serves as a useful check.

The laws of probability can be applied to a variety of calculations

The laws of probability have wide applications. For example, what are the probabilities that a family with two (and only two) children will have two girls, two boys, or one girl and one boy? For purposes of discussion we will assume that male and female births are equally probable. The probability of having a girl first is 1/2, and the probability of having a girl second is also 1/2. These are independent events, so we combine their probabilities by multiplying: 1/2 × 1/2 = 1/4. Similarly, the probability of having two boys is also 1/4.

In families with both a girl and a boy, the girl can be born first or the boy can be born first. The probability that a girl will be born first is 1/2, and the probability that a boy will be born second is also 1/2. We use the product law to combine the probabilities of these two independent events: 1/2 × 1/2 = 1/4. Similarly, the probability that a boy will be born first and a girl second is also 1/4. These two kinds of families represent mutually exclusive outcomes–that is, two different ways of obtaining a family with one boy and one girl. Having two different ways of obtaining the desired result improves our chances, so we use the sum law to combine the probabilities: 1/4 + 1/4 = 1/2. Notice that the probabilities of the three types of families (all mutually exclusive outcomes) add up to 1. This serves as a useful check that the calculations have been done correctly. You may also wish to confirm these results by making a Punnett square.

In working with probabilities, it is important to keep in mind a point that many gamblers forget. We can say that "chance has no memory." This means that if events are truly random, past events have no influence on the probability of the occurrence of independent future events. For example, if two brown-eyed people have a child, what is the probability that it will have blue eyes? If their first child has blue eyes, what is the probability that their second child will also have blue eyes? The colour of the iris of the human eye is controlled by alleles at several loci, but alleles at one locus are primarily responsible. The allele for brown eye colour B, is usually dominant to the allele for blue, b. If the two brown-eyed parents are heterozygous, there is 1 chance in 4 that any child of theirs will have blue eyes. Each fertilization is a separate, independent event; its result is not affected by the results of any previous fertilizations. If these two heterozygous brown-eyed parents have had three brown-eyed children and are expecting their fourth child, what is the probability that the child will have blue eyes? The uninformed might guess that this one must have blue eyes, but in fact there is still only 1 chance in 4 that the child will have blue eyes and 3 chances in 4 that the child will have brown eyes.

If we phrase the question differently, however, we obtain a very different answer. If two heterozygous people marry and expect to have four children, what is the probability that all four will have brown eyes? The probability of brown eyes for each child is 3/4, so we combine these independent events by the product rule: $3/4 \times 3/4 \times 3/4 \times 3/4 = 81/256$ or 0.32. Why the different answers for the two types of problems? Remember that once a brown-eyed child is born, chance (3/4) is replaced by certainty (1), so the calculation becomes $1 \times 1 \times 1 \times 3/4 = 3/4$. The chance that the fourth (as yet unborn) child will have brown eyes is therefore 3/4 (and the chance of blue eyes is 1/4).

When working probability problems, common sense is more important than blindly memorizing rules. Examine your results to see if they appear reasonable; if they do not you should reevaluate your assumptions.

Extensions of Mendelian Principles

Multiple Alleles

In a population of individuals, there can be not just two but *multiple alleles* of genes. Any given diploid individual can possess only two different alleles. Multiple alleles obey the same rule of transmission as alleles of which there are only two kinds. Dominance relationships among multiple alleles vary in that for some groups of alleles, every homozygous and heterozygous genotype produces a different phenotype, whereas in others the alleles may be arranged in a descending series in which every allele is dominant over all alleles below it. An example is the human ABO blood group system in which three alleles. I^A, I^B, and i, determine the blood group. That is, I^AI^B specifies AB blood group, I^AI^A or I^Ai specifies A, I^BI^B or $I^B i$ specifies B and ii specifies O.

Codominance

Codominance is different from incomplete dominance. In the case of codominance, the heterozygote exhibits the phenotypes of both homozygotes rather than exhibiting an intermediate phenotype as in incomplete dominance. For example, the I^A and I^B alleles are codominant since I^AI^B individuals have a phenotype that is essentially a combination of those shown by individuals with A and B blood groups, at least in term of the red blood cell surface antigens specified by the alleles involved.

Gene Interaction

Nonallelic genes may not function independently in determining the phenotypic characteristics in an organism. Interaction between gene products may occur to produce new phenotypes without modifying typical Mendelian ratios. Or, interaction between gene products may cause modifications of Mendelian ratios by one gene product interfering with the phenotypic expression of another nonallelic gene or genes. Here, the phenotype is controlled mostly by the former gene and not the latter when both genes occur together in the genotype. This type of interaction is called *epistasis.*

Lethal Alleles

A *lethal allele* is one which, when expressed, is lethal to the individual. Both recessive lethal and dominant lethal alleles are known. A recessive lethal allele may or may not have an effect in the heterozygous condition. A dominant lethal allele, to be detected, must

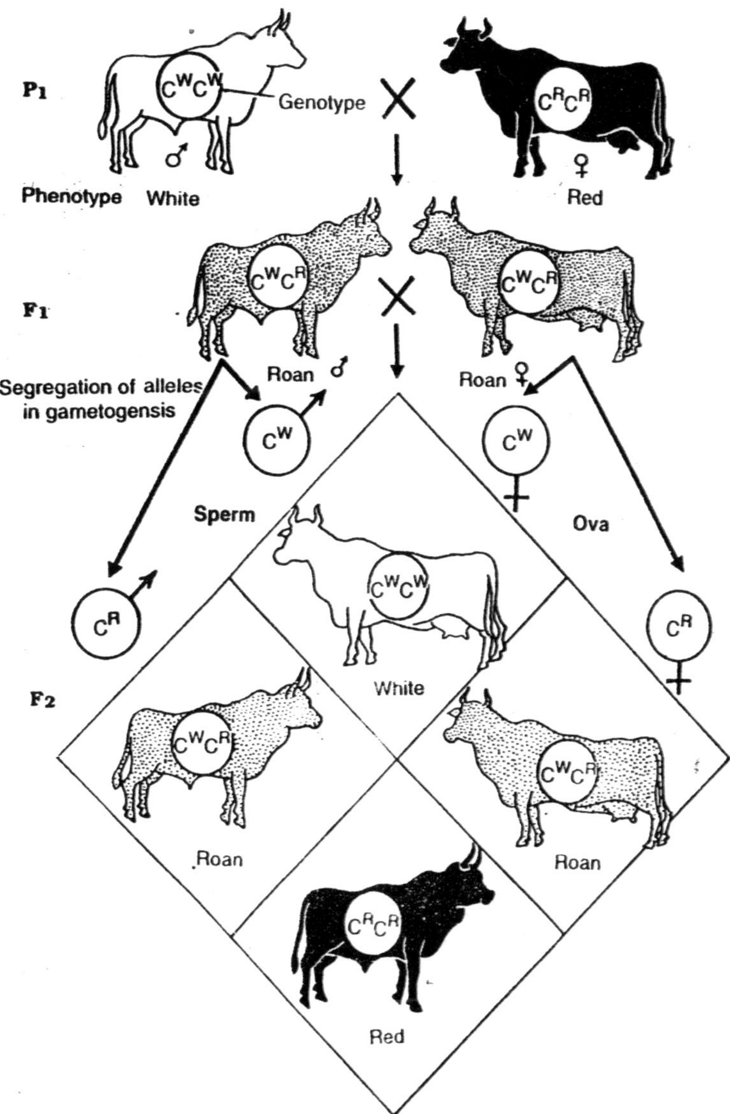

Fig. 1.11. A monohybrid cross in a white-coated and a red-coated cattle showing codominance.

exert its lethal effects at some time during development or during the lifetime of the organism. The human genetic disease Huntington's chorea, for example, is caused by a dominant lethal gene that exerts its effects generally in the adult.

Penetrance

In some cases not all individuals with a given genotype exhibit the phenotype specified by that genotype. The frequency with which a dominant or homozygous recessive gene manifests itself in the phenotype of the individuals is called the *penetrance* of the gene. For example, if 65% of the individuals with a particular gene show the corresponding phenotype, there is 65% penetrance.

Expressivity

Expressivity refers to the kind of phenotypic expression of a penetrant gene or genotype. Expressivity may be slight, intermediate, or severe.

Variations in Mendelian Phenotypic Ratios

Very soon after 1900 investigators reported inheritance patterns that gave phenotypic ratios different from those established by Mendel. In every case, once the apparent "exceptions" were understood, the result strengthened and broadened the basis of Mendelian inheritance. The observed variations included dominance and recessiveness relationships between alleles of a pair and the number of alleles of a gene. A few examples will illustrate these variations in inheritance patterns and how Mendelian principles explained each of these different phenotypic ratios.

Dominance Relationships

Mendel choose seven characters that provided unambiguous pairs of contrasting alternatives. The progeny were not intermediate between parental types; in fact, on the basis of the phenotype, the progeny could not be distinguished from their parents. One allele of each pair was dominant over its recessive alternative since all the F_1 resembled only one of the two parents and since only two phenotypic classes appeared in monohybrid F_2 progeny. All seven characters studied by Mendel yielded the same results, namely, *complete dominance* of one allele over the other allele of the pair.

In early studies of the four-o-clock and certain other flowering plants, members of the F_1 generation were intermediate between the two parental phenotypes for some traits. The heterozygous F_1 plants did not resemble either parent. When four-o'-clock with red flowers were crossed with others having white flowers, all the F_1 progeny had pink flowers. Since complete dominance was not observed, did this mean that the alleles for red and white flower colour were not transmitted to the F_1 in accordance with Mendelian predictions?

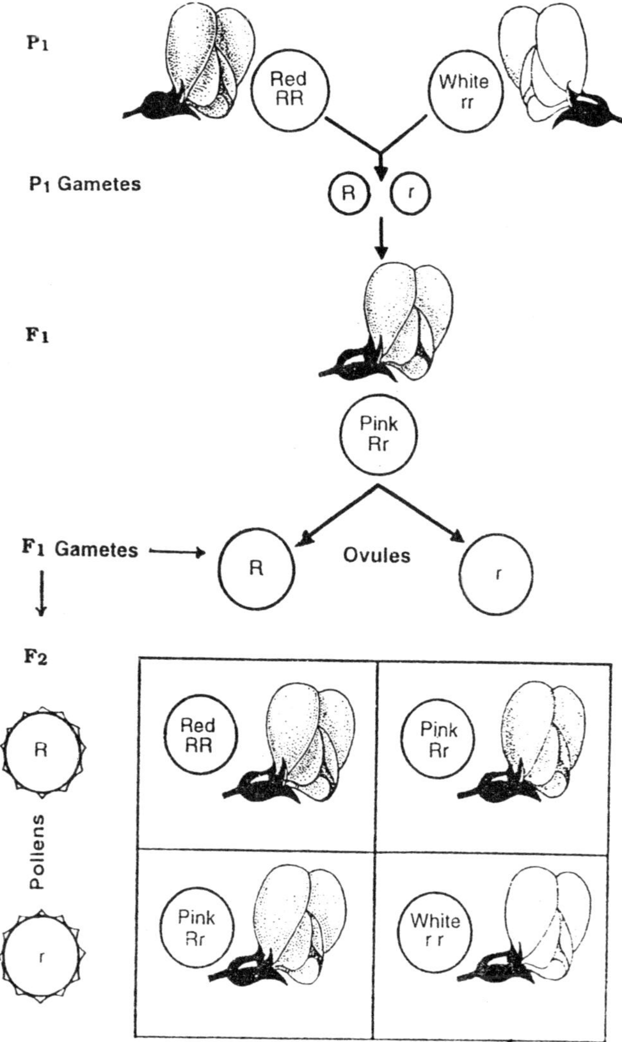

Fig. 1.12. A cross between a red flowered and a white flowered pea plant showing incomplete dominance.

The situation was clarified when pink F_1 plants were interbred to produce the F_2 generation. The F_2 progeny consisted of three phenotypic classes: 1/4 red: 2/4 pink: 1/4 white. This inheritance pattern could be interpreted as showing that the alleles for red and white flower colour were present in the F_1 plants, did segregate from each other into

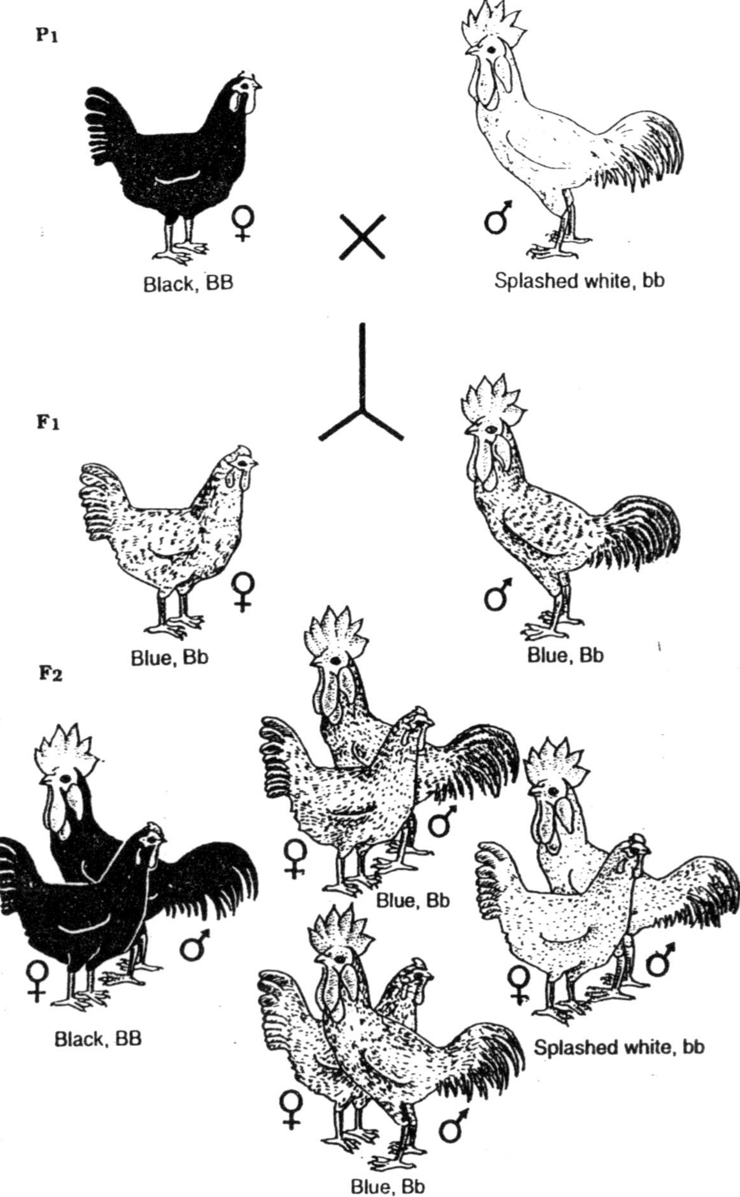

Fig. 1.13. A monohybrid cross between black and splashed white Andalusian fowl showing the incomplete dominance.

gametes, and were passed on, unchanged, to the F_2 progeny. To test whether this was indeed the case, plants having pink flowers were interbred over a number of generations. These pink × pink crosses invariably produced progeny showing the same ratio of 1 red: 2 pink: 1 white, regardless of which generation of pinks were used. If inheritance were not based on segregation of members of a pair of alleles, then flower colour should have become paler in successive generations. The consistency of the expected 1:2:1 ratio also streghtened the interpretation that only one pair of alleles was segregating since 3/4 of the progeny were coloured and 1/4 were not coloured (white).

To further test the hypothesis that only one pair of alleles of one gene governed the three flower colour types, testcrosses were made between pink-and white-flowerd plants. This kind of breeding test would produce a 1:1 ratio of pinks and whites if pinks were heterozygous for one pair of alleles and whites were recessive. If this were not the case, then flower colour would become paler as more generations of testcrosses were obtained by the continued interbreeding of pinks and whites in each of a series of test-cross generations. In testcrosses between pinks and whites, the results invariably showed 1 pink: 1 white phenotypic ratio. The two kinds of breeding tests, therefore, confirmed the hypothesis that the red and white alleles of the flower-colour gene adhered to the Mendelian rules for segregation and assortment of alleles in sexual reproduction.

The differences between the F_1 and F_2 phenotypes and ratios observed for four-o'-clocks and those observed by Mendel can easily be explained by *incomplete dominance* of the red allele over the white allele. Homozygous red- and homozygous white-flowered four-o'-clocks produce a new phenotype when crossed. The pink heterozygotes carried one allele of each kind, *R* and *r*, and the pair of alleles together in an individual cannot direct the production of either of the two homozygous parental phenotypes. Other cases of incomplete dominance are recognized by these same results, namely, the heterozygous phenotype is different from the parental phenotypes. The F_2 phenotypic ratio of 1:2:1 is characteristic of incomplete dominance of a single pair of alleles. In effect, each genotype produces a unique phenotype. The phenotypic and genotypic ratios are identical in crosses involving a pair of alleles that shows incomplete dominance. The crucial supporting evidence for such an interpretation is provided by testcross between heterozygotes and recessives. Testcross progeny will consist of only the heterozygous and recessive types in the expected 1:1 ratio.

Another kind of variation involving a pair of alleles of a single gene is the phenomenon of ***codominance***. Codominant alleles in a heterozygote are expressed equally and completely in the phenotype so that both allelic expressions can be identified in the single heterozygous individual. There is no blending of characters or diminished expression of either one of the alternative forms, as in incomplete dominance. A good example of codominance is provided by the antigens on the surface of red blood cells, phenotypically symbolized as M and N, and governed by a pair of alleles, L^M and L^N. Antigens are proteins that can be identified by a clumping, or *agglutination reaction* when mixed with the appropriate antibodies. Antibodies are produced in the serum of an individual sensitized to the specific antigen under consideration. This sensitized serum is called the antiserum. If blood from an individual is mixed with antiserum from another individual, cells may clump together or remain freely suspended in the mixture. Agglutination takes place when the antiserum contains the specific antibodies that can react with the surface antigen of the red blood cells. A person with M antigen can be identified by the agglutionation of a sample of his or her blood when mixed with serum containing anti-M antibodies. Blood cells having N antigen will remain suspended and will not undergo clumping when mixed with anti-M serum, but these cells will agglutinate in a mixture with anti-N serum. If both M and N antigens are present in the blood sample, agglutination will take place when the red blood cells are mixed with anti-M or with anti-N serum. The agglutination test, therefore, reveals the antigenic phenotype of the person in the question.

Table 1.4 The MN blood group in humans

Alleles	*Genotypes*	*Antigens produced*	*Phenotypes (blood group)*
	L^ML^M	M	M
L^M, L^N	L^ML^N	M, N	MN
	L^NL^N	N	N

When tests are conducted on a single blood sample using both anti-M and anti-N serum separately, some people's red blood cells show the agglutination reaction only with anti-M serum, some with only anti-N serum, and some with both kinds of antiserum. Persons who are homozygous for the L^M allele produce only M antigen; and those who are homozygous for L^N make only the N antigen. Heterozygotes have the genotype L^ML^N and possess M and N antigens on their red blood cell. The L^M and L^N alleles are codominant since each makes an equal and full contribution to the phenotype.

Another example of codominant alleles is provided by a globin gene governing hemoglobin synthesis. Different genes direct the production of each of the globin components that make up hemoglobin molecules in red blood cells. One globin gene, Hb_β, is responsible for the manufacture of beta-globin (β-globin), which forms part of adult hemoglobin molecules. In human populations the most common allele of the β-globin gene is Hb^A, which is responsible for synthesis of globin that forms part of *normal adult hemoglobin*. Another allele of the Hb_β gene, Hb^S, is responsible for aberrant β-globin that forms part of *sickle-cell hemoglobin*. The β-globin alleles Hb^A and Hb^S are codominant. Individuals homozygous for the Hb^A allele make only normal adult hemoglobin, homozygotes for the Hb^S allele make only sickle-cell hemoglobin, and Hb^AHb^S heterozygotes make both kinds of molecules. Each kind of hemoglobin molecule can be recognized by its behaviour when subjected to electrical forces in an electrophoresis apparatus. To test for molecule type, a sample of hemoglobin is allowed to migrate in a gel substance under the influence of an electrical field. Each kind of molecule comes to occupy a particular place in the gel and can be identified in that place after suitable treatment. Both normal adult hemoglobin and sickle-cell hemoglobin are present in blood samples of a heterozygous person; only one kind of hemoglobin is found in a homozygous individual. In heterozygous persons, sickle-cell hemoglobin causes some red blood cells to sickle. In homozygotes, the sickling is chronic and sickle-cell anemia results. Deformed red blood cells tend to clump, and these clumps cause the very narrow blood capillaries to clog. Blood flow is impeded, and various body parts may be damaged as a result of clogging episodes in the blood system of an afflicted person. These episodes, or crises, are painful as well as debilitating during the individual's relatively shortened lifetime.

This example of codominant alleles of a gene also provides an illustration of the *relative nature of dominance and recessiveness*. The Hb^S and Hb^A alleles are codominant at the molecular level of the phenotype, but not at the level of the whole organism. People enjoy the same good health whether they have two normal alleles or one normal and one Hb^S allele. At the phenotypic level of the whole organism, therefore, the normal allele behaves as a dominant over the sickle-cell allele. Dominance and recessiveness are *relative behaviors of alleles* and are not due to some special chemical or physical characteristic of the alleles themselves. This general principle applies to all genes and their alleles.

2

CHROMOSOMES

PROKARYOTIC CHROMOSOMES

Most textbooks in biology have until very recently presented a very erroneous picture of the chromosomes of prokaryotes. They have characterized prokaryotic chromosomes as "naked molecules of DNA," in contrast to eukaryotic chromosomes with their associated proteins and complex morphology. This misconception has resulted, at least in part, because (1) the pictures of prokaryotic chromosomes most often published have been autoradiographs and electron micrographs of isolated DNA molecules, *not metabolically active or functional chromosomes,* while (2) the most common pictures of eukaryotic chromosomes have been of highly condensed meiotic or mitotic chromosomes–again, *metaolically inactive chromosomal states.* We now know that functional bacterial chromosomes or "nucleoids" ("nucleoids" rather than "nuclei" since they are not bounded by a nuclear membrane) bear little resemblance to the structures seen in Cairns' autoradiographs just as the metabolically active interphase chromosomes of eukaryotes have little morphological resemblance to mitotic or meiotic metaphase chromosomes.

The contour length of the circular DNA molecule of *E.coli* is about 1100 μ. The *E.coli* cell has a diameter of only 1-2μ. Clearly, then, the chromosome must exist in a highly folded or coiled configuration within the cell. When the *E.coli* chromosome is isolated by very gentle procedures in the absence of ionic detergents (commonly used to lyse cells) and is kept in the presence of a high concentration of cations such as polyamines (small basic or positively charged proteins) or 1 M salt to neutralize the negatively charged phosphate grops of DNA, the chromosome remains in a highly condensed state comparable in size to the nucleoid *in vivo.* This structure, called the *"folded genome,"* is apparently the functional state, the single DNA molecule of *E.coli* is arranged into about 50 loops or domains, each of which is highly twisted or *"supercoiled"* (much like a tightly coiled telephone cord). This structure is dependent on RNA and protein, both of which are components of the folded genome. The folded genome can be relaxed by treatment with either deoxyribonuclease (DNase) or ribonuclease (RNase).

Supercoiling of DNA is an important feature of all chromosomes, from those of the smallest viruses to those of eukaryotes. It occurs whenever the DNA is either underwound ("negative supercoils") or overwound ("positive supercoils"). If one takes a covalently closed, circular double helix of DNA, breaks one strand, and rotates one of the ends that is produced for 360° around the complementary strand, while holding the other free end fixed, one supercoil will be introduced into the molecule. If the free end is rotated in the same direction as the DNA double helix is wound (right-handed), a negative supercoil will result. While this is probably the simplest way to visualize the phenomenon of supercoiling in DNA, it is not the mechanism used by enzymes to introduced supercoil into DNA.

Many biological functions of chromosomes can be carried out only when the participating DNA molecules are ngatively supercoiled. For example, the phage ØX174 gene A protein will nick only the ØX174 replicative form (RF) and initiate replication when the RF is in its negatively supercoiled form. All bacterial chromosomes studied to date appear to be negatively supercoiled in their functional states. When the basic proteins are carefully dissociated from the DNA of the chromosomes of *Drosophila melanogaster*, the DNA is found to contain the same amount of negative supercoiling as is found in the folded genomes of *E.coli.*

It is very likely that negative supercoiling is universally involved in certain of the biological functions of DNA molecules. Considerable evidence suggests that supercoiling is involved in recombination, gene expression, and the regulation of gene expression. In addition, negative supercoiling is almost certainly required for the replication of most, if not all, DNA molecules. An enzyme called *DNA gyrase*, which catalyzes the formation of negative supercoils in DNA, has now been isolated from several different organisms, both prokaryotes and eukaryotes. The DNA gyrase from *E.coli* has been the most extensively studied. Its activity is inhibited by the drugs novobiocin and nalidixic acid, two potent inhibitors of DNA synthesis in bacteria. This clearly inhibitors of DNA synthesis in bacteria. This clearly indicates that DNA gyrase activity is required for DNA replication.

Although the exact role (or roles) of DNA gyrase in DNA replication is not yet established, an obvious possibility is that the introduction of negative supercoils may aid in the unwinding of the strands of the double helix. Mechanistically, DNA gyrase is a most interesting enzyme. *In vitro*, it can tie knots in DNA molecules or join two circular molecules to produce interlocking rings. DNA gyrase is now known not to introduce negative supercoils into DNA molecules by cleaving one strand and rotating one of the ends as originally proposed. Rather, it produces negative supercoils two at a time by cleaving both strands of one segment of a DNA molecule, passing another segment of the molecule through the temporary gap, and then rejoining the ends. During each catalytic event, DNA gyrase "holds on" to the cut ends as the other DNA molecule or segment of the DNA molecule is passed through the transient gap, so that it can efficiently region the ends to complete the process. DNA gyrase "holds on" by means of covalent linkage of the transient ends of the cleaved DNA molecule to itself, in the same manner as the phage ØX174 gene A protein.

Morphology

Size

The size of the chromosomes varies from species to species and relatively remains constant for a particular species. The length of the chromosomes may vary from 0.2 to 50µm. The diameter of the chromosomes may be from 0.2 to 20µm. For instance, the human chromosomes are upto 6µm in length. Moreover, the organisms with less number of chromosomes contain comparatively large-sized chromosomes than the chromosomes of the organisms having many chromosomes.

The monocotyledon plants contain large-sized chromosomes than the dicotyledon plants. The plants in general have large-sized chromosomes in comparison to the animals. Further, the chromosomes in a cell are never alike in size, some may be exceptionally large and other may be too small. The largest chromosomes are lampbrush chromosomes of certain vertebrate oocytes and polytene chromosomes of certain dipteran insects.

Shape

The shape of the chromosomes is changeable from phase to phase in the continuous process of the cell growth and cell division. In the resting phase or interphase stage of the cell, the chromosomes occur in the form of thin, coiled, elastic and contractile, thread-like stainable structures, the chromatin threads. In the metaphase and the anaphase, the chromosomes become thick and filamentous. Each chromosome contains a clear zone, known as *centromere* or *kinetochore*, along their length. The centromere divides the chromosomes into two parts, each part is called *chromosome arm.* The position of centromere vary from chromosome to chromosome and it provides different shapes to the latter which are follows:

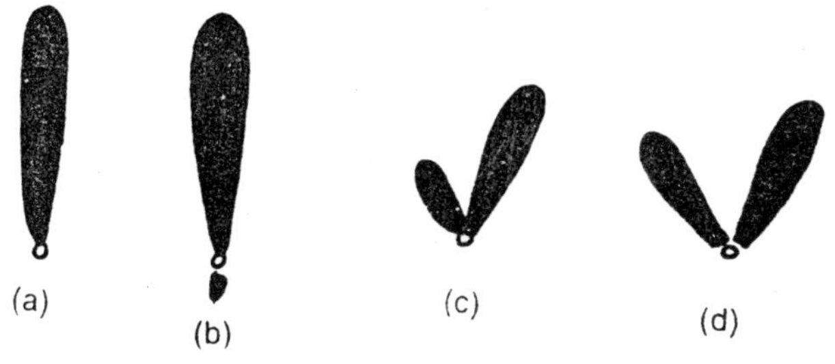

Fig. 2.1. Four morphologic types of chromosomes (a) Telocentric, (b) Acrocentric, (c) Submetacentric and (d) Metacentric.

1. *Telocentric.* The rod-like chromosomes which have the centromere on the proximal end are known as the *telocentric chromosomes.*
2. *Acrocentric.* The acrocentric chromosomes are also rod-like in shape but these have the centromere at one end and thus giving a very short arm and an exceptionally long arm. The locusts (Acridiadae) have the acrocentric chromosomes.
3. *Submetacentric.* The submetacentric chromosomes are J- or L-shaped. In these, the centromere occurs near the centre or at medium portion of the chromosome and thus forming two unequal arms.
4. *Metacentric.* The metacentric chromosomes are V-shaped and in these chromosomes the centromere occurs in the centre and forming two equal arms. The amphibians have metacentric chromosomes.

Number

The number of chromosomes in the somatic cells of higher animals and plants is known as *diploid* or *somatic* or *zygotic number*, while in the gametes (sperms and eggs) it is *haploid*, *gametic* or *reduced*. The number of chromosomes is constant in all the somatic cells of all the individuals of a species. Chromosome number is used in the identification of species and in tracing the relationship within the species. The chromosome number of some animals and plants is given in the following table:

Table 2.1. Number of chromosomes in some plants and animals.

S. No.	*Name of organism*	*Number of chromosomes*
A.	***Animals***	
1.	*Paramecium aurellia*	30-40
2.	*Hydra vulgaris*	92
3.	*Ascaris megalocephala*	2
4.	*Ascaris lumbricoides*	48 in male
5.	*Drosophila melanogaster*	8
6.	Grasshopper	24
7.	Honey bee	42, 16
8.	Mosquito-*Culex pipens*	6
9.	*Rana*	26
10.	*Felis domesticus*	38
11.	Mouse, *Mus musculus*	40
12.	*Gallus domesticus* (fowl)	78 in male
13.	*Rattus rattus* (common rat)	42
14.	*Canis familiaris*	78
15.	Chimpanzee	48
16.	Monkey-*Macaca*	42
17.	*Homo sapiens*	46
B.	***Plants***	
1.	*Alium cepa* (onion)	16
2.	*Brassica oleracea* (Cabbage)	18
3.	*Raphanus sativus* (Raddish)	18
4.	*Gossypium hirsutum* (upland cotton)	52
5.	*Zea mays* (Indian corn)	20
6.	*Triticum vulgare* (bread wheat)	42
7.	*Pisum sativum* (pea)	14
8.	*Oryza sativa* (rice)	24
9.	*Neurospora crassa* (red bread mold)	7

Chromosome Cycle and Cell Cycle

Chromosomes exhibit cyclic changes in shape and size during cell cycle. In the nondividing interphase nucleus, the chromosomes form interwoven network of fine twisted but uncoiled threads of chromatin, and are invisible. During cell division the chromatin threads condense into compact structures by helical coiling. In *prophase* of cell division the chromosomes appear as distinct threads and by *metaphase* and also in *anaphase* these become short, compact bodies having definite shapes and sizes. In anaphase these appear as rod-shaped, V-shaped, L-shaped or J-shaped. The chromosomes are studied and described at this state or at anaphase stage. In *telophase* these again uncoil to form the chromatin net.

Structure of the Chromosome

In earlier light microscopic descriptions the chromosome, or chromatid, was though to consist of a coiled thread called the chromonema lying in a matrix. The chromosome was supposed to be convered by a membranous pellicle. Electron microscopic studies later showed that there is no definite membranous pellicle surrounding the chromosome. Other structure present in the chromosome include the chromatids, centromere, secondary constrictions, nucleolar organizers, telomeres and satellites are given under the following heads:

Chromatids

During metaphase a chromosome appears to possess two threads called *chromatids*, which become interwined in the matrix of chromosome. These two chromatids are held together at a point along their length in the region of constriction of the chromosome. These chromatids are really spirally coiled *chromonemata* (sing., chromonema) at metaphase. The coiled filament was first of all observed by *Baranetzky*, in 1880, in the pollen mother cells of *tradescantia* and was called chromonema by *Vejdovsky* in 1912. The chromonema may be composed of 2, 4 or more fibrils depending upon the species. This number of fibrils in the chromonema may depend on the different phases since at one phase it may contain one fibril and at other phase it may contain two or four fibrils. These fibrils of the chromonema are coiled with each other. The coils are of two types:

1. *Paranemic coils.* When the chromonemal fibrils are easily separable from each other, such coils are called paranemic coils.
2. *Plectonemic coils.* Here the chromonemal fibrils are closely interwined and they cannot be separated easily. Such coils are called plectonemic coils.

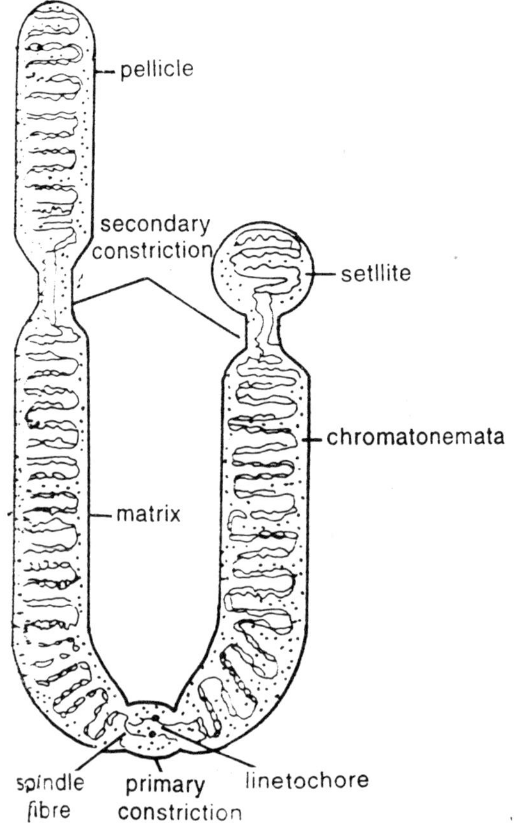

Fig. 2.2. Structure of a typical somatic chromosome at anaphase.

The degree of coiling of the chromonemal fibrils during cell divisions depend on the length of the chromosome. There are three types of coils: (i) *Major coils* of the chromonema posses 10-30 gyres, (ii) *Minor coils* of the chromonema are perpendicular to the major coils and have numerous gyres as observed in meiotic chromosomes. If splitting has not yet occurred at this stage, there will be a single chromonema, if it has already taken place there would be two chromonemata. (iii) *Standard* or *somatic coils* are found in the chromonema of mitosis where chromonemata possess helical structure, resembling the major coils of the meiotic chromosome.

Chromomeres

The chromonema of the chromosomes of mitotic and meiotic prophases have been found to contain alternating thick and thin regions and thus giving the appearance of a necklace in which several beads occur on a string. The thick or beak-like structures of the chromonema are known as the *chromomeres* and the thin region in between the chromomeres is termed as the *inter-chromomeres.* The position of the chromomeres in the chromonema is found to be constant for a given chromosome.

The cytologists have given various interpretations about the chromomers. Some consider chromomeres as condensed nucleoprotein material, while other postulated that the chromomeres are regions of the super-imposed coils. The later view has been confirmed by the electron microscopic observations. For long time most geneticists considered these chromomeres as *genes,* i.e. the units of heredity.

Centromere

The centromere is a specific region of the eukaryotic chromosome that becomes visible as a distinct morphological entity along the chromosome during condensation. It is responsible for chromosome movement in both mitosis and meiosis, functioning, at least in part, by serving as an attachment site for one or more spindle fibers. It is also the site at which the spindle fibers shorten, causing the chromosomes to move toward the poles. Electron microscopic analysis has shown that in some organisms–for example, the yeast *Saccharomyces cerevisiae*–a single spindle-protein fiber is attached to centromeric chromatin. Most other organisms have multiple spindle fibers attached to each centromeric region.

The chromatin segment of the centromeres of *Saccharomyces cerevisiae* has a unique structure in that it is exceeding resistant to the action of various DNases and has been isolated as a protein-DNA containing from 220 to 250 base pairs. The nucleosomal constitution and DNA base sequences of many of the yeast chromosomes have been determined. There are four regions, labeled I through IV. All yeast centromeres have the sequence characteristics indicated for regions I, II, and III, but the sequence of region IV varies from one centromere to another. Region II is noteworthy in that more than 90 percent of the base pairs are AT pairs. The centromeric DNA is contained in a structure (the centromeric core particle) that contains more DNA than a typical yeast nucleosome core particle (160 base pairs) and is larger. This structure is responsible for the resistance of centromeric DNA to DNase. The spindle fiber is believed to be attached directly to this particle.

The base-sequence arrangement of the yeast centromeres is not typical of other eukaryotic centromeres. In higher eukaryotes, the chromosomes are about one hundred times as large as yeast chromosomes; and several spindle fibers are usually attached to each chromosome. Furthermore, the centromeric regions of the chromosomes of many eukaryotes contain large amounts of heterochromatin, consisting of repetitive satellite DNA. For example, the centromeric regions of human chromosomes contain a tandemly repeated DNA sequence of about 170 base pairs called the alpha satellite. The number of copies in the centromeric region ranges from 5000 to 15,000, depending on the chromosome. The DNA sequences needed for spindle-fiber attachment may be interspersed among the alpha-satellite sequences. Whether the alpha-satellite sequences themselves contribute to centromere activity is unknown.

Telomeres

The sequences at the ends of eukaryotic chromosomes, called *telomeres*, play critical roles in chromosome replication and maintenance. Telomeres were initially recognized as distinct structures because broken chromosomes were highly unstable in eukaryotic cells, implying that specific sequences are required at normal chromosomal termini. This was subsequently demonstrated by experiments in which telomeres from the protozoan *Tetrahymena* were added to the ends of linear molecules of yeast plasmid DNA. The addition of these telomeric DNA sequences allowed these plasmids to replicate as linear chromosome-like molecules in yeasts, demonstrating directly that telomeres are required for the replication of linear DNA molecules.

The telomere DNA sequences of a variety of eukaryotes are similar, consisting of repeats of a simple-sequence DNA containing clusters of G residues on one strand. For example, the sequence of telomere repeats in humans and other mammals is AGGGTT, and the telomere repeat in *Tetrahymena* is GGGGTT. These sequences are repeated hundreds or thousands of times, thus spanning up to several kilobases.

Telomeres play a critical role in replication of the ends of linear DNA molecules. DNA polymerase is able to extend a growing DNA chain but cannot initiate synthesis of a new chain at the terminus of a linear chromosomes cannot be replicated by the normal action of DNA polymerase. This problem has been solved by the evolution of a special mechanism, involving reverse transcriptase activity, to replicate telomeric DNA sequences.

Euchromatin and Heterochromatin

When interphase chromosomes of metabolically active cells are chemically stained and examined under the microscope, it becomes apparent that there are two distinct types of organization of the chromatin material. One type is lightly staining and is called *euchromatin*, and the other type is darkly staining and is called *constitutive heterochromatin*. The former involves chromatin that is in a relatively uncoiled state, whereas the latter exhibits higher-order folding of the chromatin.

The distribution of constitutive heterochromatin varies from organism to organism, and examples are known where parts of chromosomes or whole chromosomes are heterochromatinized. In general, constitutive heterochromatin is found interspersed in short

Fig. 2.3. Model of constitutive heterochromatin in a mammalian metaphase chromosome. A–Constitutive heterochromatin; B–Secondary constriction I or nucleolar organizer; C–Primary constriction or centromere; D–Euchromatin; E–Secondary constriction II possible site of 5S rRNA cistrons; F–Telomere.

segments among euchromatin and is also located around the centromeres. Functionally, euchromatin contains DNA in an active or potentially active state (that is, capable, of being transcribed), whereas heterochromatin contains DNA in a transcriptionally inactive configuration. Characteristically, heterochromatin replicates later than euchromatin in the cell cycle.

Chromosome Banding

In 1969, *T.C. Hsu* and others introduced new methods for staining chromosomes by which distinct patterns of stained bands and lightly stained interbands became evident. These staining methods were enormously important since they permitted each chromosome to be identified uniquely, even if the overall morphology was identical. Distinctions can now be made among the relatively similar C group chromosomes, for example, so that we may reer to chromosome 9 or chromosome 12 instead of merely a C-group chromosome. Using the group reference is still convenient in many instances, but we now more often refer to a specific chromosome by its number as set by the karyotype conventional ordering.

The most useful chromosome banding method is *G-banding.* The earlier methods including some special steps in the staining procedure, but these were found to be unnecessary and no special conditions are really needed to visualize G-bands by staining with the Giemsa reagent. Giemsa staining had been used for many years to bring out contrast in nuclear material, and it was one of the first stains to delinete the bacterial nucleoid by microscopy.

Two main categories of chromosome banding patterns are recognized:

1. *G-bands* from Giemsa staining and *Q-bands,* which develop after staining with quinacrine and other fluorescent dyes, give relatively similar, but not identical, patterns. Fluorescent stains fade after a short time, and special microscope optics plus ultraviolet illumination are needed to see fluorescent bands. Giemsa-stained preparations are more permanent and require ordinary microscope optics and illumination. For these reasons, G-staining is used routinely.
2. *C-bands* are visualized by Giemsa staining after pretreatments using HCl and NaOH to partially denature the chromosomes in a preparation. C-bands are especially evident around the centromere and in other chromosome regions that contain substantial amounts of highly repetitive constitutive heterochromatin. The Giemsa

stain is not specific, but it binds to regions of DNA that have responded differently from non-banded regions to HCl and NaOH pretreatments.

Despite many attempts to interpret banding reactions on a molecular basis, we still know relatively little about specific interactions between DNA and any staining reagents in use Q-bands apparently result from binding between quinacrine dye and DNA regions that are rich in adenine and thymine. Since guanine and cytosine quench fluorescence GC-rich regions of DNA generally appear as unstained interbands. Since Q-bands and G-bands are relatively similar, it seems likely that a common mechanism of interaction exists between DNA and the dye. So far, this has not been shown to be true, and G-banding mechanisms remain unclear.

C-bands are very distinctively located and arise as the result of binding between Giemsa stain and residual chromatin remaining after pretreatments that extract nucleoproteins. More chromatin remains in areas of constitutive heterochromatin than in other parts of the chromosomes after denaturing steps, so more material is present to bind more of the stain and yield a band that is contrasted with lightly stained or unstained regions in between. (There is little or no nucleoprotein extraction in G- or Q-banding methods.) Locations of C-bands correlate very well with localizations of constitutive heterochromatin, according to several independent lines of evidence.

With new methods for G-banding chromosomes, investigators can unequivocally identify various structural and numerical changes in a normal complement. The greater certainty of identifying whole chromosomes or parts of chromosomes by G-bands often allows the investigator to know exactly which chromosomes are present and which chromosome parts have undergone structural rearrangements. Binding also provides a means to compare karyotypes of related species and to describe differences that apparently have an evolutionary basis.

Ultrastructure of Chromosomes

Two views have been proposed for ultra-structure of chromosomes:

(a) Multistranded View

This was proposed by *Ris* (1906). By electron microscope the smallest visible unit of the chromosome is the fibril which is 100Å in thickness. This fibril contains two DNA double helix molecules separated by a space of 25Å across and associated protein. Next largest unit is the half chromatid. The half chromatid consists of four 100Å

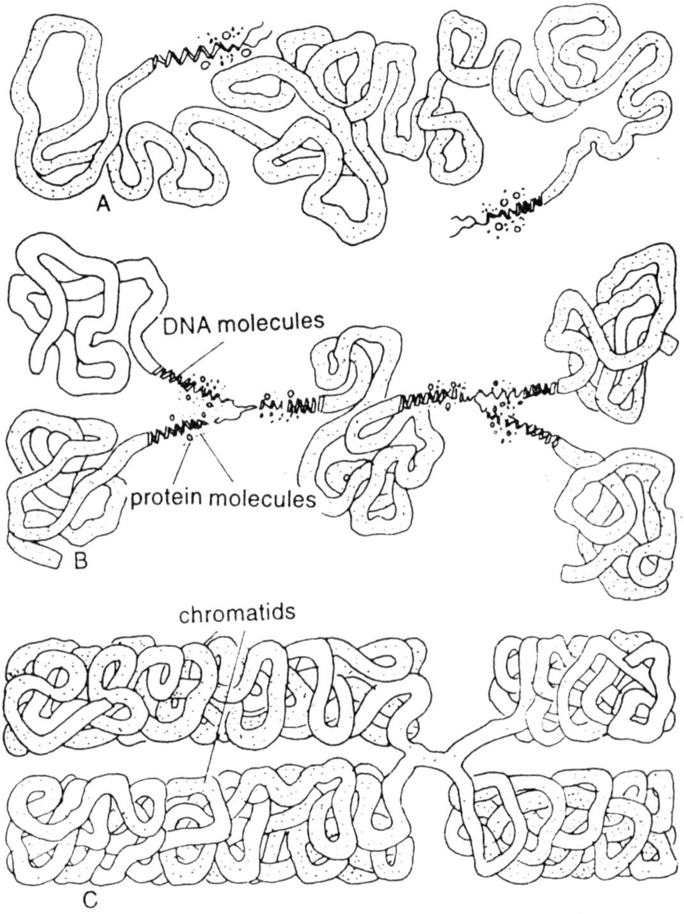

Fig. 2.4. Dupraw's folded fibre model of chromatin in interphase (A and B) and in metaphase.

fibrils so that it is 400Å in thickness and contains eight double helices of DNA and the associated protein. Two half-chromatids from a complete chromatid consisting of 16 double DNA helix molecules. As the chromosome consists of two chromatids, thus total number of helices will be 32 and diameter 1600Å thick before duplication or synthesis. After duplication chromosome has 64 double helices of DNA with corresponding diameter of 3200Å. The number of DNA helices in each unit above fibril level varies according to species. In summary, the chromosome is composed of numerous microfibrils, the smallest of which is a single nucleoprotein molecule.

(b) Folded-Fibril Model

DuPraw (1965) presented this model for the fine structure of chromosome. According to this model, a chromosome consists of single long chain of DNA and protein forming what is called as fibril. This fibril is folded many times and irregularly enterwined to form chromatid. This measure 250-300Å in thickness.

NUCLEOSOMES

Chromosomes are tight complexes between DNA and protein. It has long been known that the DNA in chromosomes is much longer than the chromosome length, and thus some means of compacting the genetic material must be available. Indeed, in chromatin the DNA is in a tightly packed state representing at least a 100-fold contraction. This is achieved by several levels of folding. The simplest level of

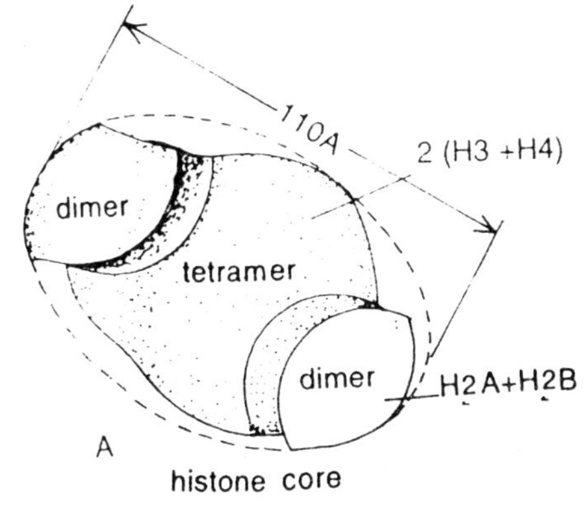

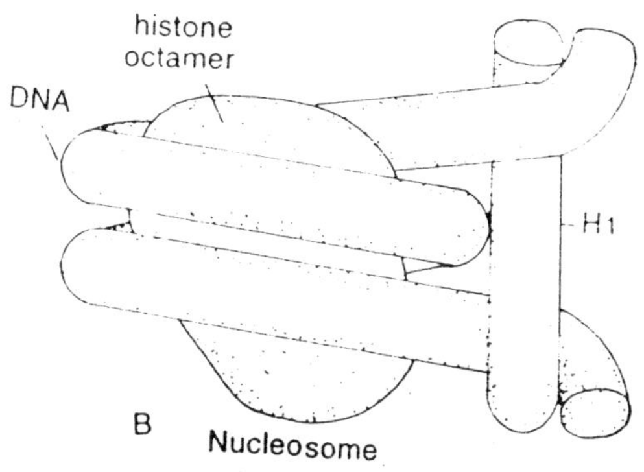

Fig. 2.5. A–Histone core; B–Nucleosome.

packing involves the coiling of DNA around a core of histones to form a structure called the *nucleosome*, and the most complex level of packing involves higher-order coiling of he DNA-histone complexes to produce chromosome morphologies exemplified by chromosomes in the division cycle.

When viewed under the microscope, the DNA–protein complex in chromosomes is seen as fibers, or nucleofilaments about 10 nm in diameter. If a nucleofilament is completely unraveled, the chromatin fiber looks like "beads on a string," where the beads are the nucleosome and the thinner thread connecting the beads is naked DNA, called linker DNA.

The nucleosome was first described in the 1970s by *A. Olins* and *D. Olins*, and by *C.L.F. Woodcock*. They described the nucleosome as 10-nm diameter, flattened spheres in which the DNA was somehow associated with a core of histones. Our current information indicates that the nucleosome is found at about 2000 base-pair intervals along the DNA molecule. If chromatin is treated for a short time with a bacterial endonuclease, microccal nuclease (which will digest any DNA not protected from its action by association with proteins), the linker DNA between nucleosomes is degraded and individual nucleosomes are related. (This procedure was pioneered by *R. Kornberg*.) Studies of isolated nucleosomes reveal them to be flat, cylindrical particles of dimensions 11 × 11 × 5.7 nm (shaped like a hockey puck or a can of tuna) with the 2-nm diameter DNA wrapped around the histone core with 80 base pairs per turn. Since there are 146 base pairs of DNA in the average nucleosome particle, the DNA is wrapped around the core approximately onc and three-fourths times. The association of the DNA with the core histone proteins is a tight one, since the DNA is well protected against enzyme attack.

By dissociating the nucleosomes with high salt concentration, it has been shown that each nucleosome consists of about 146 base pairs of DNA and a histone core (the nucleosome core particle, consisting of an octamer or two subunits each of the four histone types H2A, H2B, H3, and H4.

In the cell, chromatin exists in a more highly coiled state than the 10 nm nucleofilament. While the packaging of DNA into nucleosomes is well understood, understood less well is the higher orders of folding of the nucleosomes in the chromosomes of the cell. Above the level of the nucleosome and 10 nm nucleofilament, the next level of packing is the 30 nm chromatin fiber. Examination of isolated 30-nm chromatin

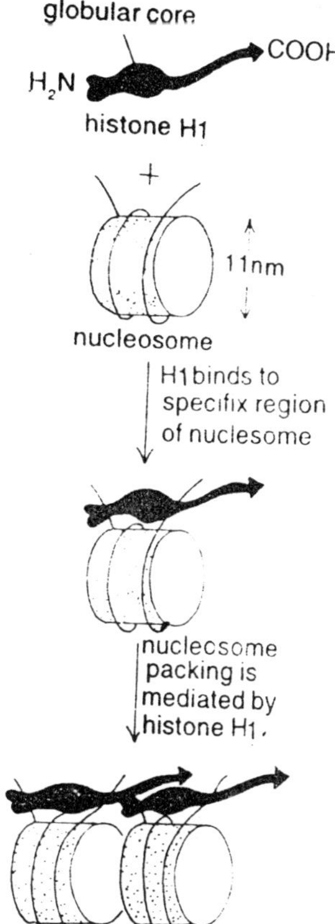

Fig. 2.6. Nucleosome packing to form solenoid.

fiber reveals the compact nature of this level of chromosome structure, making it difficult to distinguish individual nucleosomes. As a consequence, the arrangement of nucleosomes in the fiber is difficult to define. Based on biochemical and biophysical data, a recent model for the 30-nm fiber involves a zigzag ribbon of nucleosomes that is twisted to generate the 30-nm diameter chromatin fiber. In the figure, the least compact form is the relaxed zigzag ribbon, which can make the transition to the compact zigzag ribbon through no major changes in basic structure, especially the loss or addition of nucleosome-nucleosome contact. Then the 30-nm chromatin fiber is constructed by folding the compact zigzag ribbon in such a way as to minimize changes in nucleosome-nucleosome contacts. This coiling of the compact zigzag ribbon creates a double-helical ribbon with a range of possible diameters and with the nucleosomes in face-to-face contact.

What of histone H1? There is very good evidence that H1 is located at the entry/exit point of the DNA on the nucleosome. Experiments have shown that chromatin without H1 can form 10-nm nucleofilaments but not 30-nm chromatin fibers; hence, H1 must function to pack the nucleosomes together into the 30-nm fibers. Nothing is known about the regulation of this phenomenon, however.

The next levels of packing beyond the 30-nm chromatin fiber are poorly understood. The diameter of an interphase chromosome may be 300 nm while the diameter of a metaphase chromosome may be 700 nm. To achieve these levels of compaction, the simplest concept is that the 30-nm fiber becomes looped and coiled (like a rope) in various ways to give the morphologies characteristic of these chromosomes.

Polytene Chromosomes

A typical eukaryotic chromosome contains only a single DNA molecule. However, in the nuclei of cells of the salivary glands and certain other tissues of the larvae of *Drosophila* and other two-winged (dipteran) flies, there are giant chromosomes, called *polytene chromosomes*, which contain about 1000 DNA molecules laterally aligned. Each of these chromosomes has a volume many times greater than that of the corresponding chromosome at mitotic metaphase in ordinary somatic cells, and a constant and distinctive pattern of transverse banding. The polytene structures are formed by repeated replication of the DNA in a closely synapsed pair of homologous chromosomes without separation of the replicated chromatin strands or of the two chromosomes. Polytene chromosomes are typical chromosomes and are formed in

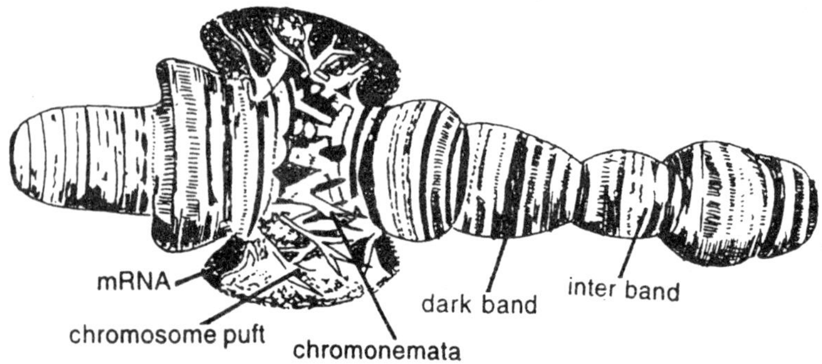

Fig. 2.7. Polytene chromosome of an insect.

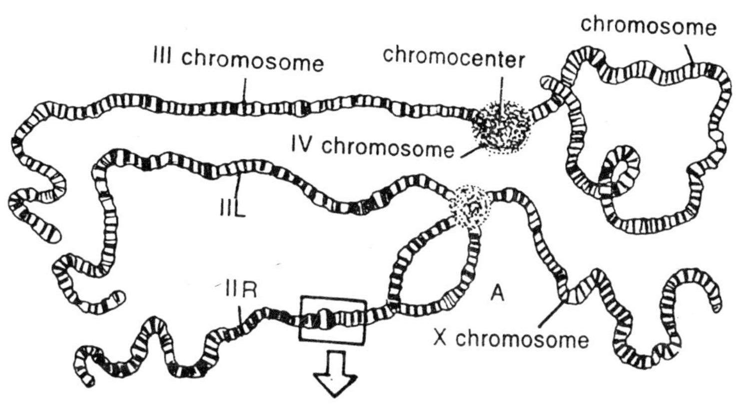

Fig. 2.8. Polytene chromosomes of Drosophila.

"terminal" cells; that is, the larval cells containing them do not divide further during development of the fly and later eliminated in the formation of the pupa. However, they have been especially valuable in the genetics of *Drosophila*.

In polytene nuclei of *D. melanogaster* and other species, large blocks of heterochromatin (a particular type of chromatin described in the following section) adjacent of the centromeres are aggregated into a single compact mass called the *chromocenter.* Because the two largest chromosomes (numbers 2 and 3) have centrally located centromeres, the chromosomes appear in the configuration, the paired X chromosomes (in female), the left and right arms of chromosomes 2 and 3, and a short chromosome (chromosome 4) project from the chromocenter. In a male, the Y chromosome, which consists almost entirely of heterochromatin, is incorporated in the chromocenter.

The darkly staining transverse bands in polytene chromosomes have about a tenfold range in width. These bands result from side-by-side alignment of tightly folded regions of the individual chromatin strands that are often visible in mitotic and meiotic prophage chromosomes as chromomeres. More DNA is present within the bands than in the interband (lightly stained) regions. About 5000 bands have been identified in the *D. melanogaster* polytene chromosomes. This linear array of bands, which has a pattern that is constant and characteristic for each species, provides a finely detailed *cytological map* of the chromosomes. The banding pattern is such that short regions in any of the chromosomes can be identified.

Because of their large size and finely detailed morphology, polytene chromosomes are exceedingly useful for *in situ* nucleic acid

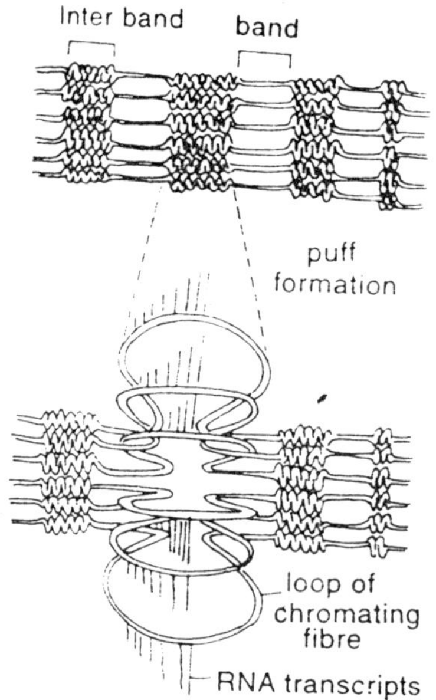

Fig. 2.9. A part of a giant polytene chromosome of Drosophila showing A-Bands and interbands; B-Puff formation.

hybridization. In the *in situ hybridization* procedure, labeled probe DNA or RNA is added to squashed polytene nuclei after denaturation of the chromosomal DNA, under conditions that favour renaturation. After washing, the only probe that remains in the chromosomes has formed hybrid duplexes with chromosomal DNA and its position can be identified cytologically.

The Organization of Nucleotide

Sequences in Eukaryotic Genomes

In bacteria, the variation of average base composition from one part of the genome to another is quite small. However, in eukaryotes, some components of the genome can be detected their base composition is quite different from the average of the rest of the genome (for example, one component of crab DNA is only 3 percent GC). These components are called *satellite DNA.* In the mouse, satellite DNA accounts for 10 percent of the genome. A striking feature of satellite DNA is that it consists of fairly short nucleotide sequences that may be repeated *as many as a million times in a haploid genome.* Other *repetitive sequences* also are present in eukaryotic DNA.

Nucleotide Sequence Composition

Eukaryotic organisms differ widely in the proportion of the genome consisting of repetitive DNA sequences and in the types of these sequences that are present. A eukaryotic genome typically consists of three components.

1. *Unique,* or *single-copy, sequences.* This is usually the major component and is typically from 30 to 75 percent of the chromosomal DNA in most organisms.
2. *Highly repetitive sequences.* This component constitutes from 5 to 45 percent of the genome. Some of these sequences are satellite DNA referred to earlier. The sequences in this class are typically from 5 to 300 base pairs per repeat and are duplicated as many as 10^5 times per genome.
3. *Middle-repetitive sequences.* This component is from 1 to 30 percent of a eukaryotic genome and includes sequences that are repeated from a few times to 10^5 times per genome.

Lampbrush Chromosomes

In the oocytic nuclei of those animals which have large yolky eggs, the prophase of first meiotic division is extremely extended. During this phase the oocyte grows and synthesizes nutrients for the future embryo. In them, the chromosomes become greatly enlarged and assume unusual configuration. A large number of loops project out from the chromatid axis, giving a lampbrush appearance. Hence, these chromosomes are called *lampbrush chromosomes.*

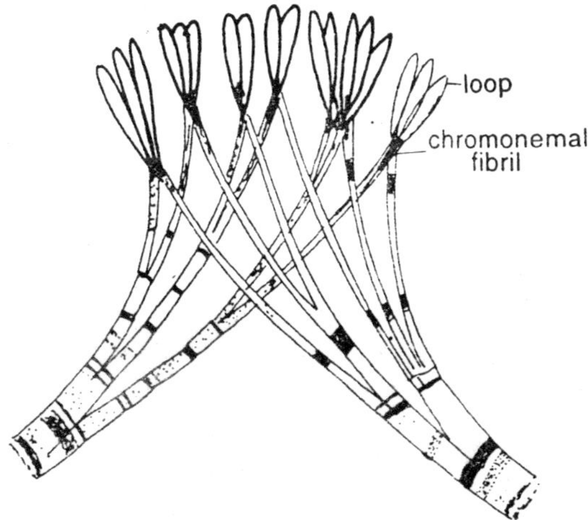

Fig. 2.10. A chromosomal loop of chromosome puff.

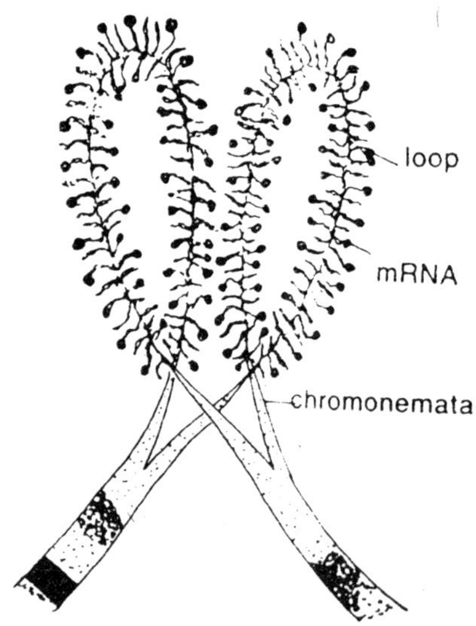

Fig. 2.11. Suggested form of the individual loops in chromosome puffs. Note the tiny fibrils with dense RNA granules attached to the loop.

The lampbrush chromosomes are bivalents each consisting of two chromatids. These persists during the prolonged diplotene phase of first meiotic prophase.

History

Lampbrush chromosomes were first observed by *Flemming* (1882) in amphibian oocyte. A detailed study was made by *J. Ruckert* (1892) in the oocytes of sharks.

Occurrence

Lampbrush chromosomes are found in the oocytes of insects, sharks, amphibians, reptiles and birds which produce large and yolky eggs. These have also been found in plants and invertebrates like *Sagitta*, *Sepia* and *Echinaster*.

Size

Lampbrush chromosomes are large enough to be seen under light microscope. These may be as long as 1000μ or more and about 20μ in width. In *Salamander* oocyte these may attain length of about 5,900μ.

Structure

A lampbrush chromosome (in diplotene stage) consists of two homologous chromosomes which are in contact only at certain points, the *chiasmata*. Each chromosome of the pair is formed of two *chromatids* which lie parallel and form the *chromosomal axis* or the *main axis*. The axis is differentiated into alternate regions of high density *loops* which are lightly coloured, and arise on both sides of the chromosomal axis.

The *chromosomal axis*, the *chromomeres* and the *loop axis* and are formed of *DNA*.

The *chromomeres* are found in pairs, one chromomere on each chromatid. These are about .25 to 2.0μm in diameter and are spaced about 2μm from centre to centre along chromatid axis. These probably represent heterochromatic regions where axial filament remains tightly coiled.

The *lateral loops* arise from the chromomeres either 2 or in multiple of two. These extend on either side of the chromosomal axis about 550μm and are about 30-50 (3.-5 nm) in diameter. Each loop consists of an *axial fibre* formed of DNA. It is surrounded with the matrix composed of *RNA* and *Proteins*. This gives fuzzy appearance of lateral loops.

ELECTRON MICROSCOPIC STRUCTURE

Electron microscopic studies by *Miller* and *Beaty* (1969) on Lampbrush chromosomes of salamander oocyte have shown the presence of dense granules on the loop axis of DNA. These dense granules represent large molecules of enzyme *RNA polymerase*. On getting attachment of DNA, these initiate RNA synthesis. Arising from these RNA polymerase molecules are seen fine fibrils of RNA which increase in length.

Each loop is considered to be one operon consisting of a series of identical copies of the same structural genes (*cistrons*) separated by spacer DNA. Each gene locus probably produces a very long RNA molecule. This interacts with protein to form *ribonucleoprotein*.

(i) According to *Callan* and *Llyod* (1960) a chromosome is the *master gene* with solenoid super coiling which produces several identical copies of its own. These extend out as a lateral loop formed of linear strand of *nucleosomes*, representing the transcriptionally active stage. These are called *salve gene copies*.

(ii) According to *spinning out and retraction hypothesis*, a chromosome is fully transcribed from end to end by spinning out a transient loop. The new loop material spins out on one side of a chromomere at the thin end of loop and returns to a condensed stage on the other side after completing the synthesis of RNA. These are associated with the rapid synthesis of yolk and protein in the maturing ovum. These disappear by the end of first prophase when chromosomes become thick and more condensed.

3

Cell Division

There are two types of organisms-acellular and multicellular. The growth and development of an individual depends exclusively on the growth and multiplication of the cells. It was *Virchow* who first of all adequately stated the cell division. In animal cell the cell division was studied in the form of segmentation division or cleavage by *Prevost* and *Dumas* in 1824. The mechanism of cell division was not precisely investigated until long afterward but *Remak* and *Kolliker* showed that the process involves a division of both the nucleus and the cytoplasm. The term *karyokinesis* was introduced by *Schleicher* (1878) to designate the changes of nucleus during division, and the term *cytokinesis* was introduced by *Whiterman* (1887) to designate the associated changes taking place in the cytoplasm.

Cell division is necessarily the avoidance of ageing, and secondly for the segregation of an individual into semi-independent units which leads to efficiency. Thus, we see that cell division is a widespread phenomenon that is essential not only for the maintenance of life but also for the development of the organism itself.

Cell division can be conveniently described as:

(i) *Direct division.* Where the nucleus and cell body undergo a simple mass division into two parts. It is also called *amitosis.*

(ii) *Indirect division.* Here the nucleus undergoes complicated changes before it is divided into two daughter nuclei.

Having seen how the DNA in the nucleus is replicated and repaired, we turn now to the process whereby the two copies of each chromosome that have been generated during the prior S phase are separated from each other and partitioned into daughter cells. These processes are mitosis and cytokinesis.

The Stages of Mitosis

Mitosis has been known and studied for a century, but only in the past 25 years has significant progress been made toward understanding the mitotic process at the molecular level. We will begin by surveying the morphological changes that occur in a cell as it undergoes mitosis; later we will examine the underlying molecular mechanisms.

Morphologically, mitosis can be described as a series of five phases, based primarily on the appearance and behaviour of the chromosomes. As with any dynamic process, we must remember that the division into phase is somewhat arbitrary and that the phase are primarily a convenience for studying and describing the process.

The five phases of mitosis are prophase, prometaphase, metaphase, anaphase, and telophase. (An alternative term for prometaphase is simply "late prophase").

Interphase

The period of metabolic activity during which cell division is not in a process, has been called the 'interphase.' This is frequently referred to as the 'resting phase' but this term is not appropriate because the cell is metabolically most active at this stage that is why *Berril* and *Huskins* (1936) referred it as the *energy phases.* The interphase is the period between the telophase of one division and the prophase of the new cell division. During this phase the cell does everything except division. It is during interphase that the genes self-duplicate and carry on their function of supervising synthesis.

The chromatin granules in the nucleus are not readily distinguishable in the living cell, but may be brought out by treatment with chemicals which kill, fix and stain them. They appear at first glance to be scattered throughout the nucleus, but careful study has produced evidence that they are arranged in a definite pattern as long coiled strands and appear as a thread like net work in ordinary stained preparation.

"Nuclear sap" or, "Karyoplasm" fills the interstics between the chromosomes. One of more rounded bodies, the *nucleoli* are usually present. Between the nucleus and the surrounding cytoplasm, is the *nuclear membrane.* In the cytoplasm adjacent to the nucleus, there is a body, the *central body,* which consists of two granules or, after each granule has replicated, of two pairs of granules, the *centrioles.*

A typical cell cycle, including interphase lasts from 20-24 hrs. Interphase in the longest period in the cell cycle, and may last for several days in cells.

Interphase can be further divided into form sub-phases:

1. G_1-phase
2. S-phase
3. G_2-phase
4. M-phase.

G_1-phase includes the synthesis and organization of the substrate and enzyme necessary for DNA synthesis. Therefore, G_1, is marked by the synthesis of RNA and protein.

G_1-phase is followed by the S-phase where the synthesis of DNA occurs.

During G_2-phase, all the metabolic activities are performed. M-phase is the period of chromosomal division.

The relative lengths of these phases differ in different organisms. A human cell in culture at 37^0C, completes the mitotic cycle in about 20 hours and the M-phase lasts for only one hour. Temperature and cell environment plays an important role in determining the rate of cell division. Even the non-meristematic cells can sometimes be made to divide by changing the environmental conditions. Those cells which are not going to divide any more, have the mitotic cycle at the G_1 phase and start differentiating.

The cells shows following changes:

1. The cell, as a whole, attains the maximum growth and possesses synthesized proteins for energy for various divisions and processes.
2. The nuclear membrane is intact and the chromosomes are found in the form of more or less loosely coiled threads, somewhat closely appressed to the membrane. In this condition of chromosomes, most of the cytologists regard them to be duplicated, while some workers are of the opinion that they are multipartite.
3. The two centrioles, which are found at right angles to each other replicate into two each. *Mazia* (1961) has described that, if the replication is checked, then division will not take place.
4. For the future spindle condensation of protoplasm into a coherent area of jelly-like consistency also takes place. The spindle also starts growing and pushes the centrioles apart.
5. DNA synthesis occurs during autosynthetic interphase, when the chromosomes and dispersed.
6. Chromocentre are also conspicuous during the interphase.

Duration of Cell Division

The time required for completion of mitosis varies from cell type to cell type and is related to the length of the G_1 phase. Temperature, within certain limits, also affects the duration of the process. In general the metaphase and anaphase stages are of short duration; most of the time of cell division is spent in prophase and telophase stages. It must be remembered that the period between cell division, interphase, is usually long. For instance, in some mammalian cell cultures 16 to 20 hours between cell divisions is a common occurrence.

Many cells of both plants and animals studied in tissue culture show the following durations in the cell cycle.

G_1	10–20 hours (usually less than 50 percent of the total)
S	6–8 hours
G_2	1–4 or more hours (usually less than 20 percent of the total)
M	1 hour
Total	18–33 hours

Formation of Mitotic Apparatus: Prophase

Toward the end of G_2, the chromosomes start to condense from the extended, highly diffuse form of interphase chromatin to the dense, coiled structures characteristic of mitosis. Although the transition from interphase to mitotic prophase is not sharply defined, a cell is considered to be in *prophase* when the chromosomes have condensed to the point of being visible as threads in the light microscope. Each chromosome has duplicated during the preceding S phase and now consists of two sister chromatids. Sister chromatids are tightly attached to each other at a constricted region, the *centromere*, which corresponds to a particular stretch of the chromosome's DNA. As the chromosomes condense, the nucleolus (or nucleoli) gradually disappears.

Meanwhile, outside the nucleus, another important organelle has sprung into action. This is the *centrosome*, an amorphous cloud of material that, in most animal cells, surrounds a pair of *centrioles*. Because they are lacking in some mitotic cells, including all plant cells and fungal cells, centrioles cannot be essential for mitosis, and their function in the centrosome remains a mystery. However, the centriole structure–a cylinder made of microtubules–is related to the centrosome's role in the cell. The centrosome is a cellular organizing center for microtubules. During interphase, the microtubules of the cytoskeleton originate there. During prophase, the cytoskeletal microtubules disassemble, and their

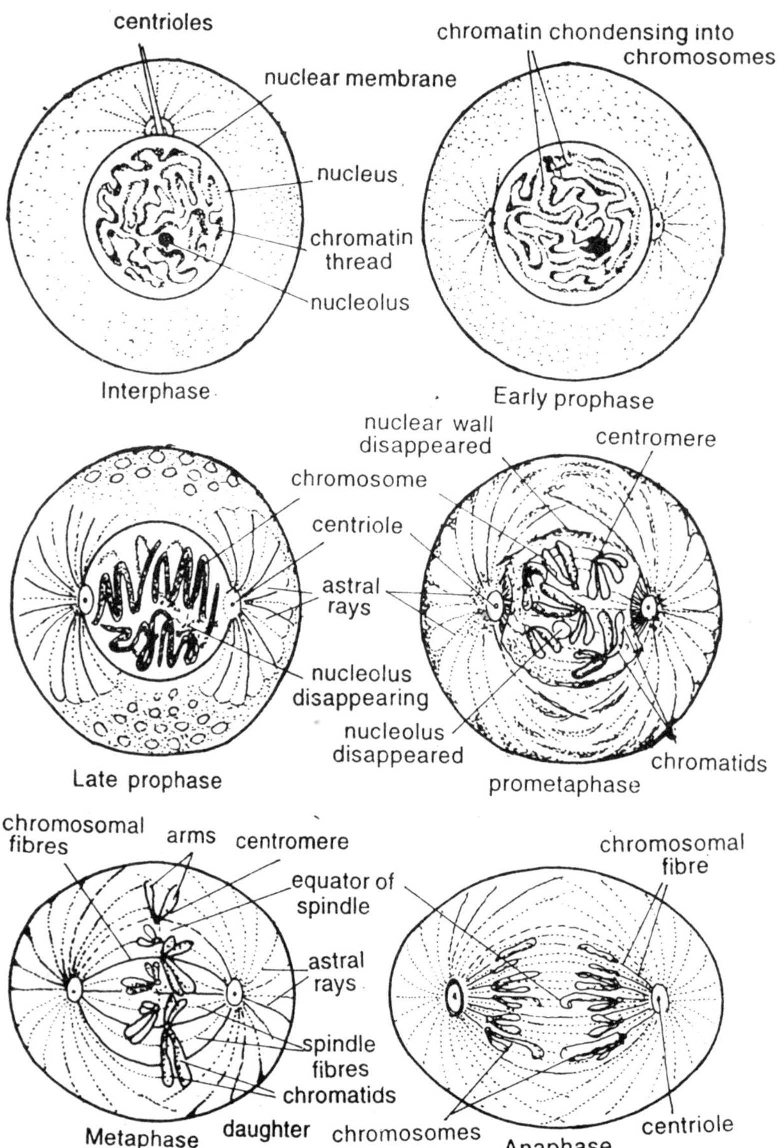

Fig. 3.1. Mitotic cell division in animal cells.

tubulin subunits start to reassemble to form the *mitotic spindle*, the apparatus that will distribute chromosomes to the daughter cells. The centrosome, including its two centrioles, has duplicated during interphase, and in prophase the two centrosomes are seen to be moving apart from each other. Radiating from them are microtubules that are growing to form the mitotic spindle. The starburst of microtubules in the immediate vicinity of a centrosome is called an aster. The terms centrosome and aster were first used for animal cells but are now used more broadly.

Prometaphase

The start of *prometaphase* is marked by the fragmentation of the nuclear envelope into membranous vesicles, allowing the mitotic spindle to enter the nuclear area. Eventually the two centrosomes are at opposite poles of the cell. On each chromosomal centromere, proteins assemble to form a protein-DNA complex called a *kinetochore*; thus there are two kinetochores on each chromosome, one on each chromatid. The two kinetochores face in opposite directions, as the diagram shows. Some of the spindle microtubules "capture" (attach to) kinetochores; others interact with microtubules coming from the other centrosome. Forces exerted within the assembly of microtubules throw the chromosomes into agitated motion and gradually move them toward the center of the cell.

Division of the Centromeres: Metaphase

The second stage of mitosis, *metaphase*, begins when the pairs of sister chromatids align in the center of the cell. When viewed with a light microscope, the chromosomes appear to be lined up in a circle along the inter circumference of the cell, as the equator girdles the earth. An imaginary plane perpendicular to the axis of the spindle that passes through this circle is called the *metaphase plate*. The metaphase plate is not an actual structures but rather an indication of where the

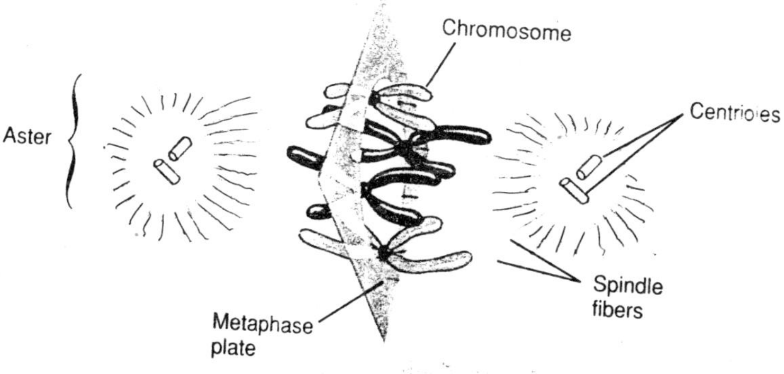

Fig. 3.2. The metaphasic plate.

future axis of cell division will occur. Positioned by the microtubules attached to the kinetochores of their centromeres, all of the chromosomes line up on the metaphase plate, their centromeres neatly arrayed in a circle, each equidistant from the two poles of the cell.

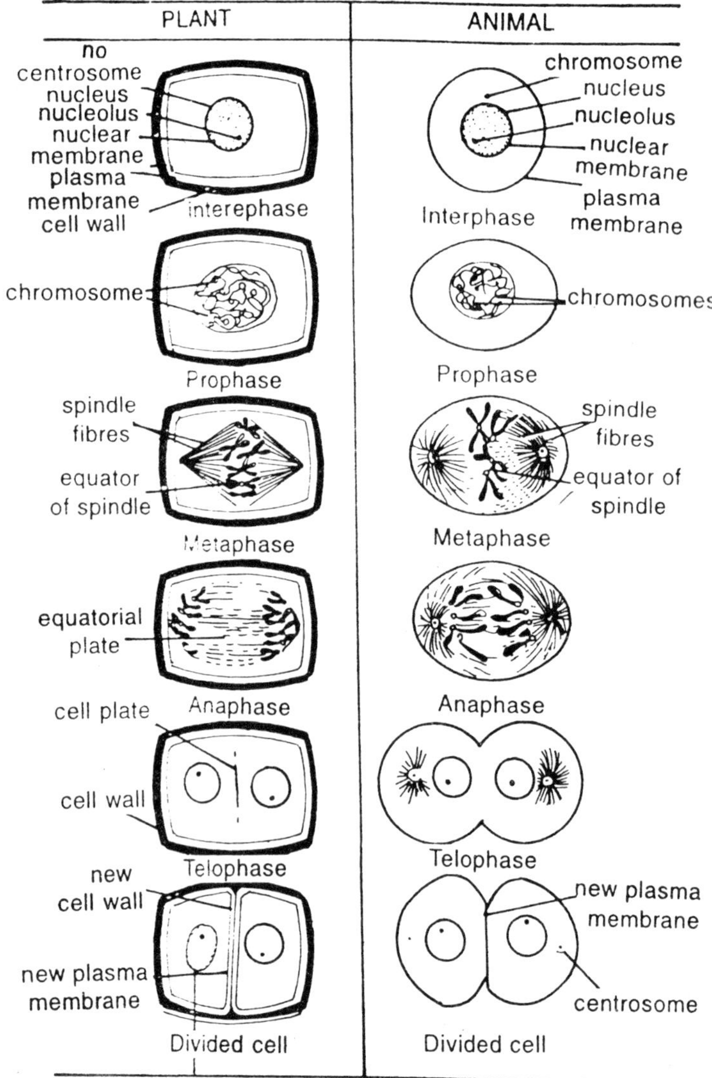

Fig. 3.3. Differences in mitosis in plants and animal cells.

At the end of metaphase, the centromeres divide. Each centromere splits in two, freeing the two sister chromatids from their attachment to one another. Centromere separation is synchronous for all the chromosomes, but the mechanism that achieves this synchrony is not known.

Separation of the Chromatids : Anaphase

Of all the stages of mitosis, anaphase is the shortest and the most beautiful to watch. Freed from each other, the sister chromatids are pulled rapidly toward the poles to which their kinetochores are attached. In the process, two forms of movement take place simultaneously, each driven by microbutules.

First, *the poles move apart*, as microtubular spindle fibers that are physically anchored to opposite poles slide past each other, away from the center of the cell. Because the chromosomes are attached to the poles by another group of microtubules, they move apart, too. If the cell is bounded by a flexible membrane, is becomes visibly elongated. This part of the anaphase is called *Anaphase B.*

Second, *the centromeres move toward the poles*, as the microtubules that connect them to the poles shorten. This shortening process is not a contraction, since the microtubules do not get any thicker. Instead, tubulin subunits are removed from the kinetochore ends of the microtubules by the organizing center. As more subunits are removed, the chromatid-bearing microtubules are progressively disassembled, and the chromatids are pulled ever closer to the poles of the cell at the rate of about 1 μm/min. This part of anaphase is called *Anaphase A.*

Reformation of Nuclei: Telophase

The separation of sister chromatids achieved in anaphase completes the accurate partitioning of the replicated genome, the essential element of mitosis. In *telophase*, the spindle apparatus is disassembled, as the microtubules are broken down into tubulin monomers that can be used to construct the cytoskeleton of the daughter cells. A nuclear envelope forms around each set of sister chromatids, which can now be called chromosomes, since each has its own centromere. The chromosomes soon begin to uncoil into the more extended form that permits gene expression. An early group of genes to be expressed are the rRNA genes, resulting in the reappearance of the nucleolus.

CYTOKINESIS

Mitosis is complete at the end of telophase. The eukaryotic cell has partitioned its replicated genome into two nuclei, which are positioned at opposite ends of the cell. While mitosis has been going on, the cytoplasmic organelles, including mitochondria and chloroplasts (if they are present), have also been reassorted to the areas that will separate and become the daughter cells. The replication of organelles takes place before cytokinesis, often in the S or G_2 phase. The process of cell division is still not complete at the end of mitosis, however, because the division of the cell proper has not yet begun. The phase of the cell cycle at which cell division occurs is called *cytokinesis*. It generally involves the cleavage of the cell into roughly equal halves.

In the cells of animals and all other eukaryotes that lack cell walls, cytokinesis is achieved by means of a constricting belt of actin filaments. The sliding of these filaments past one another decreases the diameter of the belt and pinches the cell, creating a *cleavage furrow*

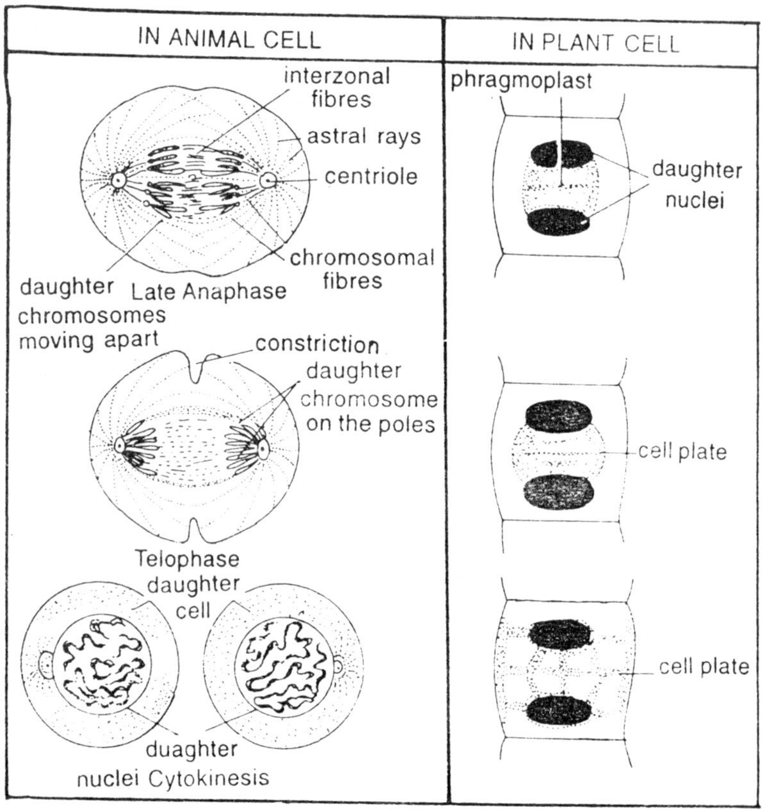

Fig. 3.4. Division of cell cytoplasm in animal and plant cell.

around the circumference of the cell. As constriction proceeds, the furrow deepens until it eventually extends all the way into the center of the cell, at which point the cell is divided in two.

Plant cells possess a cell wall that is far too rigid to be deformed by actin filament sliding. Instead, they assemble membrane components in their interior, at right angles to the spindle apparatus. This expanding membrane partition is called a *cell plate.* It continues to grow outward until it reaches the interior surface of the plasma membrane and fuses with it, effectively dividing the cell in the two. Cellulose is then laid down on the new membranes, creating two new cell walls. The space between the daughter cells becomes impregnated with pectins and is called a *middle lamella.*

In fungi and some groups of protists, the nuclear membrane does not dissolve and mitosis is confined to the nucleus. When mitosis is complete in these organisms, the nucleus divides into two daughter nuclei, one of which goes to each daughter cell during cytokinesis. This separate nuclear division phase of the cell cycle does not occur in plants, animals, or in most protists.

Control of the Cell Cycle

Research of the cell cycle has tended to focus traditionally on the events of mitosis, at least in part because the segregating chromosomes are readily visible in the light microscope. This attention to the movements of the chromosomes led biologists to conclude that the cell cycle was dissimilar in different organisms, and even in different tissues, because the chromosomes they were studying often seemed dissimilar in appearance and arrangement. The advent of sophisticated immunological and genetic engineering techniques over the last decade has led to a radical change in that conclusion. It now seems clear that the events of the cell cycle are coordinated in much the same way in all eukaryotes. The control system that human cells utilize first evolved among the protists over a billion years ago, and today it operates in essentially the same way in fungi as it does in humans. The proteins that regulate the cell cycle have been conserved so carefully that many of them function just as well when transferred from a human cell into a yeast cell.

General Strategy of Cell Cycle Control

The goal of controlling any cyclic process is to adjust the duration of the cycle so that there is sufficient time for all events to occur, without expending any more time than is necessary. In principle, there are a variety of ways to achieve this goal. An internal clock can be employed to allow adequate time for each phase of the cycle to be completed. That is how many organisms control their daily activity cycles. The disadvantage of using such a clock as a control strategy for the cell cycle is that it is not very flexible. The time required for growth or DNA replication can vary greatly, depending upon changes in the local environment of a cell. One way to achieve a more flexible and sensitive regulation of a cycle is simply to let the completion of each phase of the cycle trigger the beginning of the next phase, like a runner passing a baton to start the next leg in a relay race. Until recently, this is how biologists thought the cell division cycle was controlled. However, we now know that eukaryotic cells employ a separate centralized controller to regulate the process: at critical points in the cell cycle, further progress is dependent upon a central set of "go/no-go" switches that are regulated by feedback from the cell.

This mechanism is the same one that engineers use to control many processes. The furnace that heats a home in the winter typically goes through a daily heating cycle, as the thermostat is turned to a lower setting at night to conserve energy while we are sleeping, and then to a higher setting during the day to warm the house while we are active. When the daily cycle reaches the morning "turn on" check point, sensors report whether the house temperature is below the set point (e.g., 70°F). If it is, the thermostat triggers the furnace, which warms the house. If the house is already at least that warm, the thermostat does not start up the furnace, as no added heat is necessary. Similarly in the cell cycle, there are key check points where feedback signals from the cell about how big it is and the condition of its chromosomes can either trigger subsequent phases of the cycle or delay them to allow more time for completion of the current phase.

The cell cycle is eukaryotes is controlled at three principal check points:

Cell growth is assessed at the G_1 check point

Located near the end of G_1, just before entry into S phase, this check point makes the key decision of whether the cell will divide or not. In yeasts, where it was first studied, it is called START. If conditions are favourable for division, the cell begins to copy its DNA, initiating S phase. The G_1 check point is where more complex eukaryotes typically arrest the cell cycle if environmental conditions make cell division impossible, or if the cell passes into G_0 for an extended period.

DNA replication is assessed at the G_2 check point

The second check point occurs at the G_2 and triggers the start of M phase. If this check point is passed, the cell initiates the many molecular processes that are involved in mitosis.

Mitosis is assessed at the M check point

Occurring at metaphase, the third check point triggers the exit from mitosis and the beginning of G_1, the major growth period of the cell cycle.

Molecular Mechanism of Cell Cycle Control

How does central control of the cell cycle work? The basic mechanism is quite simple and is similar to gene control mechanisms. A set of proteins interact at the check point to trigger the next events in the cycle, and their activity is sensitive to the condition of the cell. There are two key types of proteins that participate in this interaction: cyclin-dependent protein kinases and cyclins.

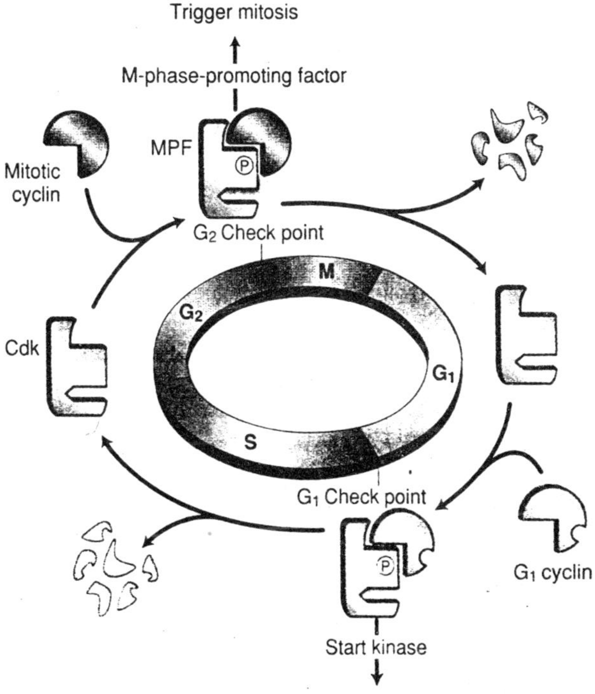

Fig. 3.5. Control of cell cycle.

Cyclin-dependent protein kinases (*Cdk's*) are enzymes that phosphorylate (that is, add phosphate groups to) the serine and threonine amino acids of important cellular enzymes and other proteins. At the G_2 check point, for example, Cdk's phosphorylate histones, nuclear membrane filaments, and the microtubule-associated proteins that form mitotic spindle. Phosphorylation of these components of the cell division machinery causes them to initiate activities that carry the cycle past the check point and into mitosis. All three check points appear to use the same Cdk molecules in yeasts; in mammals, there is a unique Cdk for each check point.

Cyclins are proteins that bind to Cdk's enabling the Cdk's to function as enzymes. Cyclins are so named because they are destroyed and resynthesized during each turn of the cell cycle. Different cyclins regulate the three check points.

Let us take a closer look at how cyclins and Cdk's interact to control the G_2 and G_1 check points of the cell cycle.

The G_2 Check Point

During G_2, the cell gradually accumulates G_2 cyclin (also called mitotic cyclin), which binds to Cdk to form a complex called MPF (mitosis promoting factor). At first, MPF is not active in carrying the cycle past the G_2 check point, but eventually a few molecules of MPF are phosphorylated and activated by other cellular enzymes. These activated MPF's in turn increase the activity of the enzymes that phosphorylate. MPF, setting up a positive feedback that leads to a very rapid increase in the cellular concentration of activated MPF. When the level of activated MPF exceeds the threshold necessary to trigger mitosis, G_2 phase ends.

MPF sows the seeds of its own destruction. The length of time the cell spends in M Phase is determined by the activity of MPF, for one of its many functions is to activate proteins that destroy cyclin. As mitosis proceeds to the end of metaphase, levels of Cdk stay relatively constant, but increasing amounts of G_2 cyclin are degraded, causing progressively more MPF to lose its activity and initiating the events that end mitosis. After mitosis, the gradual accumulation of new cyclin starts the next turn of the cell cycle.

The G_1 Check Point

The G_1 check point is thought to be regulated in a similar fashion. In unicellular eukaryotes like yeasts, the main factor controlling the start of DNA replication is cell size. Yeast cells grow and divide as rapidly as possible, and they make the START decision by comparing the volume of the cytoplasm to the size of the genome. As the cells grow their cytoplasm increases in size while their amount of DNA remains constant. Eventually a threshold ratio is reached that promotes the production of cyclins and thus triggers the next round of DNA replication and cell division.

Controlling the Cell Cycle in Multicellular Eukaryotes

The cells of multicellular eukaryotes are not free to make individual decisions about cell division in the way yeast cells are. The organization of the body cannot be maintained without severely limiting cell proliferation, so that only certain cells divide, and only at appropriate times. The inhibition of individual cell growth by other cells can be seen readily in mammalian cells growing in tissue culture: a single layer of cell expands over a culture plate until the growing border of cells comes into contact with neighbouring cells, and then the cells stop dividing. If a sector of cells is cleared away, neighbouring cells rapidly refill it and then stop dividing again. How are the cells able to sense the density of the cell culture around them? Each growing cell apparently takes up minute amounts of positive regulatory signals called growth factors (such as MPF;) that stimulate cell division. When neighbouring cells have taken up what little of the growth factor is present, there is not enough left to trigger cell division in any one cell.

Growth Factors and the Cell Cycle

It has already been observed a cell-cell interactions, growth factors work by triggering intracellular signaling systems. Fibroblasts, for example, possess numerous receptors on their plasma membranes for one of the first growth factors to be identified, platelet-derived growth factor (PDGF). Binding of PDGF to a membrane receptor initiates an amplifying chain of internal cell signals that stimulates cell division. PDGF was discovered when investigators found that fibroblasts would grow and divide in tissue culture only if the growth medium was provided with blood serum (the liquid that remains after blood clots); blood plasma (blood from which the cells have been removed without clotting) would not work. The researchers hypothesized that platelets in the blood clot were releasing into the serum one or more factors required for growth. Eventually, they isolated such a factor and named if PDGF. Growth factors like PDGF act to override cellular controls that otherwise inhibit cell division. When a tissue is injured, release of PDGF by platelets triggers neighbouring cells to divide, helping to heal the wound. Only a tiny amount of PDGF (approximately 10^{-10} M) is required to stimulate cell division.

Table 3.1. Growth Factors of Mammalian Cells

Factor	*Range of Specificity*	*Effects*
Epidermal growth factor (EGF)	Broad	Stimulates cell proliferation in many tissues; plays a key role in regulating embryonic development.
Erythropoietin	Narrow	Required for proliferation of red blood cell precursors and their maturation into erythrocytes
Fibroblast growth factor (FGF)	Broad	Initiates the proliferation of many types of stem cells; acts as a signal in embryonic development
Insulin-like growth factor	Broad	Stimulates metabolism of many cell types; potentiates the effects of other growth factors in promoting cell proliferation
Interleukin-2	Narrow	Triggers the division of activated T lymphocytes during the immune response
Mitosis-promoting factor (MPF)	Broad	Regulates entrance of the cell cycle into the M phase.
Nerve growth factor (NGF)	Narrow	Stimulates the growth of neuron processes during neural development
Platelet-derived growth factor (PDGF)	Broad	Promotes the proliferation of many connective tissues and some neuro-logical cells
Transforming growth factor b (TGF-b)	Broad	Accentuates or inhibits the responses of many cell types to other growth factors; often plays an important role in cell differentiation

Over 50 different proteins that function as growth factors have been isolated and more undoubtedly exist. Each is recognized by a specific cell surface receptor that has a shape into which the growth factor fits precisely. Binding of the growth factor to the receptor triggers events within the cell. The cellular selectivity of a particular growth factor depends upon which target cells bear its unique receptor. Some growth factors, like PDGF and epidermal growth factor (EGF), affect a broad range of cell types, while others affect only specific cell types. For example, nerve growth factor (NGF) promotes the growth of certain classes of neurons, and erythropoietin triggers cell division in red blood cell precursors. Most animal cell require a combination of several different growth factors in order to overcome the various controls that inhibit cell division.

If cells are deprived of appropriate growth factors, they stop at the G_1 check point of the cell cycle. With their growth and division arrested, they are said to be in the G_0 phase we discussed earlier. Cells like liver cells that divide only once every year or two spend most of their time in G_0 phase, and mature neurons and muscle cells usually never leave it.

Cancer and the Control of Cell Proliferation

How do growth factors influence the cell cycle? As you have seen, there are two different approaches, one positive and the other negative. PDGF and many other growth factors utilize the positive approach. They trigger passage through the G_1 check point by aiding the formation of cyclins and activating genes that promote cell division. Genes that normally stimulates cell division are sometimes called *proto-oncogenes* because mutations that cause them to be overexpressed or hyperactive convert them into oncogenes (Greek *onco*, "cancer"), leading to the excessive cell proliferation that is characteristic of cancer. Even a single such mutation (creating a heterozygote) can lead to cancer, if the other cancer-preventing genes are nonfunctional. In Mendelian terms, such mutations are said to be dominant.

Some 30 different proto-oncogenes are known. Some act very quickly after stimulation by growth factors. Among the most intensively studied of these are *myc*, *fos*, and *jun*, all of which cause unrestrained cell growth and division when overexpressed. In a normal cell, the *myc* proto-oncogene appears to be important in regulating the G_1 check point because cells in which *myc* expression is prevented will not divide even in the presence of growth factors. A critical activity of *myc* and other genes in this group of immediately responding proto-oncogenes is to stimulate a second group of "delayed response" genes, including those that produce cyclins and Cdk proteins.

Growth factors that utilize a negative approach to cell cycle control block passage through the G_1 check point by preventing the binding of cyclins to Cdk, thus inhibiting cell division. Genes that normally inhibit cell division are called *tumor-suppressor genes*. When mutated, they can also lead to unrestrained cell division, but only if both copies of the gene are mutant. Hence, these cancer-causing mutations are recessive.

The most thoroughly understood of the tumor-suppressor genes is the retinoblastoma (Rb) gene. This gene was originally cloned from children with a rare form of eye cancer that was inherited as a recessive trait, implying that the normal gene product was a "cancer suppressor" that helped keep cell division in check. The Rb gene encodes a protein that is present in ample amounts within the nucleus. This protein interacts with many key regulatory proteins of the cell cycle, but how it does so depends upon its state of phosphorylation. In G_0 phase, the Rb protein is dephosphorylated. In this state, it binds to and ties up a set of regulatory proteins like myc and fos that are needed for cell proliferation, blocking their action and so inhibiting cell division. When phosphorylated, the Rb protein releases its captive regulatory proteins, freeing them to act and so promoting cell division. Growth factors lessen the inhibition imposed by the Rb protein by activating kinases that phosphorylate it. Free of Rb protein inhibition, cells begin to produce cyclins and Cdk, pass the G_1 check point, and proceed through the cell cycle.

Variations in the Cell Cycle

As mentioned already that, eukaryotic cells do not always proceed continuously through predictable cycles of growth and division, with G_1, S, G_2 and M following one another in uninterrupted progression and with every nuclear division accompanied by cytokinesis. Such is often the case, of course, particularly in growing organisms or cultured cells that have not run out of nutrients or space. But many variations are also possible, especially in terms of the relative length of time spent in various phases of the cycle and in the immediately with which mitosis and cytokinesis are coupled.

Variations in Cell Cycle Length

Some of the most common variations in the cell cycle in vivo involve difference in generation time between different cell types. Within the same organism, some cells divide at approximately the same rate as cells in culture, but others differ greatly, depending on their role in organism. Some cells divide rapidly and continuously throughout the life of the organisms as a means of replacing cells that are lost or destroyed during the normal functioning of the organism. Included in this category are the cells that lead to sperm formation

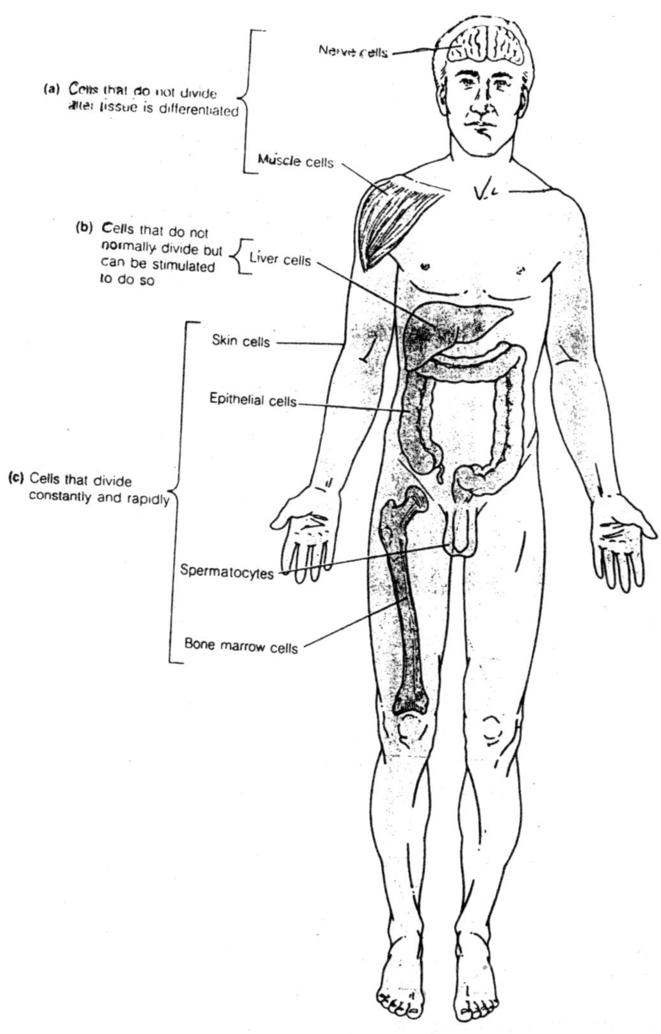

Fig. 3.6. Variations in Generation time for cells of different tissues.

and the precursor cells, called *stem cells*, that give rise to blood cells, skin cells, and the epithelial cells that line the inner surfaces of body organs such as the lungs and intestines. Human stem cells may have generations times as short as 8 hours.

Cells of slow-growing tissues, on the other hand, may have generations times of several days or more , and some cells, such as those of nerve or muscle tissue, do not divide at all. Still other cell types do not divide under normal conditions but can be induced to begin dividing again by an appropriate stimulus. Liver cells are in this category; they do not normally proliferate in the mature liver but can be induced to do so if a portion of the liver is removed surgically. Lymphocytes (white blood cells) are another example; when exposed to a foreign protein, they begin dividing as part of the immune response.

Most of these variations in generation time involve differences in G_1, although S and G_2 can also vary somewhat. Cells that divide very slowly can spend days, months, or even years in the offshoot of G_1 called G_0, whereas cells that divide very rapidly have almost no G_1 phase at all. In fact, some cells even begin DNA synthesis before mitosis is complete, eliminating G_1 entirely.

The embroys of insects, amphibians, and certain other non-mammalian animals are dramatic examples of very short cell cycles, with no G_1 phase and a very short S phase. During early embryonic development in amphibians such as the frog *Xenopus laevis*, for instance, cell division can take less than 30 minutes, even though the normal length of the cell cycle in adult tissues is about 20 hours. Under these conditions, the S phase is completed in less than 3 minutes, at least 100 times faster than in adult tissues. The incredible rate of DNA synthesis needed to sustain such a rapid cell cycle is possible because virtually all replicons are active at the same time, in contrast to the sequential activation seen in adult tissues. In addition, the average replicon length decreases, because new replicons are induced at this time.

Furthermore, these embryonic cells have little or no need to synthesize cellular components other than DNA because the fertilized egg is a very large cell with enough cytoplasm to sustain many rounds of the cell division. Each round of division subdivides the initial cytoplasm into smaller cells, until the cell size characteristic of adult tissues is reached. During the early cleavage divisions of *Xenopus* embryos, for example, not only is G_1 lacking but G_2 is unusually short, so that cells go almost directly from DNA synthesis to mitosis and back to DNA synthesis. In fact, the S phase in such cells begins even before mitosis is complete. From such examples, we know that cell growth during the G_1 and G_2 phases is not an absolute prerequisite for cell division, even though growth and division are usually coupled process.

Variations in Timing of Mitosis and Cytokinesis

For some multinucleate cells, such as the fungal and algal cells already mentioned and the skeletal muscle cells of vertebrates, the multinucleate condition is permanent. In other situations, however, the multinucleate state is only a temporary phase in the organism's development. This is the case, for example, in the development of a plant seed tissue called endosperm in the cereal grains. Here nuclear division occurs for a time unaccompanied by cytokinesis, generating many nuclei in a common cytoplasm. Successive rounds of cytokinesis then occur without mitosis, walling off the many nuclei into separate endosperm cells. A similar process occurs in developing insect eggs. The fertilized egg undergoes mitosis but not cytokinesis and soon consists of hundreds of nuclei in the same cytoplasm; later, cytokinesis catches up.

Regulation of the Cell Cycle

The variability in generation time for cells of the same organism tells us that the cell cycle must somehow be regulated. The molecular basis of this regulation is a subject of intense interest, not only for understanding the life cycles of normal cells but also for understanding how cancer cells manage to escape normal control mechanisms. Now one of the hottest areas of biological research, cell cycle regulation is beginning to reveal its underlying molecular mechanisms. We will begin our discussion with a look at the general concept of cell-cycle checkpoints and some of the early experimental evidence for the nature of their control.

Cell Cycle Checkpoints

Evidence acquired decades ago pointed to a particular point in G_1 as critical for regulation of the mammalian cell cycle. We have already seen that G_1 is the phase that varies most among cell types. Moreover, mammalian cells that have stopped dividing are almost arrested during the G_1 phase. For example, we can stop or slow down the process of cell division in cultured cells by allowing the cells to run out of either nutrients or space or by adding inhibitors of vital processes such as protein synthesis. In all such cases, the cells are arrested in G_1.

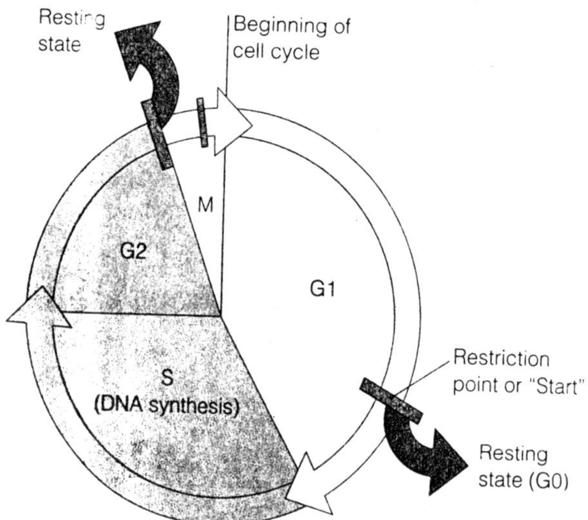

Fig. 3.7. Cell-cycle checkpoints.

These findings suggest that when a cell leaves G_1 and enters the S phase, it is committed to completing the cycle. Therefore, the release of cells from G_1 appears to be a critical control mechanism. More specifically, early researchers identified a point of no return in late G_1, which they called the *restriction point*. Cells that have passed this point are committed to division, whereas those that have not passed this point can remain in G_1 indefinitely, in the resting state called the G_0 state. As the previously mentioned experiments demonstrated, the ability to pass the restriction point can be heavily influenced by factors in the cell's environment.

Later, research with other types of cells revealed two other points of no return, and all three are now generally termed *cell cycle checkpoints*. At the end of G_2 is a major checkpoint that controls the cell's entry into mitosis (M phase). At the G_2 checkpoint, certain kinds of cells can enter a resting state analogous to G_0. Within M phase, at metaphase, is a third checkpoint, which somehow determines whether all the chromosomes are properly attached to the spindle before allowing anaphase to begin. The relative importance of the G_1 and G_2 checkpoints varies with the organism and cell type. For example, the G_1 checkpoint is the more important checkpoint in the budding yeast *Saccharomyces cerevisiae* (where it is called "Start"), as it is in most cells of multicellular organisms. However, the G_2 checkpoint is the more important one in, for example, the mitotic divisions of a fertilized frog egg and in the yeast *Schizosaccharomyces pombe* (called a *fission yeast* because it reproduces by dividing evenly in two, rather than by budding). A cell's behaviour at a checkpoint is influenced both by preceding events in the cell cycle (such as DNA replication) and by factors in the cell's environment (such as nutrients or hormones). Whatever the influence, its effects are mediated by cellular proteins, activating or inhibiting one another in chains of interactions that can be quite elaborate. However, there is an underlying unity in the molecular strategies of cell cycle regulation.

Early Evidence for Chemical Regulation of the Cell Cycle

As in many areas of scientific inquiry, early attempts to determine how the cell cycle is regulated were hampered by the difficulty of distinguishing between causal relationships and simple correlations. Just because an event usually happens at a specific point in the cell cycle does not necessarily mean it is involved in regulation of the cycle. For example, it was long thought that the critical regulatory factor was simply the ratio of cytoplasmic mass to nuclear mass and that when the cytoplasm reached a certain size, the cell would divide. It is certainly true that cell division is usually correlated with an increase in cytoplasmic mass, but there is no evidence to suggest a necessary causative relationship. Indeed, the cleavage of a fertilized egg into many smaller cells without accompanying cell growth seems to contradict such a suggestion. Moreover, there appears to be no validity to the suggestion that the transition from G_1 to S controlled by the availability of the DNA replication enzymes.

Between 1970 and 1975, it became clear that specific chemical signals present in the cytoplasm were responsible for moving the cell cycle past the G_1 and G_2 checkpoints–that is, for triggering DNA replication (S phase) and mitosis (M phase). Some of the first strong evidence for this came from experiments in which two cultured mammalian cells in different phases of the cell cycle were fused to form a single cell with two nuclei, a *heterokaryon.* If one of the original cells is in S phase and the other is in G_1, the G_1 nucleus in the heterokaryon immediately enters S phase, as though a signal present in the cytoplasm of the first cell triggers the S phase events. Similarly, if a cell undergoing mitosis is fused with another cell in any stage of its cell cycle, even G_1, the second nucleus is immediately driven into the preparatory steps for mitosis, including condensation of dispersed interphase chromatin into visible chromosomes, spindle formation, and fragmentation of the nuclear envelope. If the second cell was in G_1, the condensed chromosomes will be unduplicated.

More direct evidence for a mitosis-including chemical signal came from experiments with frog eggs. In the frog, the oocyte, an egg cell precursor, is arrested in G_2 until hormones stimulate meiosis. (Meiosis is the variation of mitosis that halves the number of chromosomes in egg or sperm production). The oocyte proceeds through most of the phase of meiosis but is arrested in M phase–in metaphase of the second of two meiotic divisions. It is now a "mature" egg cell, capable of being fertilized. Because frog oocytes and eggs are very large, about 1 mm in diameter, it is easy to transfer cytoplasm between them with a fine pipette. In a crucial experiment, it was shown that if cytoplasm taken from a mature egg cell is injected into the cytoplasm of an oocyte, the oocyte immediately begins meiosis. The hypothetical cytoplasmic chemical that induces this oocyte "maturation" was *dubbed maturation-promoting factor* (MPF). It was quickly established that MPF also induces mitosis of fertilized frog eggs (cleavage).

MPF-like activities have since been found in the cytoplasms of a broad range of eukaryotes, including yeasts, marine invertebrates, and mammals. Furthermore, the mitosis inducing factors have proven to be very similar in all these organisms. For example, in yeast cells with a defective or missing MPF gene, the human version of the gene can substitute perfectly well, despite the fact that the last ancestor common to yeasts and humans probably lived about 3 billion years ago! Through these and other kinds of experiments, investigators learned a lot about the MPF activity even before the MPF protein was purified in 1988.

The Molecular Basis of Cell Cycle Regulation

The study of cell cycle regulation entered a molecular era in 1988. This new era was brought about by the merging of results from two main lines of research, the physiological/biochemical study of developing frog eggs and the genetic study of yeasts. Their status as single-celled microbes makes yeasts particularly useful model organisms for studying many aspects of eukaryotic cell biology. Intensive research on the genetics of yeast cell cycles had begun in the late 1960s, just a few years before MPF was discovered.

Working with *S. cerevisiae*, geneticist Leland Hartwell undertook a search for mutants that were "stuck" at some point in the cell cycle. Most such mutants would be difficult or impossible to work with, because their blocked cell cycle would prevent them from reproducing. But Hartwell was able to use a powerful strategy of microbial genetics, focusing his search on *conditional mutants.* These are mutants whose defect is apparent only under certain conditions–in this case, at temperatures above the normal range for the organism. A yeast cell with such a *temperature-sensitive mutation* in a gene required for cell cycle operation reproduces normally at 20–30°C but poorly or not at all at 35–37°C. The mutant can thus be grown at the lower ("permissive") temperature for genetic and biochemical study. How can the mutant behave normally under permissive conditions? Presumably the protein encoded by the mutated gene is close enough to the normal gene product to function at the lower temperature, while the increased thermal energy at higher temperatures disrupts its active conformation (the molecular shape needed for function) more readily than that of the normal protein.

In this way, Hartwell and his colleagues identified many genes involved in the cell cycle of *S. cerevisiae* and established the points in the cell cycle at which their products functioned. Predictably, some of these genes turned out to encode DNA replication proteins, but others seemed to function in cell cycle regulation. A breakthrough discovery was made by Paul Nurse and his colleagues, who carried out similar research with the fission yeast *Schizosaccharomyces pombe.* They identified a gene called *cdc2* whose activity was essential for the initiation of mitosis–that is, for passing the G_2 checkpoint. The acronym *cdc* stands for cell division cycle. The *cdc2* gene turned out to be essentially identical to a *S. cerevisiae* gene that Hartwell's group had called *CDC28* and to have counterparts in all eukaryotic cells. (It was Nurse who showed that the human version of the gene could "rescue" mutant yeast cells.) In tribute to the importance of Nurse's discovery, the protein encoded by such a gene is often called a *Cdc2protein,* regardless of the organism where it is found.

This yeast research came together with the frog egg research when it was established that Cdc2 protein was one of two proteins making up MPF. Researchers were now primed to unravel the mysteries of the G2 checkpoint.

The Cdc2 Protein, a Protein Kinase

The Cdc2 protein is a protein kinase, an enzyme that catalyzes the transfer of a phosphate group from ATP to certain other proteins. Phosphorylation by ATP is a major theme in cell biochemistry and is the usual mechanism by which ATP functions to activate molecules, both large and small. In earlier chapters, you have seen that phosphorylation of glucose activates it for glycolysis and that phosphorylation and dephosphorylation of the protein of the sodium-potassium pump causes the shape change that allows Na^+ and K^+ to cross the plasma membrane. The sodium potassium pump protein has its own innate ability to hydrolyze ATP. However, in many other cases the phosphorylation of a protein is catalyzed by a separate protein–that is, a protein kinase. The phosphorylation of proteins by kinases, and their dephosphorylation by enzymes called *phosphatases*, is turning out to be a common cellular mechanism for regulating protein activity. And it is a mechanism that is used many times over in regulating the cell cycle.

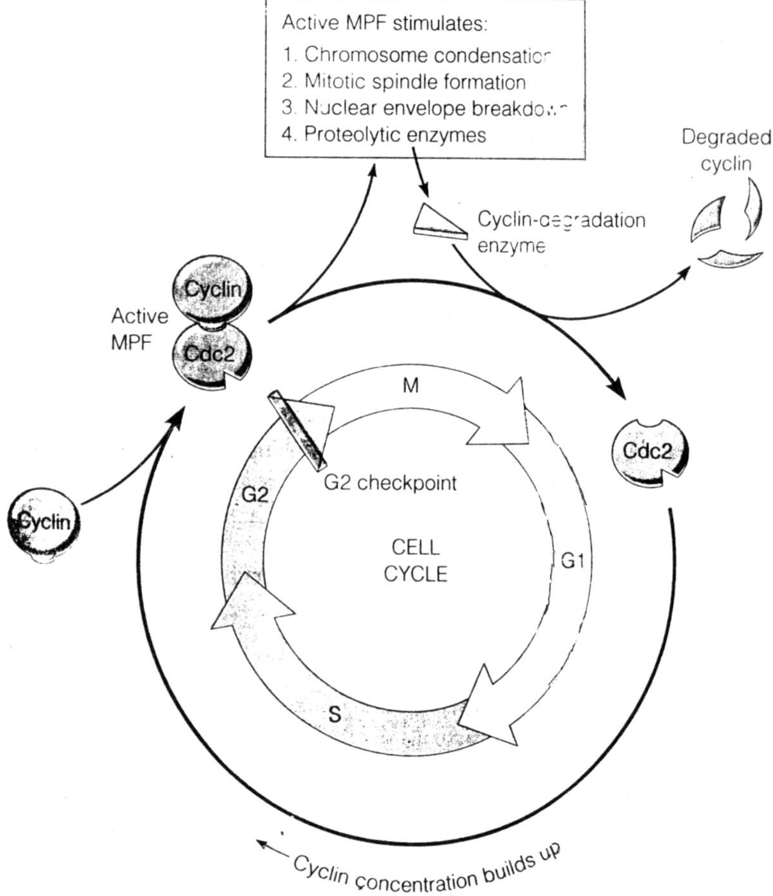

Fig. 3.8. The Cdc2 activity cycle.

The Role of Cyclins

When one looks for the Cdc2 protein at different points in the cell cycle, it is found to be present continuously at about the same concentration. However, it is not continuously active as a trigger for mitosis (or meiosis). Its MPF activity rises rapidly during G_2 phase, peaks during the first half of M phase, and then suddenly drops. MPF's activity is regulated by the second component of MPF, a protein called *cyclin.* As the name suggests, cyclins are a class of protein whose level in the cell oscillates; they are found in all eukaryotic cells. Not coincidentally, the cyclin level in a cell correlates with MPF activity level. MPF activity starts its climb when the cyclin level reaches a critical threshold.

The oscillation of cyclin level is unusual. Most other major cell proteins, like Cdc2, exist at a relatively constant concentration during the cell cycle because synthesis and any degradation occur at a constant rate as the cell grows. Unlike these proteins, cyclin is synthesized at a rate that allows cyclin accumulation to outpace the cell's growth rate– until M phase, when cyclin degradation markedly increases and destroys most of the cyclin present.

The Cdc2-cyclin complex plays a multifaced role in stimulating mitosis–it has all the activities earlier attributed to active MPF. In fact, in the scientific literature. "MPF" or "active MPF" continues to be used to mean the Cdc2-cyclin complex that stimulates mitosis. MPF (the Cdc2 cyclin complex) induces chromosome condensation, the assembly of the mitotic spindle, and the breakdown of the nuclear envelope. Only in the last of these three cases is much known about the molecular mechanism involved: The Cdc2 phosphorylates (and stimulates other kinases to phsophorylate) the *lamin* proteins of the *nuclear lamina,* to which the inner nuclear membrane is attached. phosphorylation causes the lamins to dissociate from each other, and pieces of the nuclear envelope follow suit. By the time mitosis is well under way, yet another activity of the Cdc2-cyclin complex becomes important: It activates proteolytic enzymes that cause its own demise by degrading cyclin, including both cyclin bound to Cdc2 and free cyclin. The Cdc2 protein is recycled.

What about the G_1 checkpoint, the restriction point identified decades ago in mammalian cells? In yeast, in which the G_1 checkpoint is called Start, the go-ahead signal is also given by a Cdc2-cyclin complex, although the cyclin is a different one. In cells of vertebrates, including both frogs and humans, there is a whole family of different cyclins and also a family of proteins more or less similar to the Cdc2 protein. The generic term for a member of the Cdc2 protein family is *cyclin-dependent protein kinase* (Cdk). The various types of Cdk and cyclins act in different combinations at different stages of the animal cell cycle. The details are still begins determined, but current evidence supports the involvement of the animal cell's Cdc2 along with cyclins B and A at the G_2 checkpoint, and Cdk proteins called Cdk2, Cdk4 and Cdk5, along with cyclins E and D (several kinds), at the G_1 checkpoint. Cdk2-cyclin A seems to be important during S phase. The different cyclins are made during different phases of the cell cycle.

You may be wondering about the third checkpoint mentioned earlier, the M-phase checkpoint where the decision is made whether or not to separate the metaphase chromatids and initiate anaphase. Here neither a new cyclin nor a new Cdk seems to be involved. Instead, the onset of anaphase appears to be triggered by proteolytic enzymes activated by the Cdc2-cyclin complex. However, cyclin breakdown (and the concomitant inactivation of the Cdc2-cyclin complex itself) is, surprisingly, not the anaphase-triggering event. In an experiment using an in vitro system based on from egg extracts a nondegradable from of cyclin B was added, creating a nondegradable, continuously active Cdc2-cyclin complex. Although this complex prevented mitosis from proceeding to completion, sister chromatid separation did occur. This result suggests that the proteolytic enzymes that normally attack cyclin must also attack other key proteins, perhaps including proteins required for holding sister chromatids together.

Regulation of Cdk-Cyclin Complexes by Other Kinases

Unfortunately for students of this subject, there are additional levels of complexity in the regulation of the cell cycle–making up, along with Cdk proteins and cyclins, the chains of activating and inactivating proteins we mentioned earlier. Fortunately, the reactions catalyzed by these proteins have a common theme: phosphorylation and dephosphorylation. To put it another way, most of these other proteins are protein kinases and phosphatases.

Figure indicates the main proteins and the four main reactions involved in the formation of active MPF during G_2, starting with the Cdc2 protein. This scheme is probably similar to what happens at the G1 checkpoint, also. The initial complex formed by the joining of the Cdc2 protein and the MPF (mitotic) cyclin is inactive in the cell; to trigger mitosis, the complex requires the addition of a phosphate group on a particular amino acid of Cdc2 (Thr-161). In the figure, this phosphate is highlighted with yellow. It is added by a specific kinase, which the figure calls "activating kinase." But before that enzyme acts, another *inhibiting* kinase phosphorylates the protein in two other places. (Thr-14 and Tyr-15), such a that its active site is blocked. So the last set pin the activation sequence is actually the removal of the inhibiting phosphates by a phosphatase enzyme. The extra phosphorylations and dephosphorylation steps provide other points in the pathway where the process is subject to control by other factors. In addition, a positive feedback loop is involved: The active form of Cdc2-cyclin activates more and more phosphatase.

At either checkpoint, the Cdk protein and/or the cyclin protein may need to be further modified by additional sequences of reactions before the final Cdk2-cyclin complex is fully active. The details of these process may vary with the organism and, in a multicellular organism, with the cell type. Furthermore, various environmental influences, such as nutrients and hormones, may help determine which cyclins accumulate and at what rate. Most cells have many layers of cell cycle control.

Putting It All Together: The Cell Cycle Regulation Machine

Figure is a generalized and simplified summary of the operation of the molecular machine that regulates the eukaryotic cell cycle, as currently understood. Although much of what we know to date comes from research on the G_2 checkpoint in frogs and yeasts, most cell cycle decisions are probably controlled in a similar way, with the key molecules being protein kinases and cyclins.

The cell cycle machine can be described in terms of two fundamental, interacting mechanisms. One mechanism is an autonomous clock, which on its own goes through a fixed cycle over and over again. The molecular basis of this clock is the synthesis and degradation of cyclins, which occur in a rhythmic fashion. The other mechanism adjusts the clock as needed, by providing feedback from the cell's internal and external environments. This mechanism makes use of cyclin-dependent kinases and additional proteins that, directly or indirectly, interact with cyclins. Many of the additional proteins are themselves protein kinases or phosphatases. It is this part of the cell cycle machine that transmits information about the state of the cell's metabolism–including DNA replication–and about conditions outside the cell. Energy required to activate the machine is supplied by ATP.

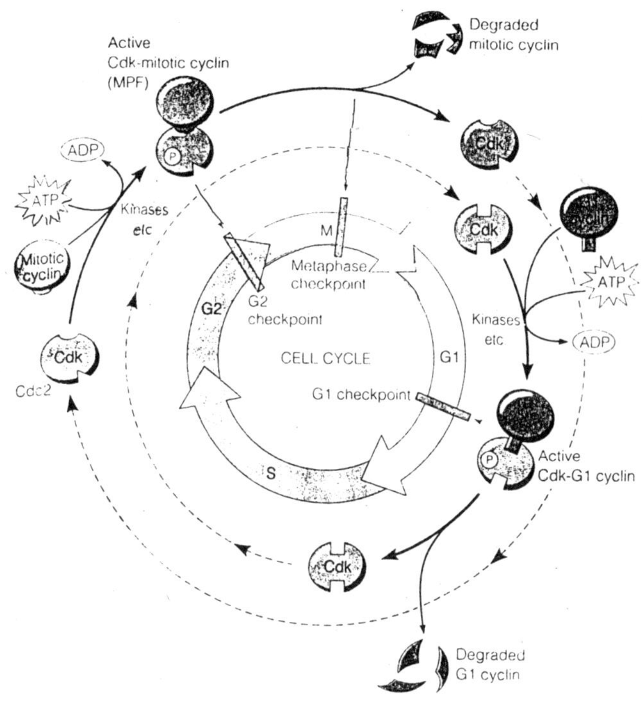

Fig. 3.9. A general model for cell cycle regulation.

Even though the Cdk-cyclin core of the cell cycle regulatory machine has been identified, we are still in the dark about many aspects of the cell cycle regulation. How exactly do the Cdk-cyclin complexes influence cell cycle events? That is, what are the actual substrates in vivo for Cdk-cyclin kinase activity? Researchers have a number of candidates, but only lamins have thus far been proven to qualify. Only with more information about substrates will we be able to determine to what extent the different Cdk-cyclins control fundamentally distinct process or, perhaps, cell-type specific versions of the same process. Not only do we need to know more about the cell cycle in the systems that are already under intensive study, such as yeasts and mammalian cells; we also need to learn more about the cells of other organisms– plants, for instance. Such studies are likely to reveal that at least some of the key cell cycle proteins are also involved in regulating other aspects of cell metabolism. Already it is known that yeast has Cdk (called PHO85) that, in combination with various cyclins, participates in both the cell cycle and phosphate metabolism.

At the other end of the chain of cell cycle control is the issue of how growth-promoting or growth-inhibiting signals coming from outside the cell connect with the cell cycle machinery. Hence the study of signal transduction pathways, which we treat intimately interconnected with the study of the cell cycle. In addition, the cancer, which can be described as a genetic disease of the cell cycle. When the normal cell cycle is disturbed by a normal molecules and events, normal cell growth may turn cancerous. More and more, therefore, there is convergence of research on the cell cycle per se and research on the cellular basis of cancer. Thus, we will return to the cell cycle at a number of points in later parts of this book.

What are the Functions of Mitotic Cell Division?

Mitotic cell division plays several roles in the lives of multicellular eukaryotic organisms. First, in conjuction with the differential expression of genes in different cells, it allows a fertilized egg eventually to become an adult consisting of perhaps trillions of individual cells.

Mitotic cell division also allows an organism to maintain its tissues, many of which require frequent replacement. For example, your skin cells live for only about 2 weeks. As dead skin flakes off imperceptibly but constantly, skin cells are continuously replaced by cell division. Your red blood cells become worn out after about 4 months. Through cell division, specialized cells in bone marrow give rise to new red blood cells, which enter the bloodstream at the rate of 3 million per second! Cells of your stomach lining, exposed to acid and digestive enzymes, survive only about 3 days before they must be replaced through the division of underlying cells.

Mitotic Cell Division Forms the Basis of Asexual Reproduction

Mitotic cell division provides the basis of asexual reproduction, in which offspring are formed from a single parent without the uniting of male and female gametes. This mode of reproduction is normal for many unicellular organisms, such as *Tetrahymena*, and yeasts. Many multicellular organisms can also reproduce asexually. Small replicas of the parent grow by means of cell division. Like its relative the sea anemone, a *Hydra* can reproduce by growing a miniature replica of itself as a bud. Eventually the bud separates from its parent, going off to live independently. Because mitosis produces genetically identical cells, these offspring are genetically identical to their parents; they are called clones.

Many plants reproduce both asexually and sexually. The beautiful aspen grooves of Colorado, Utah, and New Mexico develop asexually from shoots growing up from the root system of a single parent tree. The entire groove, although seeming to be a population of separate trees to the admiring visitor, may actually be considered to be a single individual, with its multiple trunks interconnected by a common root system. Recently, biologists at the University of Colorado reported that one of the single largest organisms yet discovered on Earth is a huge aspen groove in Utah, covering 106 acres and including about 47,000 trunks with a total mass of up to 6 million kilograms (13 million pounds). Although individual trunks age and die, the groove lives on; some grooves are thought to be hundreds of thousands of years old. If you plant an aspen tree in your backyard, who knows what you may be starting!

Mitosis also gave rise to the nucleus that produced Dolly. As you will learn in "Scientific Inquiry" Much Ado About Dolly, researchers removed the nucleus of a cell from the udder of a sheep and used it to produce a whole new lamb. This type of asexual reproduction in mammals can occur only in the laboratory!

Autoradiographic Studies on Chromosomal Duplication

Autoradiography is useful technique for cell studies that involve small amounts of materials. Radioactively labeled compounds are used. After the experiment has been performed, the radioactive atoms are localized, by putting a photographic film against the fixed (chemically killed and immobilized) cells. The β rays (electrons) emitted by the radioactive atoms expose the silver grains in the photographic emulsion. Thus by looking at the dark spots on the developed film, one can see the pattern of radioactive compounds in the biological material.

J. Herbert Taylor and his associates studied the transfer of atoms of DNA in the chromosomes of the English broad bean (*Vicia faba*) during mitosis. DNA in the root tips of the broad bean was labeled with tritium (^3H), the radioactive isotope of hydrogen, by exposing the growing tips to a solution containing ^3H-labeled thymidine. After about one third of a division cycle the seedling were transferred to a growth medium without labeled thymidine but containing colchicine. After periods equal to one or two division cycles, the root tips were fixed and pressed against the photographic film. The electrons emitted by the incorporated tritium are of such low energy that they do not penetrate deeply into the film, and they therefore produce images only at their point of entry into the film emulsion. The chromosomes and their autoradiorgrams can be viewed simultaneously with the light microscope.

The broad bean contains 12 chromosomes (2n =12, n =6). Some nuclei treated in this experiment contained 12 chromosomes, some contained 24, and some contained 48. The chromosomes in cells with 12 metaphase chromosomes have not duplicated following labeling. The chromosomes in cells with 24 and 48 metaphase chromosomes have duplicated once and twice, respectively. The chromosomes in the nuclei containing 12 chromosomes were equally radioactive in the two chromatids. Those chromosomes that had experienced a division and were in the second metaphase after labeling (24 chromosome nuclei) were labeled in one chromatid only. In those cells with 48 chromosomes, two sets of chromosomes were completely unlabeled and the other two sets were labeled like those that had undergone only one division. Although the actual arrangement of DNA in chromosomes is yet unknown, the transmission of atoms to daughter chromosomes supports the idea that a DNA double helix runs the entire length of the chromosome. For simplicity of presentation the metaphase chromatids in Figure 7.10 have been represented as containing two linear DNA molecules rather than a double helix.

Abnormalities in Mitosis

Under unfavourable conditions the mitotic divisions become defective and thus various abnormalities are formed when exposed to physical and chemical agents like temperature, radiations, necrotics and enzyme inhibitors etc. Some abnormal mitotic divisions are given under the following heads:

1. C-Mitosis (Abnormal Spindle Formation)

Brachet (1975) has described that colchicine inhibits mitosis by disorganizing spindle formation. The results are the formation of polyploid cells, and reduplication of chromosomes. Besides, during

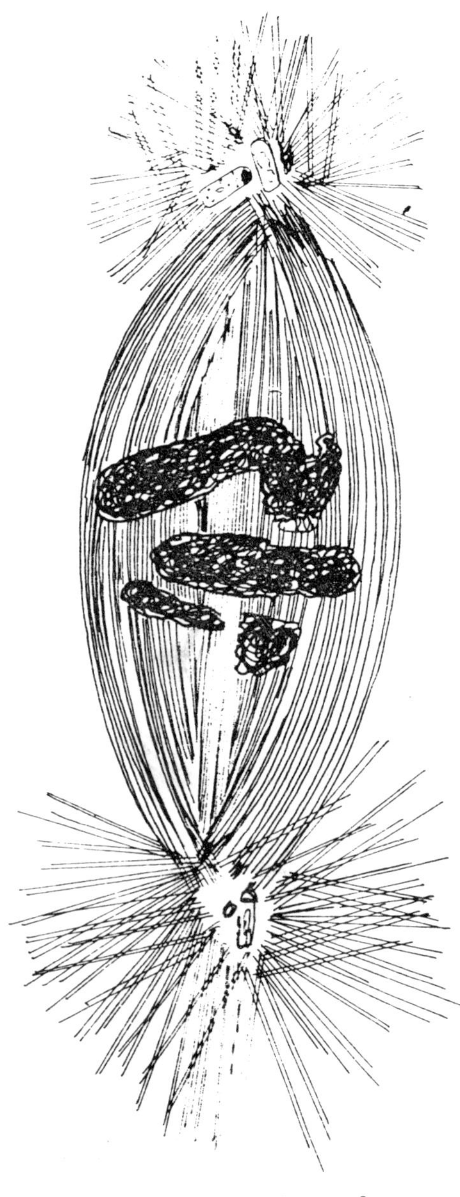

Fig. 3.10. Showing achromatic figure

mitosis, sometimes cell nucleus is either deformed into a single spheriod compact mass (pycnosis) or becomes broken down (karyorrhexis).

2. Cytastcral Mitosis

Wilson (1901) noticed the formation of numerous asters in the cytoplasm of unfertilized eggs called cytasters by placing them in hypertonic sea water. In the eggs of marine invertebrates, production of many small cytoplasmic asters is of frequent occurrence. The intermixing of nucleoplasm and hyaloplasm seems to be neccesary condition for the aster formation. *Costello* (1940) showed that asters can be produced only if germinal vesicles has broken down.

3. Multipolar and Catenar Mitosis

Mitotic division showing several spindles and centrosomes are common in many protozoans and fish eggs. Multipolarity is usually caused by uneven division of centrosomes as well as irregular distribution of chromatids of the different spindles. It results in the formation of cells as aneuploidy (uneven chromosome number). Catenar mitosis have been described by *Dalcq* and *Simon* (1932) in amphibian eggs. During this, dividing cell forms a large number of asters are capable of dividing in the absence of spindle or nucleus in an autonomous way.

4. Achrosomal Mitosis

Sometimes, mitosis in induced by various agents without use of nucleus or chromosomes. *Briggs et. al.* (1951), in fertilized eggs with heavily X-rayed sperms, removed the egg's maturation spindle by pricking and sucking. In this way, he obained very nice non-nucleate blastulae. Similarly, *Stauffer* (1945) obtained normal non-nucleated Axolotl blastula.

5. Anastral Mitosis

In plant cells and many oocytes, generally asters around the centromeres are absent, thus producing anastral mitosis. *Bataillon and Tchou Su* (1933) described in details the anastral mitosis in amphibian interspecific hybrids.

4

MEIOSIS

Since it was evident that the job of mitosis is to maintain the chromosome number in each nucleus, early investigators noted an apparent conflict in the events of gamete fertilization. They knew that during this process, two nuclei fuse, but that the chromosome number nevertheless remains constant. What prevented the doubling of chromosome number at each generation? This conflict was resolved by the prediction of a special kind of nuclear division that halved the chromosome number. This special division, which was eventually discovered in the gamete-producing tissues of plants and animals, was called meiosis.

Meiosis is also preceded by a premeiotic S phase, during which the bulk of DNA synthesis for meiosis occurs. (Some DNA synthesis also occurs during the first prophase of meiosis.) Since meiosis consists of two cell divisions, they are distinguished as meiosis I and meiosis II. The events of meiosis I are quite different from those of meiosis II, and both differ significantly from those of meiosis. Each meiotic division is formally divided into prophase, metaphase, anaphase, and telophase. Of these, the most complex and lenghty is prophase I, which has its own subdivisions: leptotene, zygotene, pachytene, diplotene, and diakinesis. Once again, try to imagine those processes as dynamic, merging into each other with no clear borders.

Prophase I

The prophase I differ from mitotic prophase in two aspects. Firstly, it is of longer duration than the mitotic prophase and secondly, most cytogenetical events such as synapsis, crossing over, etc. occur during prophase first. For the sake of convenience, the prophase first has been subdivided into five consecutive stages: leptonema, zygonema, pchynema, diplonema and diakinesis. The name of these substages of prophase first have been derived from the specific morphology or behaviour of the chromosomes within the nuclear membrane at each substage of prophase. Thus leptotene means thin thread, zygotene means yolked thread, pachytene means thick thread, diplotene means double thread and diakinesis is essentially the end of both diplotene and prophase I.

Leptotene

The chromosomes become visible at this stage as long, thin threads. No longitudinal doubleness is apparent. The process of chromosome contraction continues in leptotene and throughout the entire prophase. One other feature of leptotene is the development of small areas of thickening, called chromomeres, along the chromosome, which give it the appearance of a bead necklace.

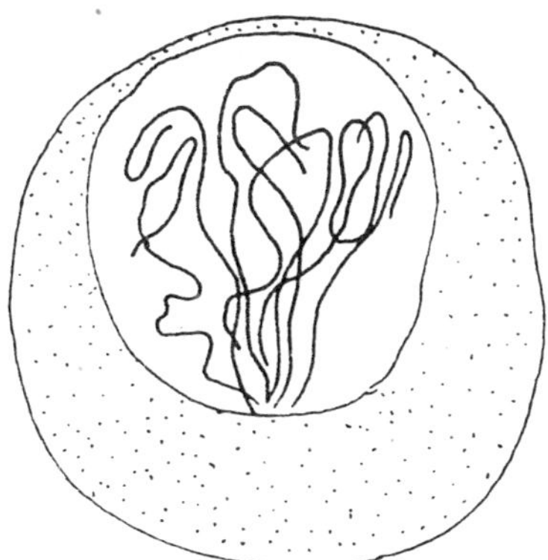

Fig. 4.1. Leptotene chromosomes arranged in a 'bouquet'.

Zygotene

This is time of active pairing at which it becomes apparent that the chromosome complement of the meiocyte is in fact two complete chromosome sets. Thus, each chromosome has a pairing partner, and the two become progressively paired, or synapsed, side by side in zipper fashion. Each pair is called a homologous pair, and the two members of a pair are called *homologous.* We see that meiocytes always contain two homologous chromosome sets, or genomes. A cell with two homologous genomes is called diploid. One genome is said to have a haploid chromosome number, usually designated *n.* Hence diploid cells are 2*n.* It should be noted that the occurrence of pairing represents a striking difference from mitosis, in which there is not such process.

What is mechanism whereby two homologous can pair so precisely along their length? First, how do they find each other in the first place? The probable answer to this is that the ends of the chromosomes, the telomeres, are anchored in the nuclear membrane, and it is likely that homologous telomeres are close, so that the zippering-up process can begin there. Second, how does the zippering-up work? Although the mechanism involved is not precisely understood, one important factor is an elaborate structure composed of protein and DNA, called a synaptonemal complex, that is always found sandwiched between homologous during synapsis.

Pachytene

This stage is characterized by thick threads representing full synapsis. Thus the number of units in the nucleus is equal to the number *n*. Nucleoli are often pronounced at this stage. The beadlike thickenings of the chromosomes, called chromomeres, are aligned precisely in the paired homologous, producing a distinctive pattern for each pair.

Diplotene

Here the DNA synthesis that had occurred in the premeiotic S phase becomes manifest as a longitudinal doubleness of each paired homologue. Once again these units, formed by longitudinal division, are called chromatids. Hence, since each member of homologous pair produces two sister chromatids, the synapsed structure now consists of a bundle of four homologous chromatids. At diplotene the pairing between homologs becomes less tight; in fact, they appear to repel each other, and as they separate slightly, cross-shaped structures called chiasmata (singular, chiasma) appear between two nonsister chromatids. One or more chiasmata are found on each chromosome pair. Chiasmata are visible manifestations of events, called crossovers, that occurred earlier, probably during zygotene or pachytene, when there is some DNA synthesis. Crossovers represent one major way in which meiosis differs from mitosis (where they occur only rarely). A crossover is a precise breakage-and-reunion event occurring between two nonsister chromatids. Studies performed on abnormal lines of organisms that undergo crossing-over very inefficiently, or not at all, show severe disruption of the orderly events that partition chromosomes into daughter cells at meiosis. Thus, crossing-over obviously plays a key

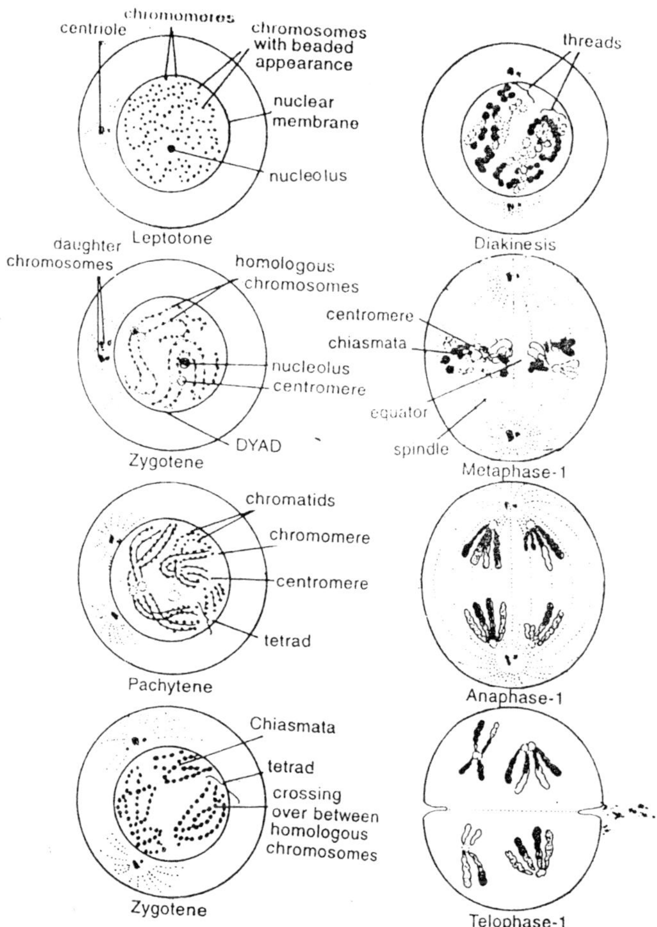

Fig. 4.2. Diagrammatic representation of different stages in the first meiotic division.

role in determining the behaviour of paired homologs, and the occurrence of at least one cross-over per pair is usually essential for proper segregation. Crossovers have another interesting role, which is to promote genetic variation by making new gene combinations.

Diakinesis

This stage does not differ appreciably from diplotene except for further chromosome contraction. By this time, the long filamentous chromosome threads of interphase are replaced by compact units far more maneuverable in the movements of the meiotic division.

Metaphase I

During the first meiotic metaphase, the bivalents orient themselves at random on the equatorial plate. The centromere of each chromosome of a terminalized tetrad is directed towards the opposite poles. The chromosomal microtubular spindle fibres remain attached with the centromeres and homologous chromosomes become ready to separate.

Anaphase I

In contrast to mitotic anaphase in which separation of sister chromatids occurs, the meiotic anaphase I is characterized by the separation of whole chromosomes of each homologous pair (tetrad), so that each pole of the dividing cell receives either a paternal or maternal longitudinally double chromosome of each tetrad. This ensures a change in chromosome number from diploid to monoploid or haploid in the resultant reorganized daughter nuclei.

Telophase I

The arrival of chromosomes at the poles of the spindle signals the end of anaphase I and the beginning of telophase I. During telophase I, the chromosomes may persist for a time in the condensed state, the nucleolus and nuclear membrane may be reconstituted and cytokinesis may also occur to produce two haploid cells. In Trillium meiocytes are reported to progress directly from anaphase I to prophase II.

II. Homotypic or 2nd Meiotic Division

Before the start of this division, most of the activities of heterotypic division cease for sometime. This resting stage is termed as the interkinesis. 2nd meiotic division involves following steps:

Prophase II

This stage superficially resembles that of the mitosis prophase with following exceptions:

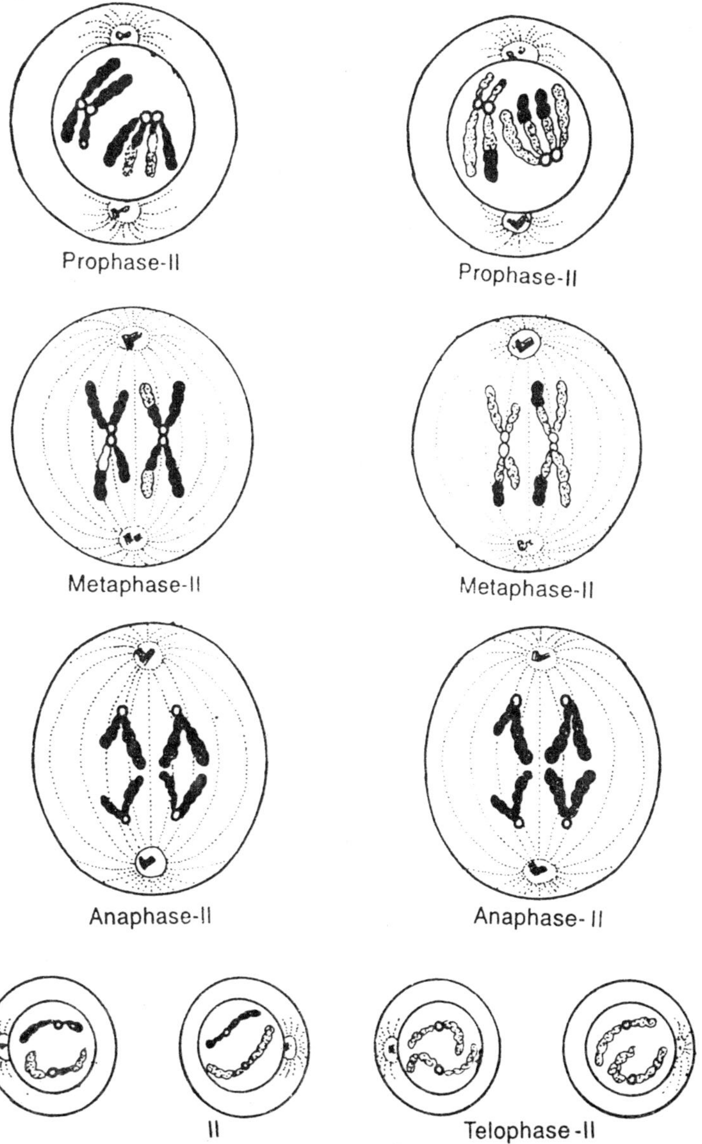

Fig. 4.3. Diagrammatic representation of homeotypic division of meiosis.

At the early prophase II, the two chromatids of each dyad looks like X, since they are conjoined by a common centromere. The four arms are widely separated. There is no rational coiling. The X-shaped dyads are quite longer than at telophase. The chromonemata are still not completely coiled. The genetic constitution of the two chromatids of each dyad depends upon the kind and number of cross overs which took place in the phase I. If there is no crossing over, the dyads would consist of two identical sister chromatids of either paternal or maternal origin. At the end of the prophase II the nucleolus disappears. The nuclear membrane disappears and the acromatic figure is developed.

Metaphase II

As usual, all the chromosomes arrange themselves on the equator for a short duration. The centromeres touch the equator but their arms radiate out in different directions. Later on the centromere in each dyad divides into two sister centromeres.

Anaphase II

Sister centromeres separate to the poles, pulling with them the chromatids to which they are attached. The chromatids, however, become much thick and stout.

Telophase II

The chromosomes at each pole uncoil and thin out to form the nuclear net. Each group gets surrounded by a nuclear membrane. Nucleolus reappears. Thus two nuclei are recognised in each cell. This is soon followed by cytokinesis and two cells are formed from each haploid daughter cell. Thus as a result of meiosis four cells are produced, each with a haploid set of chromosomes, i.e., each contains just one members of each homologous pair.

Cytokinesis

Sometimes meiosis I is followed by cytokinesis and sometimes it is deferred until the end of meiosis II. The details of the process are same as during mitosis. Cell plate formation results into a tetrad of cells with reduced chromosome numbers. The arrangement of cells in a tetrad is different in different organisms, but is characteristic of the species.

Like mitosis, meiosis too is a dynamic process and its different stages merge into subsequent ones. There is no sharp demarcation between different stages. DNA synthesis takes place during interphase, prior to prophase I. Duplication of chromosomes takes place during interphase itself and the chromosomes that appear during leptotene are double and not single. Number of chromosomes is reduced during meiosis I. If the reduction process is examined more closely, it becomes

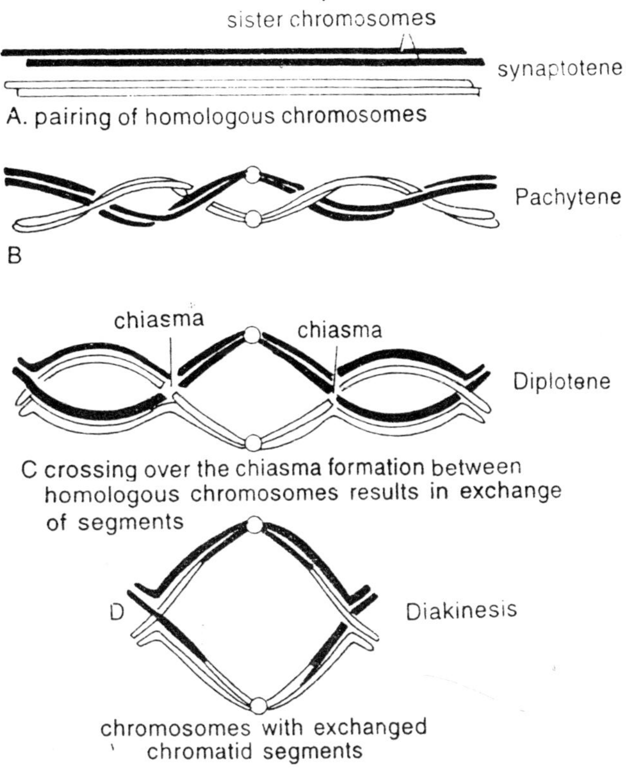

Fig. 4.4. Mechanism of chiasmata formation and crossing over.

clear that there is qualitative reduction during meiosis I of those part of chromosomes which are not involved in exchange. The homologous chromosomes or the maternal and paternal chromosomes segregate or undergo qualitative reduction during meiosis I. Their quantitative reduction takes place during meiosis II, when their centromeres divide and the sister chromatids separate. But the reverse is true of those parts of chromatids which are exchange due to the formation of chiasmata. For these parts quantitative reduction takes place during meiosis I and qualitative reduction during meiosis II. Thus, complete of both qualitative and quantitative reduction takes place as a result of meiosis I and II and none of them alone can bring about complete reduction. It is, therefore, not proper to say that meiosis I is reductional and meiosis II equational. Duration of meiosis varies a great deal in different organisms. So does the duration of different stages in an organism. Prophase I lasts for the longer duration.

Meiosis Leads to Genetic Diversity

What are the consequences of the synapsis and separation of homologous chromosomes during meiosis? In mitosis, each chromosome behaves independently of its homologous; its two chromatids are sent to opposite poles at anaphase. If we start a mitotic division with chromosomes, we end up with x chromosomes in each daughter nucleus, and each chromosome consists of one chromatid. In meiosis, things are very different.

In meiosis, synapsis organizes things so that chromosomes of maternal origin pair with the paternal homologs. Then the separation during meiotic anaphase I ensures that each pole receives one member from each pair of homologous chromosomes. (Remember that each chromosome still consists of two chromatids.) For example, at the end of meiosis I in humans, each daughter nucleus contains 23 out of the original 46 chromosomes–one member of each homologous pair. In this way, the chromosome number is decreased from diploid to haploid. Furthermore, meiosis I guarantees that each daughter nucleus gets one full set of chromosomes, for it must have one of each pair of homologous chromosomes.

The products of meiosis I are genetically diverse for two reasons. First, synapsis during prophase I allows the maternal chromosome to interact with the paternal one; if there is crossing over, the recombinant chromatids contain some genetic material from each chromosome. Second, which member of a pair of chromosomes goes to which daughter cell at anaphase I is a matter of pure chance. If there are two pairs of chromosomes in the diploid parent nucleus, a particular daughter nucleus could get paternal chromosome 1 and maternal chromosome 2, or paternal 2 and maternal 1, or both maternals, or both paternals. It all depends on the random way in which the homologous pairs line up at metaphase I.

Note that of the four possible chromosome combinations just described, two produce daughter nuclei that are the same as one of the parental types (except for any material exchanged by crossing over). The greater the number of chromosomes, the less probable that the original parental combinations will be reestablished. Most species of diploid organisms do, indeed, have more than two pairs. In humans, with 23 chromosome pairs, 2^{23} different combinations can be produced.

Meiotic Errors

The Source of Chromosomal Disorders

A pair of homologous chromosomes may fail to separate during meiosis I, or sister chromatids may fail to separate during meiosis II or during mitosis. This phenomenon is called nondisjunction, and it results in the production of aneuploid cells. Aneuploidy is a condition in which one or more chromosomes or pieces of chromosomes are either lacking or are present in excess.

If, for example, the chromosome 21 pairs fails to separate during the formation of a human egg (and thus both members of the pair go to one pole during anaphase I), the resulting egg contains either two copies of chromosome 21 or none at all. If an egg with two of these chromosomes is fertilized by a normal sperm, the resulting zygote and infant has three copies of the chromosome: He or she is trisomic for chromosome 21. As a result of carrying an extra chromosome 21, such a child demonstrates the symptoms of Down syndrome: impaired intelligence; characteristic abnormalities of the hands, tongue, and eyelids; and an increased susceptibility to cardiac abnormalities and diseases such as leukemia.

Other abnormal events can also lead to aneuploidy. In a process called translocation, a piece of a chromosome may break away and become attached to another chromosome. For example, a particular large part of one chromosome. For example, a particular large part of one chromosome 21 may be translocated to another chromosome. Individuals who inherit this translocated piece along with two normal chromosomes 21 have Down syndromed.

Other human disorders result from particular chromosomal abnormalities. Sex chromosome aneuploidy causes disorders such as Turner syndrome and Klinefelter syndrome, discussed in Chapter 10 in connection with sex determination. Deletion of a portion of chromosome 5 results in cri du chat (French for "cat's cry") syndrome, so named because an afflicted infant's cry sounds like that of a cat. Symptoms of this syndrome include severe mental retardation.

Trisomies (and the corresponding monosomies) are surprisingly common in human zygotes, but most of the embryos that develop from such zygotes do not survive to birth. Trisomies for chromosomes 13, 15, and 18 greatly reduce the probability that an embryo will survive to birth, and virtually all infants who are born with such trisomies die before the age of 1 year.

Trisomies and monosomies for other chromosomes are lethal to the embryo. At least one-fifth of all recognized pregnancies spontaneously terminate during the first two months, largely because of such trisomies and monosomies. (The actual proportion of spontaneously terminated pregnancies is certainly higher, because the earliest ones often go unrecognized).

Polyploids can have Difficulty in Cell Division

Both diploid and haploid nuclei divide by mitosis. Multicellular diploid and multicellular haploid individuals develop from single-celled beginnings by mitotic divisions. Mitosis may proceed in diploid organisms even when a chromosome from one of the haploid sets is missing or when there is an extra copy of one of the chromosomes (as in Down syndrome).

Under some circumstances triploid ($3n$), tetraploid ($4n$), and higher-order polyploid nuclei from. Each of these ploidy levels represents an increase in the number of complete sets of chromosomes present. If, through accident, the nucleus has one or more extra full sets of chromosomes–that is, if it is triploid, tetraploid, or of still higher ploidy–this abnormally high ploidy in itself does not prevent mitosis. In mitosis, each chromosome behaves independently of the others.

In meiosis, by contrast, chromosomes synapse to begin division. If even one chromosome has no homologous, anaphase I cannot send representatives of that chromosome to both poles. A diploid nucleus can undergo normal meiosis; a haploid one cannot. A tetraploid nucleus has an even number of each kind of chromosome, so each chromosome can pair with its homolog. But a triploid nucleus cannot undergo normal meiosis, because one-third of the chromosomes would lack partners.

This limitation has important consequences for the fertility of triploid, tetraploid, and other chromosomally unusual organisms that may be produced by plant breeding or by natural accidents. Modern bread wheat plants are hexaploids, the result of the accidental crossing of three different grasses, each having its own diploid set of 14 chromosomes.

Cell Death

As we mentioned at the start of the chapter, an essential role of cell division in complex eukaryotes is to replace cells that die. In humans, billions of cells die each day, mainly in blood and epithelia lining organs such as the intestine. Cells die in one of two ways. The first, necrosis, occurs when cells either are damaged by poisons or are starved of essential nutrients. The scab that forms around a wound is a familiar example of necrotic tissue. More typical in an organism is apoptosis, a series of events that constitute genetically programmed cell death.

Why would a cell initiate apoptosis, which is essentially "cell suicide"? One reason is that the cell in question is no longer needed by the organism. For example, before birth a human fetus has weblike hands, with connective tissue between the fingers. As development proceeds, this unneeded tissue disappears as the cells undergo apoptosis.

A second reason for apoptosis is that the longer cells live, the more prone they are to damage that could lead to cancer. This is especially true of cells in the blood and intestine, which are exposed to high levels of toxic substances. In these cases, cells sacrifice their lives to the good of the organism.

Like the cell cycle, the events of apoptosis are very similar in most organisms. The cell becomes isolated from its neighbours, chops up its chromatin into nucleosome-sized pieces, and then fragments itself. In a remarkable example of the economy of nature, the surrounding living cells ingest the remains of the dead cell. The genetic signals that lead to apoptosis are common to many organisms.

5

Nucleic Acids as the Genetic Material

Earth originated about 5 billion years ago and life originated about 3 billion years ago and that too in water on this earth. All life that has appeared on earth is descended from organisms that were originally very simple. The physical and chemical evolution on earth that preceded the appearance of organisms (prebiotic evolution) was followed by an organismal evolution over the last 4 billion years which culminated in organisms of extraordinary complexity, such as ourselves. Note that present-day organisms–those in existence the past few thousand years–would occupy much less horizontal space in figure than the vertical line marking the interval at 0, for even if 1000 vertical lines could be drawn between 0 and –1, each thin line would represent 1000 thousand (that is, 1 million) years!

It has been reported in this latest thin line of time a great variety of organisms–plant and animal, microscopic to mammoth, incredibly diverse–making up some 2 million kinds, or species. These organisms range from viruses and bacteria through protozoans, sponges, corals and jellyfish, flat, round, and segmented worms, shellfish and starfish, spiders and insects, finned fishes, amphibians, reptiles, birds, mammals, algae and fungi, mosses, ferns, and seed plants in a bewildering tangle of forms of life. By collecting information about these various organisms and by organizing such information–that is, by developing the science of biology–we can group facts about organisms, and establish principles and generalities that apply to many species. Such common threads among species are revealed by studying their structue, form, function and composition.

All organisms have a common origin, and all of them require one or more *cells* and many cell products for their structure and function. Cells vary in size and complexity from the tiny and relatively simple cell of a bacterium, about 100 of which can fit across the dot of an *i*, to the giant and relatively complex yolk, which is the single cell of a chicken or an ostrich egg.

Features of a typical cell can be suggested by imagining a plastic bag containing a very porous sponge which is saturated with a thick vegetable soap. This sponge, in turn, surrounds a smaller plastic bag containing noodle soup. The outer plastic bag represents the *cell membrane*, or the outside limit of the cell. The inner plastic bag represents the *nuclear membrane*, the outside limit o the *nucleus*. The sponge represents the *endoplasmic reticulum*, which is a network of membranes (forming channels often interconnected) that also connect the other two membranes. The soups and membranes make up *protoplasm*, which is called *cytoplasm* outside the nucleus. The vegetables in the cytoplasm include various types of bodies (membrane-bound bodies are called *organelles*), such as *ribosomes*, *Mitochondria* and, in green cells, *chloroplasts*. The noodles of the nucleus represent the *chromosomes*. One or a few chromosomes are also present in each mitochondrion and chloroplast. Nucleus-containing cells are *eukaryotic* cells, and organisms composed of such cells are *eukaryotes*. The cells of bacteria and blue-green algae, on the other hand, can be likened to a single plastic bag, the size of a mitochondrion or smaller, containing soup with relatively few vegetables and only a noodle or two. Such cells consists largely of a small mass of protoplasm bounded by a cell membrane, containing ribosomes and one or a few chromosomes. Lacking a nucleus the cells are said to be *prokaryotic*; the cells also lack an endoplasmic reticulum, mitochondria, and chloroplasts. Although all *prokaryotes* are single-celled organisms, eukaryotes may also be composed of single cells, as are protozoans such as amebae or of as many as trillions of cells, as in each human being.

The common features of cell structure and form are accompanied by common features of cell function. The thousands of chemical reactions and physical changes that occur in protoplasm comprise the cell's *metabolism*. Metabolism occurs primarily at sites and surfaces provided by membranes, organelles, and large molecules (including chromosomes). The reactions that synthesize more energy-containing substances or protoplasm are *anabolic;* the reactions which degrade energy-providing substances or protoplasm are *catabolic*. By drawing raw materials and energy from the environment and processing them through metabolism, cells are able to maintain themselves, grow, and divide to produce more cells. All of today's organisms are characterized functionally, therefore, by their capacity for two basic functions; (1) *self-maintenance*, which involves growth, replacement, and/or repair of parts of an organism; and (2) *self-reproduction*, which involves making more of the same kind of organisms.

Requirements for the Genetic Material

The genetic material is of central importance to cell function and therefore must fulfil a number of basic requirements:

1. It must contain the information for cell structure, function, and reproduction in a stable form. This information is encoded in the sequence of basic building blocks of the genetic material.
2. It must be possible to replicate the genetic material accurately such that the same genetic information is present in descendant cells and in successive generations.
3. The information coded in the genetic material must be able to be decoded to produce the molecules essential for the structure and function of cells.
4. The genetic material must be capable of (infrequent) variation. Specifically, mutation and recombination of the genetic material are the foundations for the evolutionary process.

The nucleic acids, *deoxyribonucleic acid* (DNA) and ribonucleic acid (RNA), meet all these requirements.

Presence of Genetic Material

Organisms have unique requirements at the structural level, at the functional level and at the macromolecular level. Each organism must possess a facility or factor that (1) persists during the entire existence of the organism, (2) is repeated in each of its progeny, (3) is different in different organisms, and (4) contains the information needed for synthesizing characteristic proteins and nucleic acids. For purposes of scientific investigation, we assume that we are dealing with a material factor rather than a spiritual one. Since this material factor must contain the instructions for the creation, or genesis, of an organism, we can call it *genetic material*. All the different organisms that have appeared on earth are likely to have had a single ancestor, (1) there is only one basic tye of genetic material, and (2) the genetic material can be changed yet still be reproduced. The single most important feature of all organism is, with this view, genetic material. This material is characterized by being preserved, replicated, capable of replicating its modifications, and contains the information for unique protein and nucleic acid. The study of the properties and functions of genetic material thus comprises the core of the study of all organisms, that is, of the science of biology.

Some mature viruses, or *virions*, are composed only of nucleic acid and protein in combination. Clearly, their genetic material must be one or the other or some combination of these two types of macromolecule. A determination can be made through the infection experiments described in the next section.

A Genetic View

The concept of the gene has been the focus of some hundred years of work to establish the basis of heredity. A gene is a sequence of DNA that carries the information representing a protein. Until very recently this would have been an adequate (if incomplete) bio-chemical description. The sequence of DNA could be identified as a continuous stretch of nucleotides, related to the protein sequence by a colinear read-out. But now it is clear that the sequence representing protein is not always continuous; it may be interrupted by sequences not concerned with specifying the protein. So genes may be in pieces that are put together during the process of gene expression. The large number or genes that make up the genome of any species are organizd into a comparatively small number of chromosomes. The genetic material of each chromosome consists of an extremely long stretch of DNA, containing many genes in a linear order. How many genes are present in total has been a puzzle for a long time.

Recently the view that each gene may reside by itself as a unique entity has been superseded by the realization that, in many cases, there may be clusters of related genes that constitute small families. The genome generally has been viewed as rather stable, subject to changes in overall constitution and organization only on an extraordinarily slow evolutionary time scale. This contrasts with recent evidence that in some instances there may be rearrangements that occur regularly; and there may be components of the genome that are relatively mobile. The concept of the gene has therefore undergone an evolution in which, although many of its traditional properties remain, exceptions have been found to show that none constitutes an absolute rule. Starting from the discovery of the gene as a fixed unit of inheritance, its properties were defined in terms of its residence at a definite position on the chromosome, this in turn leading to the view that the genetic material of the chromosome is a continuous length of DNA representing many genes.

A gene may exist in alternative forms that result in the expression of a different characteristic (such as red versus white flower colour). These forms are called *alleles*. The law of independent segregation states that these alleles do not affect each other when present in the same plant, but segregate unchanged by passing into different gametes when the next generation forms. In a true-breeding organism, a *homozygote*, both alleles are the same. But a mating between two parents each of which is homozygous for a different allele generates a hybrid or *heterozygote*. If one allele is *dominant* and the other is *recessive*, the organism will have the appearance or *phenotype* only of the dominant type (so the heterozygote is indistinguishable from the true-breeding dominant parent). But Mendel's first law recognizes that the genetic constitution or *genotype* of the hybrid comprises the presence of both alleles. This is revealed, when the hybrid is crossed with another hybrid to form the second generation. The critical point is that the alleles do not mix, but are physical entities whose interaction is at the level of expression.

Alleles may exhibit *incomplete* (*partial*) *dominance* or no dominance (sometimes known as *codominance*). In the latter case the heterozygote is distinguished from the homozygotes by properties that are intermediate between them. In the snap-dragon, for example, a cross between red and white generates hybrids with pink flowers. Although there is no dominance, the same rule is observed that the first hybrid (F_1) generation is uniform; and the same ratios are generated in a further hybrid cross, although three phenotypes can be distinguished instead of two.

The Independence of Different Genes

Mendel's second law which is the independent assortment summarizes of different genes. When a plant that is dominant for two different characters is crossed with a parent that is recessive for both, as before the F_1 consists of plants uniformly of the dominant type. But in the next hybrid cross, two classes of plants are seen. One consists of the two *parental types*. The other consists of new phenotypes, representing plants with the dominant feature of one parent and the recessive feature of the other. These are called *recombinant types*; and they occur in both possible (*reciprocal*) combinations. The ratios in which these occur can be explained by supposing that gamete formation involves an entirely random association of one of the alleles for the first character with one of the alleles for the other.

All four possible types of gamete are formed in equal proportion, and associate at random in forming the zygotes of the next generation. Once again, the characteristic ratio of phenotypes conceals a greater variety of genotypes, which can be confirmed by the *backcross* to the recessive parent. Essentially the backcross provides a method for looking directly at the genotype of the organism that is being examined.

The law of independent assortment therefore says that the behaviour of any pair (or greater number) of genes can be predicted overall by the rules of mathematical combination. Thus the assortment of one gene does not influence the assortment of the other. Implicit in this concept is the view that assortment is a matter of statistical probability and not an exact result. The ratios of progeny types will approximate increasingly closely to the predicted numbers as a greater number of crosses are performed.

Chromosomes in Heredity

When the chromosomal theory of inheritance was simultaneously proposed by Sutton and by Boveri, it resolved the discussion that had been continuing for some time on the role of chromosomes. They had already been implicated in heredity, although in a somewhat hazy manner, in 1903 it was realized that their properties corresponded exactly with those ascribed to Mendel's particulates units of inheritance. The cell theory established in the middle of the nineteenth century proposed that all organisms are composed of cells, and that these can arise only from preexisting cells.

Early cytology showed that a "typical" cell consists of a dense nucleus separated by a membrane from the less-dense surrounding cytoplasm. Within the nucleus, the granular region of *chromatin* could be recognized by its reaction with certain stains. Not long after Mendel's work, it was found that the chromatin consists of a discrete number of thread-like particles, the *chromosomes*. Chromosomes can be visualized in most cells only during the process of cell division.

The two types of division in sexually reproducing organisms explain both the perpetuation of the genetic material and the process of inheritance as predicted by Mendel's laws. During growth of the organism, the *cell cycle* falls into two parts. The long period of *interphase* represents the time during which the cell engages in its synthetic activities and reproduces its components. Then the short period of *mitosis* is an interlude during which the actual process of division into two daughter cells is accomplished. The products of the series of mitotic divisions that generate the entire organism are called the *somatic cells*. At the end of mitosis, each daughter cell can be seen to start its life with two copies of each chromosome. These are called *homologous.*

The total number of chromosomes in what is known as the *diploid set* can therefore be described as *2n*. The typical somatic cell exists in the diploid state. During interphase, a growing cell duplicates its chromosomal material. This is not evident at the time and becomes apparent only during the subsequent mitosis. During mitosis, each chromosome appears to split longitudinally to generate two copies. These are called *sister chromatids*. At this point, the cell contains *4n* chromosomes, organized as *2n* pairs of sister chromatids. In other words, there are two (homologous) copies of each sister chromatid pair. The series of events, by which mitotic division is accomplished. Essentially the sister chromatids are pulled toward opposite poles of the cell, so that each daughter cell receives one member of each sister chromatid pair. Now these are chromosoems in their own right. The *4n* chromosomes present at the start of division have been divided into two sets of *2n* chromosomes. This process is repeated in the next cell cycle. Thus mitotic division ensures the constancy of the chromosomal complement in the somatic cells.

Meiosis generates cells that contain the *haploid* chromosome number, *n*. This involves two, successive divisions. Once again, the chromosomes have been duplicated previously, so that the cell enters division with *4n*. At the time of first division, the homologous pairs of sister chromatids *synapse* or *pair* to form *bivalents*. Each bivalent contains all four of the cell's copies of one homologue. The first division causes each bivalent to segregate into its two component sister chromatid pairs. This generates two sets of *2n* chromosomes, each set consisting of *n* sister chromatid pairs.

Now the second meiotic division follows, in which both of the sets of *2n* divide again. This division resembles the mitotic division, since one member of each sister chromatid pair segregates to a different daughter cell. The overall result of meiosis is to divide the starting number of *4n* chromosomes into four haploid cells. These may then give rise to mature eggs or sperm. In forming these gametes, homologous of paternal and maternal origin are separated, so that each gametes gains only one of the two homologues of its parent.

A critical feature in relating this process to the predictions of Mendel's laws was the realization that nonhomologous chromosomes undergo segregation independently, so that either member of one homologous pair enters the gamete at random with either member of a different homologous pair.

Genes occur in allelic pairs, one member of each pair having been contributed by each parent; the diploid set of chromosomes results from the contribution of a haploid set by each parent. The assortment of nonallelic genes into gametes should be independent of origin, nonhomologous chromosomes undergo independent segregation. The critical provision is that each gamete obtains a complete haploid set, and this is fulfilled whether viewed in terms of factors or chromosomes.

ARRANGEMENT OF GENES

Linear

The proof that genes reside on chromosomes required a demonstration that a particular gene is always present on a particular chromosome. This was provided by the properties displayed by a variant of the fruit fly *Drosophila melanogaster* obtained by Morgan in 1910. This white-eyed male appeared spontaneously in a line of flies of the usual red eye colour.

The red eye colour is the *wild type* of character. The white eye is the *mutant* phenotype. The event responsible for generating the mutant phenotype is a genetic change called a *mutation.* Most mutations prevent the gene in which they occur from functioning properly or at all; and so the study of mutation is for the most part the study of inactive alleles.

Sometimes it is possible for a mutation to be reversed by a genetic change that restores the original state of the genetic material. This is called a *reversion* to the wild type. The white mutation could be located on a particular chromosome because of its association with sexual type.

In many sexually reproducing organisms there is an exception to the rule that chromosomes occur in homologous pairs whose separation (disjunction) at meiosis produces identical haploid sets. Male and female sets may differ visibly in chromosome constitution, the most common form of difference being the replacement of one member of a homologue pair with a different chromosome in one of the sexes. This pair is referred to as the *sex chromosomes*, and the remaining homologous pairs are called the *autosomes.*

The chromosome complements of the two sexes can be described as 2A + XX and 2A + XY, where the haploid set of autosomes is denoted A and the two sex chromosomes are X and Y. The sex with the complement of 2A + XX is called *homogametic*; if forms gametes only of the type A + X. The sex with the complement of 2A + XY is called *heterogametic*; it forms equal proportions of gametes of the types A + X and A + Y. The random union of gametes from one sex with gametes from the other sex perpetuates the equal sex ratio at zygote formation. In *Drosophila* the females are homogametic.

A critical prediction of Mendel's laws is that the results of a genetic cross should be the same regardless of orientation– that is, irrespective of which parent introduces which allele. But the reciprocal crosses with white eye in *Drosophila* give different results.

The cross of white male × red female gives the entirely red F_1 expected if red is dominant and white is recessive. But in the F_2 all the white-eye flies that reappear are males. In the reciprocal cross of red male × white female, all the F_1 males are white-eyed and all the females are red eyed. Crossing these gives an F_2 with equal proportions of white and red eyes in each sex. This pattern of inheritance exactly follows that of the sex chromosomes. If the alleles for red and white eyes are carried on the X chromosome, with no locus for eye colour present on the Y chromosome, the phenotype of a male will be determined by the single allele present on its X chromosome. This allele will be transmitted to all of its daughters and none of its sons. This is the typical pattern of *sex linkage.*

The separation of chromosomes seen at meiosis explains the independent assortment of genes that are carried on different chromosomes. But the number of genetic factors is much greater than the number of chromosomes. Each chromosome may bear as many as 1000 to 10000 genes. The quickening pace of genetics after the turn of the century was shown by the passage of only three years between the report of the first eye-colour mutant and Stretevant's study in 1913 of the pattern of inheritance of six sex-linked mutations. Morgan had shown in 1911 that each of these factors shows the same sex-linked pattern as white/red eye colour. By the same logic each must therefore be carried by the X chromosome.

Morgan proposed that the cause of genetic linkage is the simple mechanical result of the location of the factors in the chromosomes. He suggested that the production of genetic recombinant classes can be equated with the process of *crossing over* that is visible during meiosis. Early in meiosis, at a stage when there are four copies of each chromosome organized in a bivalent, pairwise exchanges of material occur between the closely associated (synapsed) homologue pairs. This exchange is called a *chiasma.*

In maize and in *Drosophila,* suitable *translocations* had occurred in which a part of one chromosome had broken off and become attached to another. This allows the translocation chromosome to be distinguished by its appearance from the normal chromosome. In suitable crosses, it is then possible to show that the formation of genetic recombinants occurs only when there has been a physical crossing-over between the appropriate chromosome regions. As the distance between the genes increases, the probability of crossing-over between them will increase. Thus if crossing-over is responsible for recombination, genes lying near each other will be tightly linked, and genetic linkage will decrease with physical distance apart. Reversing the argument, genetic linkage can be taken to be measure of physical distance.

The concept that genes on the same chromosome are linked was extended by Sturtevant with the proposal that the extent of recombination between them can be used as a *map distance* to measure their relative locations. This is expressed as the percent recombination; that is,

$$\text{Map distance} = \frac{\text{Number of recombinants} \times 100}{\text{Total number of progeny}}$$

Distance is given in terms of *map units,* defined by 1 map unit (or centiMorgan) equal 1% recombination.

Thus if two genes A and B are 10 units apart, and it is 5 units from B to a further gene C, the direct measure of distance between A and C will be close to 15. The genes can therefore be placed in *linear order.* A crucial point in the construction of maps is that the distance between genes does not depend on the *alleles* that are used, but only on the gene *loci.* The locus defines the *position* on the chromosome at which the gene for a particular trait resides; the various alternative forms of the gene–that is, the alleles used in mapping–all reside at the *same* location.

The genetic mapping is concerned with identifying the locations of gene loci, which are fixed and in a linear order. In a mapping experiment the same result is obtained irrespective of the particular combination of alleles. Sturtevant concluded that his results "form a new argument in favour of the chromosome view of inheritance, since they strongly indicate that the factors investigated are arranged in a linear series, at least mathematically."

This last qualification is interesting. Although a genetic map can be constructed that represents a chromosome as a linear array of genes, this does not prove that the genetic content of the chromosome is *physically* a continuous array of genes. Many other models were considered before it became clear that a chromosome contains a single thread of genetic material. Linkage is not displayed between all pairs of genes located on a single chromosome.

The maximum recombination that occurs in the 50% predicted by Mendel's second law. (Although there is a high probability that recombination will occur between two genes lying far apart on a chromosome, each individual recombination event involves only two of the four associated chromosomes, thus generating 50% crossover between the genetic factors).

The factors that are well separated may show no direct linkage, in spite of their presence on the same chromosome. However, each can be linked via intermediate factors, and so a genetic map can be extended beyond the limit of 50 map units directly measurable between any pair of genes.

In fact, because of the distortion of linkage measured directly between markers that are not closely linked, map distances usually are based on measurements between fairly close factors, and may be subject to corrections from the simple percent recombination. The *linkage group* includes all those genes that can be connected either directly or indirectly by linkage relationships. Those close together show direct linkage; those more than 50 map units apart in practice assort independently.

As linkage relationships are extended, each gene identified in an organism can be placed into linkage with a group of previously mapped genes. Thus the genes fall into a discrete number of linkage groups. Genes in one linkage group always show independent assortment with regard to genes located in other linkage groups. The number of linkage groups is the same as the number of chromosomes.

The relative lengths of the linkage groups are similar to the actual relative sizes of the chromosomes. The example of *Drosophila* (where it happens to be particularly easy to measure the chromosome lengths). Mendel's concept of the gene as a discrete particulate factor can therefore be extended into the concept that the chromosome constitutes a linear unit divided into many genes, whose physical arrangement may underlie their genetic behaviour.

IDENTIFICATION OF DNA AS THE GENETIC MATERIAL

Two classic experiments led to the identification of DNA as the genetic material and, in so doing, laid the foundation for molecular genetics. These experiments are described in the following section.

The Transformation Experiments

The development of the current idea that DNA is the genetic material began with an observation in 1928 by Fred Griffith, who was studying the bacterium responsible for human pneumonia–i.e., *Streptococcus pneumoniae* or *Pneumonoccus.* The virulence of this bacterium

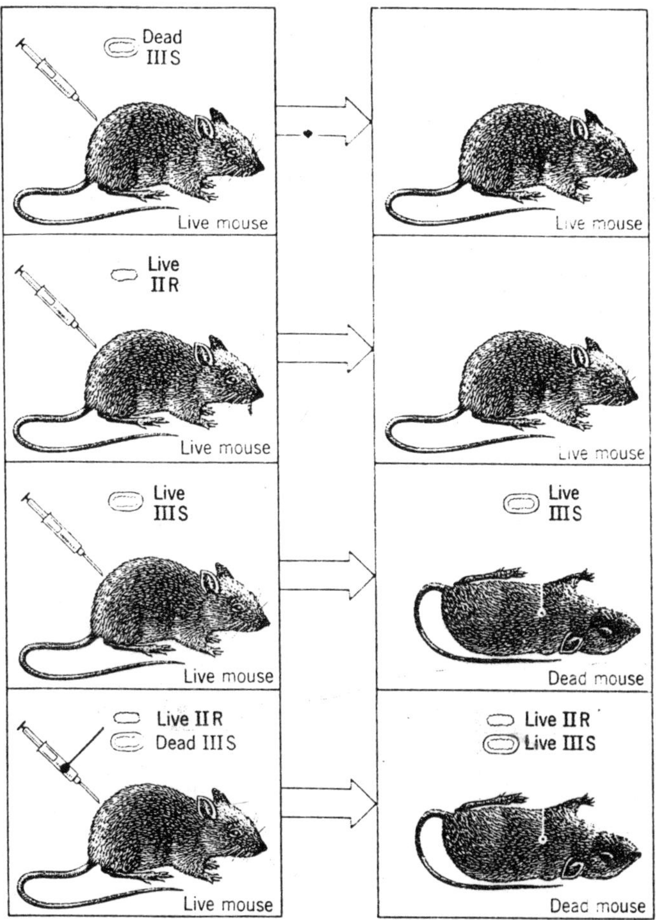

Fig. 5.1. Griffith's demonstration of transformation in pneumococcus.

was known to be dependent on a surrounding polysaccharide capsule that protects it from the dense systems of the body. This capsule also causes the bacterium to produce smooth-edged (S) colonies on an agar surface. It was known that mice were normally killed by S bacteria. Griffith then isolated a rough-edged (R) colony mutant, which proved to be both nonencapsulated and nonlethal. He subsequently made a significant observation–namely, whereas both R and heat-killed S were nonlethal, a mixture of live R and heat-killed S was lethal. Furthermore, the bacteria isolated from a mouse that had died from such a mixed infection were only S–i.e., the live R had somehow been replaced by or *transformed* to S bacteria. Several years later it was shown that the mouse itself was not needed to mediate this transformation because when a mixture of R and heat-killed S was grown in a culture fluid, living S cells were produced. A possible explanation for this surprising phenomenon was that the R cells restored the viability of the dead S cells; but this idea was eliminated by the observation that living S cells grew even when the heat-killed S culture in the mixture was replaced by a cell extract prepared from broken S cells, which had been freed from both intact cells and the capsular polysaccharide by centrifugation. Hence, it was concluded that the cell extract contained a *transforming principle*, the nature of which was unknown.

15 years later when Oswald Avery, Colin MacLeod, and Maclyn McCarty partially purified the transforming principle from the cell extract and demonstrated that it was DNA. These workers modified known schemes for isolating DNA and prepared samples of DNA from S bacteria. They added this DNA to a live R bacterial culture; after a period of time they placed a sample of the S-containing R bacterial culture on an agar surface and allowed it to grow to form colonies. Some of the colonies (about 1 in 10^4) that grew were S type.

To show that this was a permanent genetic change, they dispersed many of the newly formed S colonies and placed them on a second agar surface. The resulting colonies were again S type. If an R colony arising from the original mixture was dispersed, only R bacteria grew in subsequent generations. Hence the R colonies retained the R character, whereas the transformed S colonies bred true as S. Because S and R colonies differed by a polysaccharide coat around each S bacterium, the ability of purified polysaccharide to transform was also tested, but no transformation was observed. Since the procedures for isolating DNA then in use produced DNA containing many impurities, it was necessary to provide evidence that the transformation was actually caused by the DNA alone.

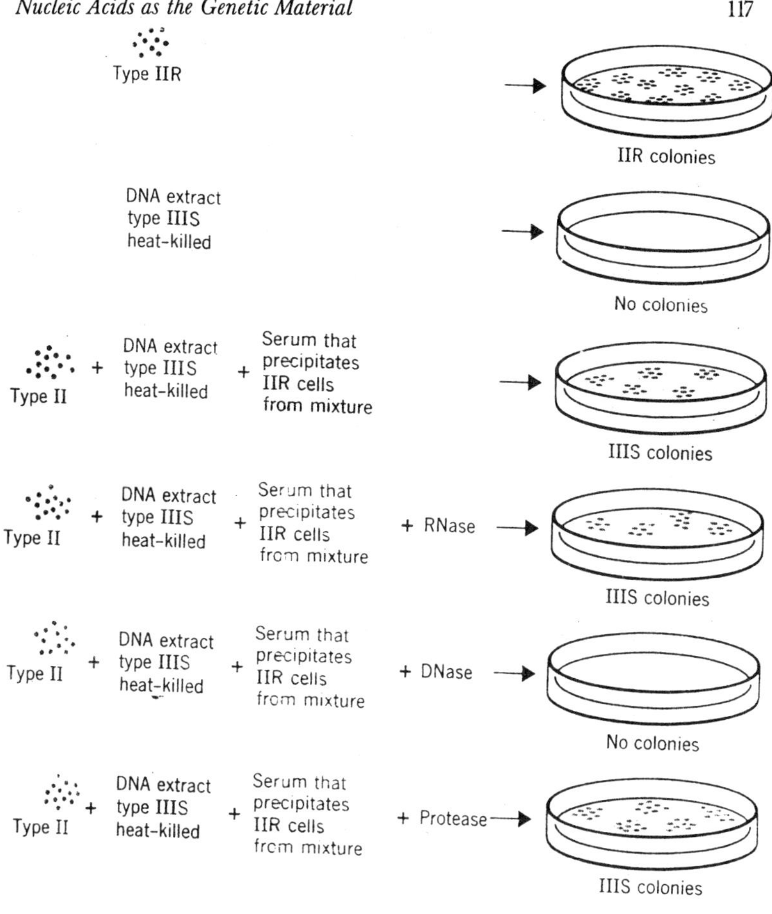

Fig. 5.2. Avery, MacLeod, and McCarty's proof that the "transforming principle" is DNA.

This evidence was provided by the following four procedures.

1. Chemical analysis showed that the major component was a deoxyribose-containing nucleic acid.
2. Physical measurements showed that the sample contained a highly viscous substance having the properties of DNA.
3. Experiments demonstrated that transforming activity is not lost by reaction with either (a) purified proteolytic (protein-hydrolyzing) enzymes–trypsin, chymotrypsin, or a mixture of both–or (b) ribonuclease (an enzyme that depolymerizes RNA).
4. It was demonstrated that treatment with materials known to contain DNA-depolymerizing activity (DNase) inactivated the transforming principle.

The transformation experiment was not accepted by the scientific community as proof that DNA is the genetic material, because it was widely believed that DNA was a simple tetranucleotide incapable of carrying the information required of a genetic substance.

The tetranucleotide hypothesis was based on chemical analyses that indicated that DNA consists of equimolar amounts of the four bases; this conclusion was based on inadequate chemical procedures for analyzing base composition and on the use of higher organisms as

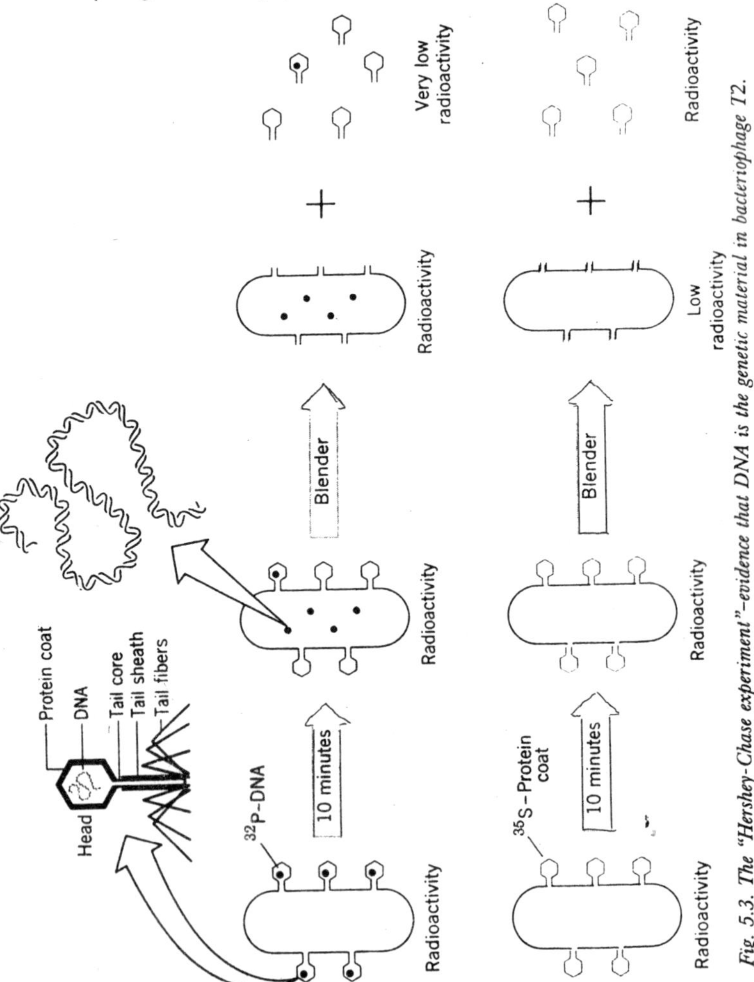

Fig. 5.3. The "Hershey-Chase experiment"—evidence that DNA is the genetic material in bacteriophage T2.

sources of DNA; in these organisms the base composition is not far from being equimolar. However, as techniques improved and a greater range of organisms was examined, it was found that the base composition varies widely from one species to the next and that DNA is not a simple small molecule. With these results, the idea that DNA is the genetic material became acceptable. The following experiment provided the final proof.

The Blendor Experiment

An elegant confirmation of the genetic nature of DNA came from an experiment with *E.coli* phage T2. This experiment, known as the *blendor experiment* because a kitchen blendor was used as a major piece of apparatus, was performed by Alfred Hershey and Martha Chase, who demonstrated that the DNA injected by a phage particle into a bacterium contains all of the information required to synthesize progeny phage particles. A single particle of phage T2 consists of DNA (now known to be a single molecule) encased in a protein shell.

The DNA is the only phosphorus-containing substance in the phage particle; the proteins of the shell, which contain the amino acids methionine and cysteine, have the only sulfur atoms. Thus, by growing phage in a nutrient medium in which radioactive phosphate ($^{32}PO_4^{3-}$) is the sole source of phosphorus, phage containing radioactive DNA can be prepared. If instead the growth medium contains radioactive sulfur as $^{35}SO_4^{3-}$, phage containing radioactive proteins are obtained. If these two kinds of labeled phage are used in an infection of a bacterial host, the phage DNA and the protein molecules can always be located by their radioactivity. Hershey and Chase used these phage to show that ^{32}P but not ^{35}S is injected into the bacterium. Each phage T2 particle has a long tail by which it attaches to sensitive bacteria.

Hershey and Chase showed that an attached phage can be torn from a bacterial cell wall by violent agitation of the infected cells in a kitchen blendor. Thus, it was possible to separate an adsorbed phage from a bacterium and determine the component(s) of the phage that could not be shaken free by agitation–presumably, those components had been injected into the bacterium.

In the first experiment ^{35}S-labeled phage particles were adsorbed to bacteria for a few minutes. The bacteria were separated from unadsorbed phage and phage fragments by centrifuging the mixture and collecting the sediment (the pellet), which consisted of the phage-bacterium complexes. These complexes were resuspended in liquid

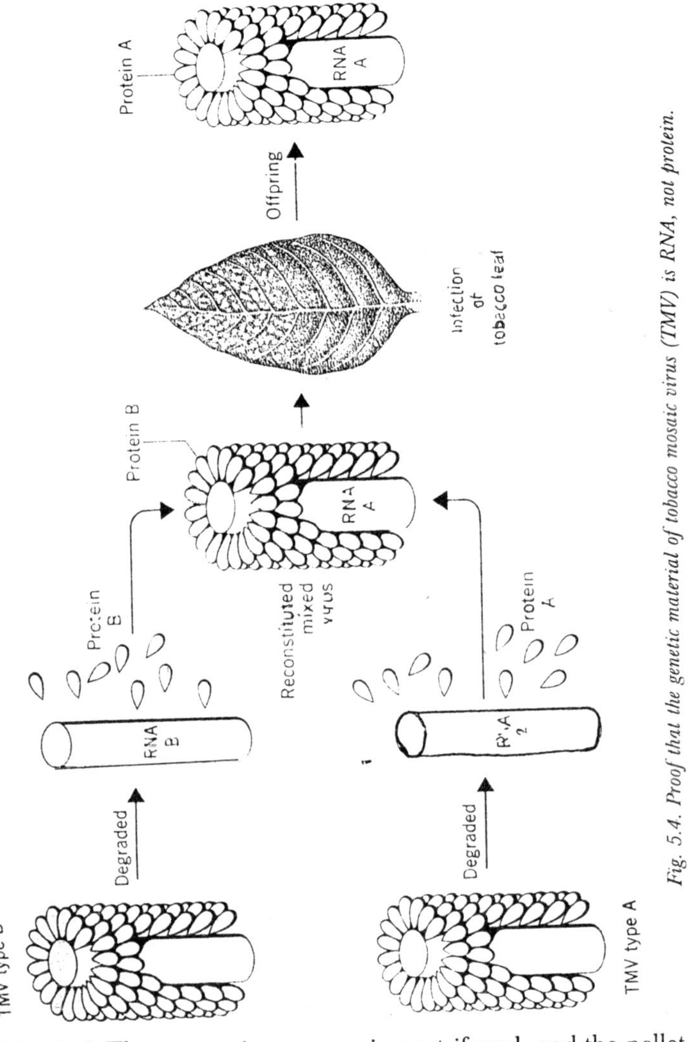

Fig. 5.4. Proof that the genetic material of tobacco mosaic virus (TMV) is RNA, not protein.

and blended. The suspension was again centrifuged, and the pellet, which now consisted almost entirely of bacteria, and the supernatant were collected. It was found that 80 percent of the ^{35}S label was in the supernatant and 20 percent was in the pellet. The 20 percent of the ^{35}S that remains associated with the bacteria was shown some years later to consist mostly of phage tail fragments that adhered too tightly to the bacterial surface to be removed by the blending. A very different result was observed when the phage population was labeled with ^{32}P.

In this case 70 percent of the ^{32}P remained associated with the bacteria in the pellet after blending and only 30 percent was in the supernatant.

Of the radioactivity in the supernatant roughly one-third could be accounted for by breakage of the bacteria during the blending. (The remainder was shown some years later to be a result of defective phage particles that could not inject their DNA.) when the pellet material was resuspended in growth medium and reincubated, it was found to be capable of phage production. Thus, the ability of a bacterium to synthesize progeny phage is associated with transfer of ^{32}P, and hence of DNA, from parental phage to the bacteria.

Transfer Experiments

Another series of experiments, known as *transfer experiments*, supported the interpretation that genetic material contains ^{32}P but not ^{35}S. In these experiments progeny phage were isolated, after blending, from cells that had been infected with either ^{35}S- or ^{35}P-containing phage and the progeny were then assayed for radioactivity, the idea being that some parental genetic material should be found in the progeny. It was found that no ^{35}S but about half of the injected ^{32}P was transferred to the progeny. This result indicated that though ^{35}S might be residually associated with the phage-infected bacteria, it was not part of the phage genetic material. The interpretation (now known to be correct) of the transfer of only half of the ^{32}P was that progeny DNA is selected at random for packaging into protein coats and that all progeny DNA is not successfully packaged.

Properties of a Genetic Material

To comprise the genes, DNA must carry the information to control the synthesis of the enzymes and proteins within a cell or organism, self-replicate with high fidelity yet show a low level of mutation, and be located in the chromosomes.

Control of the Enzymes

The growth, development, and functioning of a cell are controlled by the proteins within it, primarily its enzymes. Thus the nature of a cell's phenotype is controlled by the protein synthesis within that cell. The genetic material must determine the presence and effective amounts of the enzymes in a cell. For example, given inorganic salts and glucose, an *E.coli* cell can synthesize, through its enzyme-controlled biochemical pathways, all of the compounds it needs for growth, survival, and reproduction.

An enzyme is a protein that acts as a catalyst of a specific metabolic process without itself being markedly altered by the reaction.

Most reactions that are catalyzed by enzymes could occur anyway, but only under conditions too extreme for them to take place within living systems. For example, many oxidations occur naturally at high temperatures. Enzymes allow these reactions to occur within the cell by lowering what is called the *activation energy* (ΔG) of a particular reaction. Most metabolic processes, such as the biosynthesis or degradation of molecules, occur in pathways in which an enzyme facilitates each step in the pathway.

Each reaction product in the pathway is altered by an enzyme that converts it to the next product. The enzyme threonine dehydratase, for example, converts threonine into a-ketobutyric acid. Enzymes are composed of folded polymers of amino acids; the sequence of amino acids determines the final structure of an enzyme. The genetic material determines the sequence of the amino acids.

The three-dimensional structure of enzymes permits them to perform their function. An enzyme combines with its substrate or substrates (the molecules it works on) at a part of the enzyme called the *active site*. The substrates "fit" into the active site, which has a shape that allows only the specific substrates to enter. This view of the way an enzyme interacts with its substrates is called the *lock-and-key model* of enzyme functioning. When the substrates are in their proper position in the active site of the enzyme, the particular reaction that the enzyme catalyzes takes place. The reaction products then separate from the enzyme and leave it free to repeat the process.

Enzymes can work at phenomenal speeds. Some can catalyze as many as a million reactions per minute. Not all of the cell's proteins function as catalysts. Some are structural proteins, such as keratin, the main component of hair. Other proteins are regulatory–they control the rate of production of other enzymes. Still others are involved in different functions; albumins, for example, help regulate the osmotic pressure of blood.

Replication

The genetic material must replicate itself precisely so that every daughter cell receives an exact copy. Some *mutability*, or the ability to change, is also required because we know that the genetic material has changed, or evolved, in the history of life on earth. In their 1953 paper, Watson and Crick had already worked out the replication process based on the structure of DNA. Mutability is also a natural derivative of this process.

Location

It has been known since the turn of the century that genes, the discrete functional units of genetic material, are located in chromosomes within the nuclei of eukaryotic cells: The behaviour of chromosomes during the cellular division stages of mitosis and meiosis mimics the behaviour of genes. Thus the genetic material in eukaryotes must be a part of the chromosomes.

For a long time, proteins were considered the most probable cell components to be genetic material because they have the necessary molecular complexity. Amino acids can be combined in an almost unlimited variety, creating thousands and thousands of different proteins. The first proof that the genetic material is deoxyribonucleic acid (DNA) was provided in 1944 by Oswald Avery and his colleagues. The Watson and Crick model in 1953 ended a period when DNA was thought to be the genetic material, but its structure was unknown.

Storage and Transmission of Genetic Information by DNA

The information possessed and conveyed by the DNA of a cell is of several types:

1. The sequence of amino acids in every protein synthesized by the cell.
2. A start and a stop signal for the synthesis of each protein.
3. A set of signals that determine whether a particular protein is to be made and how many molecules are to be made per unit time.

This information is contained in the sequence of DNA bases. The amino acid sequence and the start and stop signals are not obtained directly from the DNA base sequence but via RNA intermediates. That is, DNA serves as a template for synthesis of specific RNA molecules called messenger RNA. The base sequence of the mRNA is complementary to that of one of the DNA strands and it is form the mRNA sequence that the amino acid sequence is translated. In particular, the base sequence is "read" in order in groups of three bases called *triplets* or *codons* and each group corresponds to a particular amino acid or to a start or stop codon.

The two-stage process has the advantage that the DNA molecule neither has to be used very often nor has to enter the protein-synthetic mechanism. Since protein synthesis occurs continually, one can appreciate why DNA evolved with bases having groups capable of forming hydrogen bonds. First, by specific patterns of hydrogen-bonding–that is, base-pairing–the base sequence of DNA is easily transcribed into an RNA molecule by means of an enzymatic system that adds a base to the polymerizing end of an RNA molecule only if that base can hydrogen-bond with the base being copied. The use of hydrogen-bonding enables the cell to use less genetic material to hold information than if van der Waals forces had been selected in the course of evolution; this is because van der Waals forces are so weak that the error frequency in transmitting information would be much greater than with hydrogen bonds unless more bases (or other molecules) were used for complementary binding.

Transmission from Parent to Progeny

When a cell divides, each daughter cell must contain identical genetic information; that is, each DNA molecule must become two identical molecules, each carrying the information that was contained in the parent molecule. This duplication process is called *replication.* Once again, the value of bases capable of hydrogen-bonding is apparent. That is, to make a replica, the replication system need only require that the base being added to the growing end of a new chain be capable of hydrogen-bonding to the base being copied.

Chemical Stability of DNA

In long-lived organisms a single DNA molecule may have to last one hundred years or more. Furthermore, the information contained in the molecule is passed on to successive generations for millions of years with only small changes. Thus, DNA molecules must have great stability. The sugar-phosphate backbone of DNA is extremely stable. The C–C bonds in the sugar are resistant to chemical attack under all conditions other than strong acid at very high temperatures.

The phosphodiester bond is a little less stable and can be hydrolyzed at room temperature at pH 2, but this is not a physiological condition. In considering the stability of the phosphodiester bond, one can see immediately why 2´-deoxyribose rather than ribose is the constituent sugar in DNA. The phosphodiester bond in RNA is rapidly broken in alkali. The chemical mechanism for this breakage requires the OH group on the 2´ carbon of the sugar. In DNA, because deoxyribose is used, there is no 2´-OH group, so the molecule is exceedingly resistant to alkaline hydrolysis. The N-glycosylic bond is also very stable, though it would not be so if the bases were not rings. An alteration in the chemical structure of a base definitely means loss of genetic information.

In a cell there are certainly a large number of chemical compounds that can attack a free base. In considering this problem one can see immediately the value of the double helix. One aspect of this is that the molecule isredundant in the sense that identical information is contained in both strands; that is, the base sequence of one strand is complementary to that of the other strand. In fact, there exist in cells elegant repair systems that can remove an altered base and then by reading the sequence on the complementary strand, can replace the correct base. The more important aspect of this double-helical structure is that the nature of the bases and the duplex structure of DNA provides extreme protection against chemical attack.

The bases are hydrophobic rings having charged groups that contain the genetic information. It is therefore the charged groups that need protection. The hydrophobic nature of the base causes the bases to stack so extensively that water is almost completely excluded from the stacked array. This has the effect that water-soluble compounds are often unable to come into close contact with the "dry" stack of bases. The bases themselves, with the exception of cytosine, are very stable. At a very low rate, however, cytosine is deaminated to form uracil.

Deamination is a disastrous change because the deamination product, uracil, pairs with adenine rather than with guanine. This has two effects: (1) an incorrect base will appear in mRNA and (2) an adenine instead of a guanine will occur in newly replicated DNA strands. There exists an intracellular system, for removing uracil from DNA and replacing it with thymine. The necessity for eliminating a uracil formed by deamination explains why DNA utilizes thymine and not uracil as the base complementary to adenine. If uracil were the normal DNA base, there would be no way to distinguish a correct uracil from an incorrect uracil produced by deamination of cytosine. By using thymine, the cell follows the rule: Always remove a uracil from DNA because it is unwanted.

It is not obvious why RNA uses uracil and not thymine nor why DNA evolved with cytosine rather than with some base that would not deaminate. This may have been an evolutionary accident. It is possible, though, that the original DNA and RNA molecules both contained cytosine and uracil because these were the only pyrimidines available in the primordial sea. Cells then gained the ability to methylate uracil to form thymine because this provides a cell with a criterion for eliminating the result of a cytosine→uracil conversion. It is significant that the final step in the synthesis of thymidylic acid is the methylation of deoxyuridylic acid. The thymidylic acid is then converted to the triphosphate needed for incorporation into DNA.

Mutation

The process by which a base sequence changes is called *mutation.* There are two main mechanisms of mutation:

1. A *chemical alteration* of the base that gives it new hydrogen-bonding properties and causes a new base to be present in a newly replicated daughter molecule.
2. A *replication error* by which an incorrect base or an extra base is accidentally inserted in the daughter molecules.

On the average, mutational changes are deleterious and lead to cell death. Therefore, it is important that not too many mutations occur in a single DNA molecule because otherwise the rare advantageous alteration would always occur in a cell destined to die by virtue of lethal mutations. Thus, the mutation rate must be low or controlled. This is accomplished in two ways. First, the hydrophobic, water-free core of the DNA molecule reduces the accessibility of the DNA to attacking molecules, as we discussed earlier. Second, the cell has evolved several repair mechanisms for correcting alterations and replication errors. These repair systems are not entirely efficient though and allow mutations to occur at a rate that is very low but useful in the long run.

The mutations are usually deleterious, so that it is important that the parental information is not lost. Such loss is prevented in two ways: (1) The other members of the species retain the parental base sequence and (2) a double-stranded molecule is redundant. Normally only one strand is altered, and DNA replicates in such a way that after cell division the DNA molecule in each daughter cell receives only one of the parental single strands. Thus, it is possible for one of the daughter DNA molecules to be normal and the other to be mutant; the mutant may not be able to survive but the cell with the parental base sequence will. If the mutant is better equipped to survive and multiply than the parent organisms or any other member of the species, then after a great many generations, Darwin's principle of survival of the fitest will lead to ultimate replacement of the parental genotype by the mutant phenotype in nature.

RNA as Genetic Material in Small Viruses

As more and more viruses were identified and studied, it became clear that many of them contain RNA and proteins, but no DNA. In all cases so far studied, it is clear that these "RNA viruses" store their genetic information in nucleic acids rather than in proteins just like all other organisms, although in these viruses the nucleic acid in RNA. One of the first experiments that established RNA as the genetic material in RNA viruses was the so-called reconstitution experiment of H. Fraenkel-Conrat and B. Singer, published in 1957. Fraenkel-Conrat and Singer's simple, but definitive, experiment was done with tobacco mosaic virus (TMV), a small virus composed of a single molecule of RNA encapsulated in a protein coat. Different strains of TMV can be identified on the basis of differences in the chemical composition of their protein coats.

By using the appropriate chemical treatments, one can separate the protein coats of TMV from the RNA. Moreover, this process is reversible; by mixing the proteins and the RNA under appropriate conditions, "reconstitution" wall occur, yielding complete, infective TMV particles. Raenkel-Contrat and Singer took two different strains of TMV, separated the RNAs from the protein coats, and reconstituted "mixed" viruses by mixing the proteins of one strain with the RNA of the second strain, and vice versa. When these mixed viruses were used to infect tobacco leaves, the progeny viruses produced were always found to be phenotypically and genotypically identical to the parent strain from which the RNA had been obtained. Thus, the genetic information of TMV is stored in RNA, not protein.

6

Nucleic Acids

The life in a cell is due to nucleic acid and the most interesting thing is that the nucleic acid itself is non-living. *Friedrich Miescher* in (1868) isolated a material from pus cells by digesting them for weeks with dilute HCl and called this material as *nuclein. Hoppe Segler* and his co-workers confirmed the work of *Miescher* and also proved the presence of nuclein in yeast and the erythrocytes of the birds and reptiles and various other tissues. The term *nucleic acid* was introduced by *Altmann. Albrecht Kossel, P.A. Levine* and *Walter Jones* described the chemical nature of nucleic acid. According to them the nucleic acid is composed of phosphoric acid, a sugar and nitrogenous bases. *Franklin W. Stahl* presented first evidence that nucleic acid forms the genetic material. *Chargaff* (1951) described the occurrence of nitrogenous bases in equal proportions. *Dotty* (1961) has emphasized much on the physical properties of DNA. In 1962 *Willkins, Klatson* and *Crick* proposed a model to explain the double helicular structure of DNA. They also shared the Nobel Prize for the same.

Biological Role

The functions of DNA may be summarized below:

1. It contains the genetic information that is transmitted from generation to generation, which is achieved by self-replication of DNA during cell growth and division so that two daughter double helical molecules of DNA are obtained, each identical to parent DNA.
2. It expresses its encoded genetic information for the synthesis of RNA and proteins for metabolic function and control of all cellular

activities. The genetic information is carried in the form of genes and expressed at appropriate times.

A gene is defined as the sequence of bases in DNA which specifies the complete amino acid sequence of a polypeptide chain or the base sequence of an RNA molecule (rRNA, tRNA). The sequence that specifies a polypeptide chain is commonly called a structural gene.

Most genes of eukaryotic organisms do not consist of a single continuous sequence which is transcribed into mRNA. Rather, they are interrupted by regions called introns which do not specify the protein product. The regions which are transcribed and specify the final protein product are called as *exons*.

Characteristics and Properties

1. *Genetic material.* DNA is the molecule of heredity and is responsible for the progeny to have the same characteristics as their parents.
2. *DNA content.* The DNA content of a cell is remarkably constant for each species (except in germ cells and when chromosomal variations occurs) and cannot be altered by environmental circumstances, with change in age or nutritional status.
3. *Base composition.* DNA isolated from different tissues of the same organisms has the same base composition. The base composition of DNA varies from one species to the other; while the DNA from closely related species has more or less similar base composition.

   In nearly all DNAs, the number of adenine residues is equal to the number of thymine residues i.e., A=T, and the number of guanine residues is equal to the number of cytosine residues, i.e., G=C or A+G = T+C or

$$\frac{A+G}{T+C} = 1$$

4. *Effect of pH.* DNA is a polybasic acid due to the presence of phosphate groups which are fully ionized at physiological pH. Because of negative charges present DNA binds strongly to histones and cations like Na^+ and Mg^{2+}. pH also affects the stability of the double helical structure of DNA. The hydrogen bonded base pairs are stable between pH 4.0 and pH 10.0. Outside these limits, their hydrogen bonds break and the complementary strands separate from each other, a process known as *denaturation.*
5. *Effect of temperature.* When highly polymerized double-stranded DNA is slowly heated, the double helix 'melts', as a result the

double-stranded structure is converted to a random coil over a range of a few degrees of temperature. This transition from a helix to a coil results in increase in absorbance. The midpoint temperature (T_m) is the melting temperature of the helix of a specific DNA polymer. The T_m's of different DNA's increase linearly as a function of the percentage of G-C base pairs.

6. *Absorbance.* The purine and pyramidine bases found in the DNA and also RNA, strongly absorb ultraviolet radiation of wavelength at 260 nm. This property is used to identify and estimate nucleic acids. The high molecular weight DNA typically has an optical density at 260 nm which is about 35-40 percent less than the optical density expected from adding up the individual absorbances of bases in the DNA. This phenomenon is called the hypochromic effect which is explained by the fact that in a helical structure, the bases are stacked one above the other. Interaction of π electrons between the bases then results in a decrease in absorbancy.
7. *Hydrolysis.* Gentle acid hydrolysis of DNA at pH 3.0 causes selective hydrolytic removal of all its purine bases without affecting the pyrimidine-deoxyribose bonds or the phosphodiester bonds of the backbone. The resulting DNA derivative devoid of the purine bases is called an apurine acid. Selective removal of the pyrimidine bases by hydrazine produces apyrimidinic acid. DNA is not hydrolyzed by dilute alkali unlike RNA because of no 2′-hydroxyl groups.

Enzymes that hydrolyze the phosphodiester bonds of nucleic acids are collectively known as nucleases. Nucleases can be classified by their point of attack upon the polynucleotide chain. Those that attack the polymer at either its 3′ or 5′ terminus and sequentially remove nucleotide residues one at a time or as small oligonucleotides are known as *exonucleases*; those that attack within the chain are called *endonucleases.*

Chemistry of Nucleic Acids

Having identified the genetic material as the nucleic acid DNA (or RNA), we need to examine the chemical structure of these molecules. Their structure will tell us a good deal about how they function.

Nucleic acids are made by joining *nucleotides* in a repetitive way into long, chainlike polymers. Nucleotides are made of three components: phosphate, sugar, and a nitrogenous base. When incorporated into a nucleic acid, a nucleotide contains one each of the

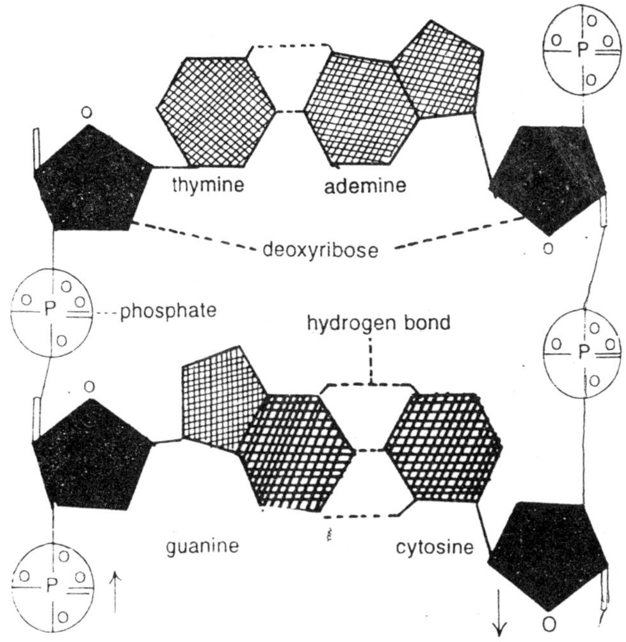

Fig. 6.1. Showing antiparallel nature of backbones of DNA.

three components. But, when free in the cell pool, nucleotides usually occur as triphosphates. The energy held in the extra phosphates is used, among other purposes, to synthesize the polymer. A *nucleoside* is a sugar-base compound. Nucleotides are therefore nucleoside phosphates. The sugars differ only in the presence (ribose in RNA) or absence (deoxyribose in DNA) of an oxygen in the 2′ position. The carbons of the sugars are numbered 1′ to 5′. The primes are used to avoid confusion with the numbering system of the bases. DNA and RNA both have four types of bases (two *purines* and *pyrimidines*) in their nucleotide chains. Both molecules have the purines *adenine* and *guanine* and the pyrimidine *cytosine.* DNA has the pyrimidine *thymine*; RNA has the pyrimidine *uracil.* Thus three of the nitrogenous bases are found in both DNA and RNA, whereas thymine is unique to DNA and uracil is unique to RNA.

A nucleotide is formed in the cell by attachment of a base to the 1′ carbon of the sugar and attachment of a phosphate to the 5′ carbon of the same sugar, the nucleotide takes its name from the base. Nucleotides are linked together (*polymerized*) by the formation of a bond between the phosphate of one nucleotide and the hydroxyl (OH) group at the 3′ carbon of an adjacent molecule. Very long strings of nucleotides can be polymerized by this *phosphodiester bonding.*

Table 6.1. Components of Nucleic Acids

			Base	
	Phosphate	*Sugar*	*Purines*	*Pyrimidines*
DNA	Present	Deoxyribose	Guanine	Cytosine
			Adenine	Thymine
RNA	Present	Ribose	Guanine	Cytosine
			Adenine	Uracil

Functional Structure

Although the identity of the nucleotides that polymerized to form a strand of DNA or RNA was known, the actual structure of these nucleic acids when they function as the genetic material remained unknown until 1953. The general feeling was that the biologically active structure of DNA was more complex than a single string of nucleotides linked together by phosphodiester bonds and that several interacting strands were involved.

In 1953, Linus Pauling, a Nobel laureate who had discovered the a-helical structure of proteins, was investigating a three-stranded structure for the genetic material, whereas Watson and Crick decided that a two-stranded structure was more consistent with available evidence. Three lines of evidence directed Watson and Crick: the chemical nature of the components of DNA, X-ray crystallography, and Chargaff's ratios.

DNA X-Ray Crystallography

Maurice Wilkins, Rosalind Franklin, and their colleagues were using *X-ray crystallography* to analyze the structure of DNA. The molecules in a crystal are arranged in an orderly fashion, such that when a beam of X rays is passed through the crystal, the beam will be scattered. The pattern of the scatter can be recorded on photographic film. The nature of this pattern depends on the structure of the crystal. The cross in the center of the photograph indicates that the molecule is a helix; the dark areas at the top and bottom come from the bases, stacked perpendicularly to the main axis of the molecule.

Chargaff's Rule

Until Erwin Chargaff's work, scientists had laboured under the erroneous *tetranucleotide hypothesis* in which it was believed that

DNA was made up of equal quantities of the four bases; therefore, a subunit of this DNA consisted of one copy of each base. Chargaff carefully analyzed the base composition of DNA in various species. He found that although the relative amount of a given nucleotide differs among species, the amount of adenine equaled that of thymine and the amount of guanine equaled that of cytosine. That is, in the DNA of all the organisms studied, there is a 1:1 correspondence between the purine and pyrimidine bases. This is known as *Chargaff's rule.* Chargaff's observations disproved the tetranucleotide hypothesis; the four bases of DNA were not in a 1:1:1:1 ratio. His results were extremely important to Watson and Crick in the development of their model.

Table 6.2. Nucleotide Nomenclature

Base	Nucleotide (Nucleoside monophosphate)	*Abbreviation*					
		Mono-phosphate		Dipho-sphate		Tripho-sphate	
		Ribose	*Deoxy-ribose*	*Ribose*	*Deoxy-ribose*	*Ribose*	*Deoxy-ribose*
Guanine	Guanosine monophosphate	GMP		GDP		GTP	
	Deoxyguanosine monophosphate		dGMP		dGDP		dGTP
Adenine	Adenosine monophosphate	AMP		ADP		ATP	
	Deoxyadenosine monophosphate		dAMP		dADP		dATP
Cytosine	Cytidine monophosphate	CMP		CDP		CTP	
	Deoxycytidine monophosphate		dCMP		dCDP		dCTP
Thymine	Deoxythymidine monophosphate		dTMP		dTDP		dTTP
Uracil	Uridine monophosphate	UMP		UDP		UTP	

Table 6.3. Percentage Base Composition of Some DNAs

Species	*Adenine*	*Thymine*	*Guanine*	*Cytosine*
Human Being (Liver)	30.3	30.3	19.5	19.9
Mycobacterium tuberculosis	15.1	14.6	34.9	35.4
Sea Urchin	32.8	32.1	17.7	18.4

The Watson-Crick Model

With the information available, Watson and Crick began making molecular models. They found that a possible structure was one in which two helices coiled around one another (a *double helix*) with the sugar-phosphate back-bones on the outside and the bases on the inside. This structure would fit the dimension established for DNA by X-ray crystallography if the bases from the two strands were opposite each other and formed rungs in a helical ladder.

The diameter of the helix could only be kept constant (about 20Å–angstrom units) if there were one purine and one pyrimidine base per rung. Two purines per rung would be too big and two pyrimidines would be too small. After further experimentation with models of the bases, Watson and Crick found that the hydrogen bonding necessary to form the rungs of their helical ladder could occur readily between certain base pairs, the pairs that Chargaff found in equal frequencies. Thermodynamically stable hydrogen bonding occurs

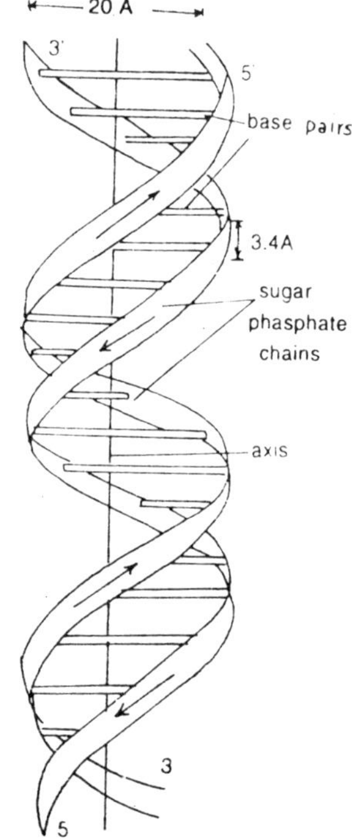

Fig. 6.2. Watson and Crick's model of DNA showing double helical structure.

between thymine and adenine and between cytosine and guanine. The relation is called *complementarity.* There are two hydrogen bonds between adenine and thymine and three between cytosine and guanine.

Another point about DNA structure relates to the fact that *polarity* exists in each strand. That is, one end of a DNA strand will have a 5′ phosphate and the other end will have a 3′ hydroxyl group. Watson and Crick found that hydrogen bonding could only occur if the polarity of the two strands ran in opposite directions; that is, the two strands were *antiparallel.*

Molecular Structure of DNA

The DNA molecule is a polymer consisting of several thousand pairs of nucleotide monomers. Each nucleotide consists of the pentose sugar–*deoxyribose*, a *phosphate* group, and a *nitrogenous base*, which may be either a *purine* or a *pyrimidine.*

Pentose Sugar

Levine (1909) identified ribose sugar as a main constituent of nucleic acid structure. *Mori* (1929) isolated sugar from the guanine nucleoside and showed that it is deoxyribose sugar. It is pentose type which in DNA molecule shows the absence of one oxygen molecule from carbon-2 position of ribose sugar. Both deoxyribose and ribose (pentose sugars of nucleic acids) have a pentagonal ring with five carbons, among which two (i.e., 3′ and 5′) are attached to phosphoric acid and three (1) to the base This sugar is called *deoxyribose* and it simply acts as a support column to which bases are attached.

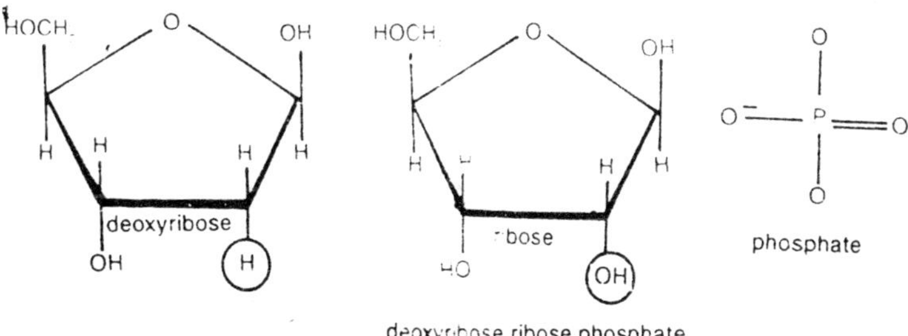

Fig. 6.3. Pentose sugars and phosphate.

Phosphate

In the DNA strand the phosphate groups alternate with deoxyribose. Each phosphate group is joined to carbon atom 3′ of one deoxyribose and to carbon atom 5′ of another. Thus each strnd has a 3′ end and a

5′ end. The two strands are oriented in opposite directions. The 3′ end of one strand corresponds to the 5′ end of the other. Consequently the oxygen atoms of deoxyribose point in opposite directions in the two strands. *Phosphodiester bonds* are formed between the sugars of two different nucleotides and a phosphate group.

Nitrogenous Bases

The nitrogenous bases of nucleic acid are of two types: *Purine* and *Pyrimidine.* Purine bases comprise mainly adenine (A) and guanine (G) which are common to both DNA and RNA while pyrimidine bases comprise cytosine (C) and thymine (T) in DNA and in the cases of RNA, thymine (T) is replaced uracil (U). These bases are arranged linearly as the back bones of chains which (back bones) are composed of alternating sugar and phosphate. The base have specific way of combining, if C is found in one chain G will be opposite to it and in other if A is on one T must be on the other. Or in other words adenine (A) is always paired with thymine (T) and guanine (G) with cytosine (C).

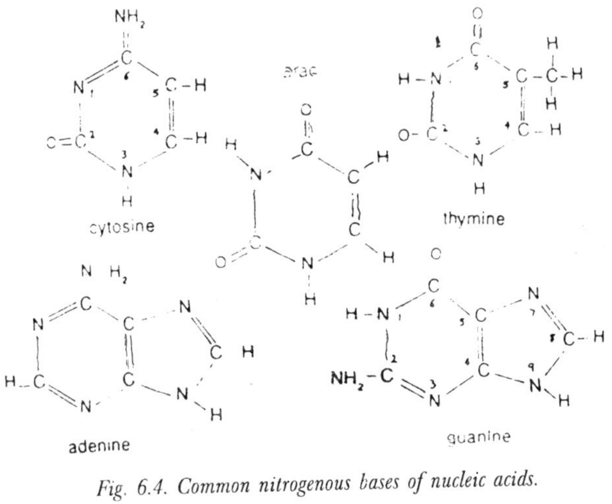

Fig. 6.4. Common nitrogenous bases of nucleic acids.

Among bases A-T and G-C, two hydrogen bonds are formed between A and three hydrogen bonds are formed between C and G. This hydrogen bond formation precludes A-C or G-T pairs. Only these arrangements A-T and G-C are possible, for two purines would occupy too much space to allow a regular helix and correspondingly, two pyrimidines would occupy too little. The stickness of these pairing rules result in a complementary relation between the sequences of bases on the two interwined chains. For example, if we have a sequence ATGTC on one chain, the opposite chain must have a sequence TACAG.

The hydrogen atom with its positive charge is shared between an oxygen atom and a nitrogen atom, both with slight negative charges. Although hydrogen bonds are weak, the fact that there are so many gives stability to the DNA molecule. The weak hydrogen bonding enables the two strands of the DNA to separate during replication.

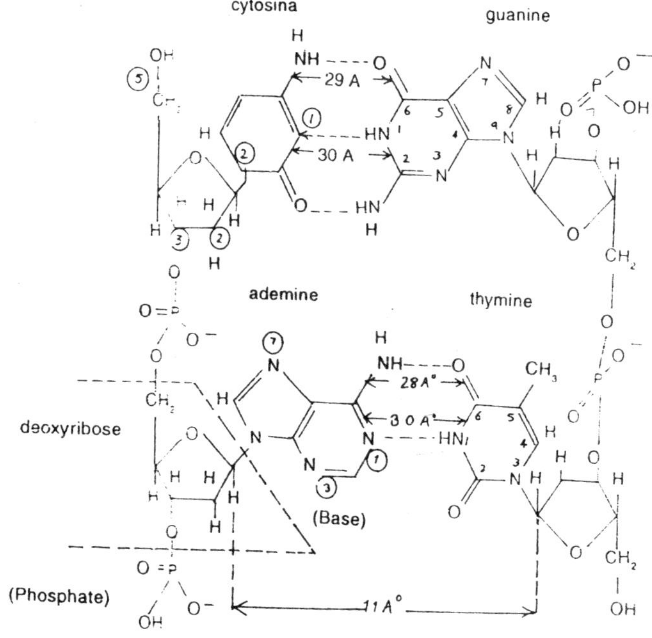

Fig. 6.5. Segment of a DNA molecule.

The bases are joined to the pentoses by N-C *glycosidic bonds.* For the purine, the glycosidic bond is between the C_1 position of the pentose sugar and the N_9 position of the bases. For the pyrimidines the linkage joins the C_1 and N_2 positions.

Nucleosides and Nucleotides

A sugar molecule and a nitrogenous base form a *nucleoside,* and a nucleoside plus a phosphate group form a *nucleotide.* In other words a nucleoside is a base-sugar combination and a nucleotide is a nucleoside phosphate. The nucleotides of RNA are called ribonucleotides, and those of DNA deoxyribonucleotides. Ribonucleotides contain the sugar ribose, and deoxyribonucleotides the sugar deoxyribose. The nucleosides are usually abbreviated as A, T, G or C, with *d*-as a prefix for deoxynucleosides (e.g., dG-is the abbreviation for deoxy guanosine).

Table 6.4. Four Nitrogen Bases, Nucleosides and Nucleotides of DNA Molecule.

Nitrogen base	Base deoxyribose =deoxyribo-nucleoside	Deoxyribonucleoside + Phosphoric acid = Deoxyribonucleotide	Abbreviation for nucleotide
1. Adenine (A)	Deoxyadenosine	Deoxyadenylic acid (Deoxyadenosine monophosphate)	3´-dAMP
2. Guanine (G)	Deoxyguanosine	Deoxyguanylic acid (Deoxyguanosine monophosphate)	3´-dGMP
3. Cytosine (C)	Deoxycytidine	Deoxycytidylic acid (Deoxycytidine monophosphate)	5´-dCMP
4. Thymine (T)	Deoxythymidine	Thymidylic acid (Deoxythymidine monophosphate)	5´-dTMP

Nucleosides

The nucleosides are compounds formed by linking purine and pyrimidine bases to either D-ribose or 2-deoxy-D-ribose in a N-β-glycosidic bond. The point of attachment of the base to the sugar is N-9 of the purines or N-1 of the pyrimidines to C-1 of D-ribose or 2-deoxy-D-ribose. The carbon atoms of the sugar are designated by prime numbers (i.e., C-1´, C-5´), while the atoms in the bases lack the prime sign.

Table. 6.5. Lists the trivial names of the purine and pyrimidine nucleosides which are related to the bases that occur in RNA and DNA.

Table 6.5. Names of nucleosides.

Base	*Ribonucleoside*	*Deoxyribonucleoside*
Adenine	Adenosine	2´-Deoxyadenosine
Guanine	Guanosine	2´-Deoxyguanosine
Uracil	Uridine	2´-Deoxyuridine
Cytosine	Cytidine	2´-Deoxycytidine
Thymine	Thymine ribonucleoside	2´-Deoxythymidine

Nucleotides

These are phosphoric acid esters of nucleosides in which phosphoric acid is esterified with one of the hydroxyl groups of D-ribose (2´, 3´ and 5´ hydroxyl group) or 2´-deoxy-D-ribose (3´ and 5´ hydroxyl group). Therefore, 2´, 3´ or 5´ ribonucleoside monophosphate and 3´ or 5´ deoxyribonucleoside monophosphates can be formed. However, the 5´ position is most commonly phosphorylated.

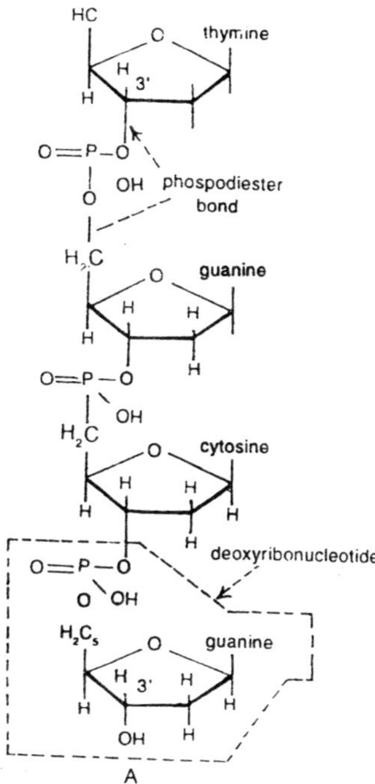

Fig. 6.6. Polynucleotide chain of DNA.

Nucleoside monophosphates can be linked to a phosphate or a pyrophosphate group through anhydride bonds to give nucleoside di- and triphosphate, respectively. The compounds are called acids or mono-, di- or triphosphate accordingly. The structure of the mono-, di- and triphosphates of adenosine are shown in figure as examples.

In the absence of oxygen atom at C-2´ of the above sugar ring, the structures are correspondingly known as deoxyadenosine-5´-monophosphate (dAMP), deoxyadenosine-5´-diphosphate (dADP) and deoxyadenosine-5´-triphosphate (dATP), respectively. A list of biologically important nucleotides is present in table.

Table 6.6. Biologicaly important nucleotides* and their nomenclature.

Ribonucleotides	*Deoxyribonucleotides*
Adenosine-5´-monophosphate (adenylic acid; AMP)	Deoxyadenosine-5´-monophosphate (deoxyadenylic acid; dAMP)
Guanosine-5´-monophosphate (guanylic acid; GMP)	Deoxyguanosine-5´-monophosphate (deoxyguanylic acid; dGMP)
Cytidine-5´-monophosphate (cytidylic acid; CMP)	Deoxycytidine-5´-monophosphate (deoxycytidylic acid; dCMP)
Uridine-5´-monophosphate (uridylic acid; UMP)	Deoxythymidine-5´-monophosphate (deoxythymidylic acid; dTMP)

* Each of the 5´-monophosphates exists as the 5´-diphosphate and 5´-triphosphate. Thus, as an example, there occurs GMP, GDP, GTP, dGMP, dGDP and dGTP.

An important discovery has been the identification of cyclic nucleotides. An important cyclic nucleotide is 3´5´-cyclic adenosine monophosphate (3´,5´-cyclic AMP or cAMP) which is called a second messenger plays a key role in the biochemical action of a number of hormones.

Metabolic Function of Nucleotides

All types of cells contain a wide variety of nucleotides and their derivatives. Nucleotides serve many functions some of which are as follows:

(i) Role in energy metabolism

ATP is the main form of chemical energy available to the cells. It is generated in cells by oxidative phosphorylation and substrate-level phosphorylation. ATP is utilized to drive metabolic reactions, as a phosphorylating agent, and is involved in such processes as muscle contraction, active transport and maintenance of cell membrane integrity. As a phosphorylating agent, ATP serves as the phosphate donor for the generation of the other nucleoside-5´-triphosphates (e.g. GTP, UTP, CTP).

(ii) Monomeric units of nucleic acids

The nucleic acids, DNA and RNA are composed of monomeric units of the nucleotides as building-blocks.

(iii) Physiological mediators

cAMP plays an important role as a ‘second messenger’ in epinephrine- and glucagon-mediated control of glycogenolysis and glycogenesis. cGMP acts as a mediator of cellular events. ADP is important for normal platelet aggregation and hence blood coagulation.

(iv) Components of coenzymes

Coenzymes such as NAD^+, FAD and coenzyme A are important metabolic constituents of cells and are involved in many metabolic pathways.

(v) Activated intermediates

The nucleotides also serve as carriers of activated intermediates required for a variety of reactions. UDP-glucose is a key intermediate in the synthesis of glycogen and glycoproteins. CTP is utilized to generate CDP-choline, CDP-ethanolamine which are involved in phospholipid metabolism.

(vi) Allosteric effectors

Many of the regulated steps of the metabolic pathways are controlled by the intracellular concentrations of nucleotides.

Table 6.7. Showing Amount of Various Bases in Different Tissues.

S.No.	*Source*	*Adenine*	*Guanine*	*Cytosine*	*Thymine*	$\frac{A+T}{G+C}$
1.	Human liver	30.3	19.5	19.9	30.3	1.53
2.	Human sperm	30.7	19.3	18.8	31.2	1.62
3.	Hen red cells	28.8	20.5	21.5	29.2	1.38
4.	Rat bone marrow	28.6	26.4	21.5	28.4	1.33
5.	Herring sperm	27.8	22.2	22.6	27.5	1.23
6.	*Paracentrotus lividus* (sea urchin) sperm	32.8	17.7	18.4	32.1	1.85
7.	Salmon	29.7	20.8	20.4	29.1	1.43
8.	Wheat germ	26.5	23.5	23.0	27.0	1.19
9.	Yeast	31.3	18.7	17.1	32.9	1.79
10.	*Diplococcus pneumoniae*	29.8	20.5	18.0	31.6	1.59
11.	K-12 *Escherichia coli*	26.0	24.9	25.2	23.9	1.00
12.	*Mycobacterium tuberculosis*	15.1	34.9	35.4	14.6	0.42
13.	Bacteriophage T_2	32.5	18.2	16.7	32.6	1.86

Structural Variation in DNA

Following three types of DNA have been recognized.

1. Double stranded DNA
2. Single Stranded DNA
3. Circular DNA

Double Stranded DNA

This is the common DNA present in all the living organisms. These contain two polynucleotide chain running antiparallel and twisted around each other in the form of regular double helix. The detailed structure of this type of DNA has already been explained.

Single Stranded DNA

At first it was thought that all DNA molecules are double stranded except during replication, when a small region around the replicating fork is temporarily in a non-hydrogen-bonded, single-stranded form. It therefore came as quite a surprise when experiments revealed that the DNA of several groups of small bacterial viruses exists as single-stranded molecules in which the amount of A is not equal to the amount of T and the amount of G does not equal the amount of C. Among these single-stranded phages are the spherically shaped φX174 and S13 viruses and the rod-shaped f1 and M13 viruses.

The chromosomes of the parvoviruses, minute viruses that infect many vertebrate organisms, are also single stranded DNA molecules. When these single-stranded viral chromosomes enter their host cells, they serve as templates for the formation of complementary strands. The resulting double helices in turn serve as templates for new single strands that then become incorporated into new virus particles. Thus, the fundamental mechanism for orderning nucleotides during the synthesis of single-stranded DNA is basically the same as that used for double-helical DNA. Nucleotide selection always occurs by attaction of the complementary base. Single-stranded DNA replication usually differs from double-helical replication in that it uses only one of the two complementary strans (the–strand) as a template for the progeny (+) strand.

Although single-stranded DNA can serve as the chromosome of a small virus, it would not be very effective in storing genetic information within the chromosomes of cells. In bacteria and in many eukaryotic cells (e.g., the haploid phase of yeast cells), there usually is only one copy of a given gene. If interphase chromosomes contained single-stranded DNA, with the double helix existing only briefly during mitosis, then the daughter cells would contain two completely different sets of genetic informations. This follows from the fact that the two complementary chains do not have identical base sequences and hence would code for entirely different amino acid sequences.

The double-strand ensures that each daughter cell will contain the same genetic information. The double-stranded form also permits effective DNA repair mechanisms to exist. For example, when one strand is damaged by exposure to X-rays, the remaining strand can provide the genetic information to rebuild the damaged section. It is probably no accident that only very small viral chromosomes are single-stranded. If they were to become, say, the size of T2, they would present too large a target for chain-damaging events.

Structure of Single-stranded DNA

Single-stranded DNA molecule have a strong tendency to fold back on themselves to form irregular double-helical hairpin loops whenever their sequences permit significant numbers of nucleotides to base-pair. Imperfect as these loops are, they nonetheless are energetically favoured under physiological salt conditions, since they allow much more effective stacking of the flat hydrophobic surfaces of the bases than is possible in any fully extended single-stranded structure.

Often, as many as half the bases of single-stranded DNA are so hydrogen-bonded, and their molecules are highly compacted. If, however, there are only minimal numbers of, neutralizing cations (e.g., less than or equal to 0.01 M Na^+) then the electrostatic repulsion between the partially unneutralized phosphates will keep the single-stranded chains highly extended, and there will be fewer hairpin loops. Single-stranded DNA is always denser than double-helical DNA. Much more complete hydrogen bonding to water is possible for the atoms of the highly regular double helix than for the atoms of the inherently irregularly shaped single-stranded DNA. In the absence of base-pairing hydrogen bonds, Van der Waal's interactions take over and bring the atoms of DNA closer together than would be the case if they formed regular hydrogen bonds with each other. An analogy can be made with behaviour of water, which becomes less dense as it forms the regular, hydrogen-bonded lattice of crystalline ice.

Rigorous Crystallography

X-ray diffraction patterns from parallel-oriented fibers of DNA provided the first clues that the polynucleotide chains of DNA have helical conformations. They also told us that DNA was a multichained molecule composed of two or three polynucleotide chains. However, the double-helical nature of DNA did not directly emerge from X-ray diffraction analysis alone. Equally important was the use of model building in which the question was asked. Given the chemical features of a single DNA chain, into what three-dimensional helical conformations could it fold? The simplest models were those in which the sugar-phosphate backbones were on the outside with the bases stacking in the center.

The double helix emerged when it was realised that thymine and guanine had keto, not enol, configurations and that by base-pairing adenine to thymine and guanine to cytosine, a stereochemically pleasing, regular helical molecule resulted. They were originally discovered through their antimicrobial and anticancer attributes. Each time these intercalating agents become inserted into a DNA molecule, they necessarily *increase* the spacing of successive base pairs along the helical axes to roughly 7Å, almost the distance between phosphate atoms in a fully extended polynucleotide chain. In this situation, very little rotation is possible around the helical axis, and so intercalating agents not only extend double helices, but extensively unwind them. For example, when a molecule of either ethidium bromide (an inhibitor of DNA synthesis) or actinomycin D (a powerful inhibitor of RNA synthesis) intercalates, the rotation angle between the two adjacent base pairs is reduced from 36° to 10°. The fact that intercalation occurs of readily indictes that it must be energetically favoured, with the van der Waals bonds holding the inserted molecules to the base pairs being stronger than those found between conventionally stacked base pairs. Intercalation is additional evidence for the *metastability* of the double-helical structure–its ability to temporarily assume many inherently unstable configurations that normally quickly revert back to the standard B conformation. One such metastable variant must be short stretches of extended chains with gaps separating adjacent base pairs, into which an intercalating agent may bind.

The Chromosomes of Viruses, E.coli, and Yeast are Single DNA Molecules

The first estimates of the average molecular weights of DNA centered at about a million, the size needed to encompass an average-size bacterial gene coding for some 300 to 400 amino acid. Therefore, it seemed natural to equate single DNA molecules with single genes. However, these early reports were inaccurate owing to DNA breakage during its isolation and study. Now it is clear that virtually all undegraded DNA molecules contain the information of at least several genes.

The most certain molecular weight values come from DNA-containing viruses. Regardless of whether the DNA content is relatively small or large, each virus particle contains a single DNA molecule. For example, all the DNA of the small monkey (simian) virus SV40 is present within a single molecule of molecular weight ~ 3×10^6 daltons (5×10^3 has pairs, or 5 kbp), while the DNA molecule of the large bacterial virus T2 has a molecular weight of 1.2×10^8 daltons (2×10^3 kbp). In these cases, the entire viral chromosome, rather than separate genes, corresponds to a single DNA molecules. Likewise, the chromosome of an *E.coli* cells is a single DNA molecule whose molecular weight is about 2.5×10^9 daltons (about 4×10^3 kbp) and whose extended length is, roughly 1 mm. DNA molecules of the yeast *Saccharomyces cerevistiae* range in size between the T2 and *E.coli* DNAs, with their average size being that expected if each of the yeast's 17 chromosomes contain one DNA molecule.

The centromeres of yeast chromosomes (if not those of all eukaryotic chromosomes) do not represent discontinuities between separate DNA molecules, but instead are specialized regions of DNA evolved to interact with the spindle bodies upon which chromosomal segreation occurs during cells division. As of yet, there is no direct proof that the very much larger chromosomes of higher plants and animals (often 50 times larger than the *E.coli* chromosomes) also contain only one DNA molecule. However, we now expect that the one chromosome–one DNA molecule rule will hold for the chromosomes of all organisms. In any case, some DNA molecules are larger than any other biological molecules by several powers of ten.

Circular Versus Linear DNA Molecules

Autoradiography and electron microscopy, initially suggested that all DNA molecules are linear and have two free ends. But when it became possible to take a better look at undergraded DNA, many DNA molecules were found to be circular. For instance, the small monkey DNA virus SV40 has a 5000-base pair circular double-helical chromosome; the similarly short chromosomes of single-stranded phages are likewise circular, as are almost all autonomously replicating plasmid DNAs.

Most, if not all, bacterial chromosomes are also circular We suspect that the DNA found in the rare circular chromosomes of higher cells are likewise circular molecules. Circular shapes were initially puzzling, since they seemed to present obstacles to the untwisting of double helices during DNA replication. Only when specific enzymes that first snip and then join DNA chains were discovered did the untwisting dilemma disappear. Equall important is the realization that linear DNA molecules do not have the uncomplicated structures first envisioned for them. As we see in the next chapter, replicating the ends of DNA molecules is not a straightforward process.

All linear DNA molecules have evolved special tricks to replicate their ends, as well as to prevent their free ends from either being nibbled back by cellular enzymes or being enzymatically ligated together to form even larger chromosomes. For example, the ends of the linear adenovirus chromosomes have specialized proteins covalently attached to the 5′ ends of strands; and poxvirus chromosomes have closed hairpin loops at their ends, with their basic structures being that of largely self-complementary circular single strands. Still other linear viral chromosomes (e.g., T2 and T7) have the same sequence at both ends, allowing the ends of the chromosome to recombine together during DNA replication to form very long, end-to-end aggregates (concatamers). Moreover, some DNA molecules that are linear when isolated from a virus particle (e.g., phage λ) are found as circles inside the host cell. This tells us that the linear and circular forms of such DNA molecules are interconvertible. Starting with a closed circle, a specific enzyme introduces two breaks, one in the + strand and one in the – strand. As these breaks are very close to each other, the intervening hydrogen bonds occasionally are broken by thermal agitation, and then the circle unfolds. Most importantly, the resulting linear form contains single-stranded ends that have complementary nucleotide sequences (“*sticklly ends*”).

At a later time the linear form can base-pair and resume a circular configuration. If the mission phosphodiester bonds are then reformed, the covalent circle is regenerated. There is solid evidence that the phage λ DNA interconverts between its circular and linear forms using precisely this mechanism. Later, we shall see that the duplication of such “sticky” DNAs involves circular replicative intermediates. Thus, the linear form is most likely an adaptation for injection of the viral chromosome through the narrow phase tail.

Supercoiling of Circular DNA

DNA is a very flexible structure, its exact molecular parameters are a function of both the surrounding ionic environment and the nature of the DNA-binding proteins with which it is complexed. Because their ends are free, linear DNA molecules can freely rotate to accommodate changes in the number of times the two chains of the double helix twist about each other. But once the two ends become covalently linked to form circular DNA molecules, the absolute number of times the chains twist about each other (the *linkage number*) cannot change. For the most part, changes in the average number of base pairs per turn of the double helix will necessarily be accommodated by the formation of the appropriate number of supercoils in the opposite direction.

Supercoiling refers to the further twisting of double-helical DNA molecules. Untwisting of the double helix usually leads to supercoiling in the negative (left-handed) direction while overtwisting leads to positive (right-handed) supercoiling. The supercoiled state is inherently less stable than uncoiled DNA, and the cutting, or *nicking*, of a single strand instantly converts a supercoiled molecule into its simple (*relaxed*) circular state. Because circular DNA molecules become increasingly compacted as they become more supercoiled, relaxed DNA is easily distinguished from supercoiled DNA by its slower sedimentation in a centrifugal field and by the longer time it takes to move through an agarose gel. At the present time gel electrophoresis affords the most direct way to measure supercoiling, since it can separate molecules that differ by only single turns.

Just because isolated DNA molecules would energetically prefer to be in the relaxed state does not mean that the "natural state" of DNA is relaxed. In fact, virtually all DNA within both prokaryotic and eukaryotic cells exists in the negative supercoiled state. As such it seldom, if ever, occurs as free DNA, but is complexed with specific DNA-binding proteins to form compacted molecules called *chromatin.* How this compaction occurs is best understood for eucaryotic chromatin, whose most prominent DNA-binding components are the histones. *Histones* are relatively small, positively charged, arginine- and lysine rich proteins that aggregate together to form discrete ellipsoid-shaped packets (histone cores) around which the DNA supercoils. The resulting *nucleosomes* give to chromatin its beaded appearance.

Since the same collection of histones binds to nearly all sections of DNA, histones are thought to play essentially structural roles, as opposed to enzymatic or regulatory roles. Only about half the mass of chromatin proteins is histone. The remaining proteins are not nearly as well characterized with their exact functions yet to be determined. Viral DNA chromosomes are also complexed with proteins, both when multiplying within cells and when packaged into virus particles. The SV 40 viral chromosome is at all time compacted into 24 histone-containing nucleosomes identical it structure to those existing in the chromosomes of their host cells. When freed from its protein components, SV 40 DNA is found as a negative supercoil. In contrast, no histones are complexed with the DNA found in adenovirus particles.

The vertebrate viruses encode a unique DNA binding protein to help package their DNA within their protein capsules. When, however, an adenovirus DNA functions within a cell, its own DNA-binding protein is released and replaced with the same histone components that bind to all other cellular DNA.

DNA Supercoiling Around Nucleosome

In forming the nucleosomes of eukaryotic cells, 200-base-pair-long segments of DNA negatively (left-handedly) supercoil twice around the histone cores. The energy lost by supercoiling may be partially compensated by the energy gained from the ionic and hydrogen bonds formed between the histones and the DNA. The number of supercoils in solution need not be the same as found in chromatin. In chromatin, the double helix is slightly more twisted than in solution, with 10 base pairs per helical turn in chromatin versus 10.5 base pairs per helical turn in naked DNA in solution. As a result, there are fewer superhelices found in solution than found in chromatin.

Histone-like Proteins in Prokaryotes

For many decades, it was believed that bacterial DNA, unlike eucaryotic DNA, occurs naked and does not have a compacted chromatin-like arrangement. But recently, using more genetle preparative procedures, condensed *E.coli* chromosomal sections have been seen that are organized into bead-like packets from which small, basic proteins analogous to the histones can be extracted. That bacterial DNA is complexed with proteins that have histone-like properties undoubtedly reflects its need to be regularly compacted in order to function.

Crystallization of these proteins, DNA-binding protein II, reveals that its 9500-dalton chains associate in pairs as dimers containing extended arginine containing arms able to interact with the phosphates of one turn of the DNA backbone. So it is likely that dimers sited adjacently along prokaryotic DNA are oriented into a helical arrangement with the DNA bound on the outside. Interestingly, the average *degree of supercoiling* (i.e., the number of super-helicle twists per ten base pairs) is about 0.05 for all naturally occuring DNA supercoils, regardless of their source.

Topoisomerases of Supercoiled DNAs

That supercoiling not only occurs but is biologically very important is affirmed by the existence of the enzyme topoisomerase II, which specifically generates the negative supercoils that characterize so much of chromosomal DNA. *Topoisomerases* are a group of enzymes that convert (isomerize) one topological version of DNA into another. They do so by changing the linkage number, which is the number of times two DNA chains twist around each other.

Prokaryotic topoisomerase II (sometimes called *gyrase*) uses the energy of ATP to generate negative supercoils by untwisting DNA in the left handed direction. Its eukaryotic counterpart, however, may be capable of using ATP to generate negative supercoils only in the presence of other, yet to be described proteins. The supercoiling action of topoisomerase II is counterbalanced by a second enzyme, topoisomerase I, which converts supercoiled DNA to the unstrained, energetically more favourable relaxed state. The relative amounts of topoisomerase I and topoisomerase II in cells are finely tuned, tending to create just the right amount of negative supercoiling. Thus, mutations that lower the number of topoisomerase I molecules are viable only if the number of topoisomerase II molecules also decrease.

Both topoisomerase I and II work by catalyzing the breakage and rejoining of DNA phosphodiester bonds. Unlike virtually all other enzymes, their action does not lead to new patterns of covalent bonds. Rather, their role is to create temporary gaps in polynucleotide chains.

When acting topoisomerase do not create free ends but instead become themselves covalently attached to one of the two broken ends (depending on the specific enzymes, either the 3′ and 5′ end). Through such catalysis, a tyrosine group on the topoisomerase becomes linked to the terminal phosphate group of the cut polynucleotide chain. The free energy present in the original phosphodiester bond is thus preserved so that it can be used to rejoin the broken chain. When a DNA chain is broken by a topoisomerase, its broken ends do not fall apart but are held together by the topoisomerase. At no time do the broken ends have the capacity to rotate freely. If that happened, topoisomerase would be limited to only completely relaxing double helices. Instead, individual topoisomerases function to make discrete changes of either one positive turn (topoisomerase I) or two negative turns (topoisomerase II) in the linkage number. Such discrete steps result from DNA chains passing through either transient single-stranded breaks (topoisomerase I) or transient double-stranded breaks (topoisomerase

Table 6.8. DNA Topoisomerases

	Procaryotic (E.coli)		*Eucaryotic (Hela)*	
	Type I	*Type II*	*Type I*	*Type II*
Approximate MW	100,000	400,000	100,000	309,000
Subunits	Monomer	Tetramer (A_2B_2)	?	Dimer (subunit MW=172,000)
Gene	top A	gyr A Polypeptide MW 105,000; activity inhibited by nalidixic acid and oxolinic acid gyr B Polypeptide MW 95,000; activity inhibited by coumermycin and novobiocin	?	?
Covalent intermediate with DNA	At 5´ P	Subunit A Tyrosine covalent bond to 5´ P	At 3´ P	At 5´ P
Relaxation	Mg^{2+} required; ATP-independent	ATP-independent	No Mg^{2+} required; ATP-independent	ATP-dependent
Negative supercoiling	None	ATP-dependent	None	None

II). Topoisomerases are constructed in such a way that they can undergo reversible conformational changes that create cavities through which DNA chains can pass.

The ATP requirement for bacterial topoisomerase II (gyrase)-induced supercoiling most likely reflects the use of ATP in mediating the necessary conformational changes that lead to chain passage. Topoisomerase I, however, relaxes supercoiled DNA without requiring a cyclical input of energy. Complete understanding of how topoisomerase I and II work can occur only when their precise structures are determined through X-ray crystallographic procedures. Even now, however, the prediction can be made that double helices wrap themselves around topoisomerase II in positive supercoils prior to the start of the cutting-rejoining cycle. Without this restraint on the orientation of the DNA, topoisomerase II would break as well as make negative supercoils, and the net result of its action would be the eventual relaxation of the double helical substrate to the uncoiled state.

Looped and Supercoiled Structure

The fact that linear DNA molecules have free ends would seem to preclude their possession of supercoiled segments. Yet, examination of gently isolated linear eukaryotic chromosomes shows that their chromatin is organized into a large number of successive looped domains organized along a scaffold containing two major DNA-binding proteins. Somehow, attachment to the scaffold prevents the rotation of one domain from being transmitted to adjacent sections, since the DNA within each of these loops independently supercoils and may be under different torsional strain.

There is much evidence both from *Drosophila's* giant salivary chromosomes and from the looped lamp brush chromosomes of the salamander that the individual loops represent functional units of chromatin, with given loops usually being transcribed into one or more very long RNA molecules. How the protein scaffold might maintain independently supercoiled loops was a complete mystery until recently. Now it is less so, following the discovery that a major component of the scaffold is none other than topoisomere II.

The data presented illustrate several points:

1. The sixteen possible nearest-neighbour frequencies occur with a large number of frequencies.
2. The sums of the vertical columns show that the amount of A is equal to the amount of T and that G equals C, thus indicating the DNA was probably replicated correctly.

3. The two DNA strands are of opposite polarity. This is borne out by the frequency equivalence of the pairs of sequences: CpT and ApG, GpT and ApC, GpA and TpC, and CpA and TpG, as predicted by antiparallel DNA strands. Were the two strands of the same polarity, different matching sequences would have been predicted, for example, TpA and ApT, GpA and CpT, CpA and GpT, etc.

To show that newly synthesized DNA is made using a DNA template requires two rounds of nearest neighbour analysis. In the first round the originally isolated DNA is used in the reaction and a set of 16 nearest neighbour frequencies are obtained as described. In the second round the enzymatically synthesized DNA is used in the reaction and a second set of frequencies is produced.

The results show good agreement between the two sets of frequencies, thus indicating that DNA plays a template role in DNA replications in vitro. Since the early, comparatively, crude in vitro DNA synthesis experiments, a large number of refinements have been made. For example, the reaction mixtures now duplicate the conditions that are present in growing cells such that it is possible to synthesize DNA in vitro that is identical in all respects with DNA produced in vivo. The reaction mixtures required for efficient DNA synthesis vary, depending on the DNA used as the template. In general, there is a basic set of enzymes and proteins required for the synthesis of all DNA sources. Beyond that, each DNA requires a number of specific proteins for new synthesis to occur, and the actual set of proteins needed depends on the DNA in question.

Types of DNA's

Prokaryotic DNA

Also called the *circular a superhelicular DNA.* The DNA molecules of prokaryotes and most viruses are circular. A circular molecule may be a covelently closed circle, which consists of two unbroken complementary, single strands, or it may be a *nicked circle*, which has one or more interruption (nicks) in one or both strands. With few exceptions, covalently closed circles are twisted. Such a circle is said to be a *superhelix* or a *supercoil.* It has also been discussed already.

Eukaryotic DNA

It is the filamentous type of DNA found in all eukaryotic cells of animals and plants and has already been discussed.

Extranuclear DNA

Also known as non-chromosomal or cytoplasmic DNA. In cytoplasm this DNA is associated with some of cell organelles as given belows:

(a) Mitochondrial DNA

Mitochondrial DNA is usually found in cyclic double stranded, supercoiled molecules, the exceptions being the linear mitochondrial DNA molecules from *Tetrahymena* and *Parameciui.* Marmialian mitochondrial DNA molecules are not packaged into nucleosomes. They are about 15 kbp long and can therefore code for 15 to 20 proteins. However, some yeast ones are considerably larger and *Saccharomyces* mitochondrial DNA is 75 kbp long.

Plant mitochondrial DNA is much longer still. Yeast mitochondrial DNA molecules have been widely studied since they are readily amenable to genetic analysis. Mitochondrial DNA can only codes for a small proportion of mitochondrial proteins. Mitochondrial DNA from several mammals, *Xenopus* and *Drosophila*, has been sequenced and the sequence analysed. There are genes for two ribosomal RNAs, 22 tRNAs and for 13 protein most or all of which are involved in electron transport. The only region of mammalian mitochondrial DNA which is non-coding is the *D-loop region* involved in the initiation of DNA replication. This is also the region at which transcription of both strands is initiated.

Transcription continues uninterrupted around the cyclic molecule and the transcripts are the processed to give individual messenger, ribosomal and tRNAs. The tRNA forms the puncturation between the various protein coding regions and provides the signals for the processing enzymes. Plant mitochondrial DNA is apparently much more complex than animal mitochondrial DNA. It consists of per mutation of basic structure related to each other by recombination.

(b) Chloroplast DNA

Chloroplast DNA is in general much larger than mitochondrial DNA, being in the molecular weight range 100 million (150 kbp). In contrast to mitochondria which have from one to ten molecules of DNA per organelle, chloroplasts tend to have a very large number of copies of the DNA molecules in each organelle, in some cases greater than one hundred. Like mitochondrial DNA, the DNA in chloroplasts carries the coding information for essential membrane components, tRNA and rRNA. All known chloroplast DNA molecules are cyclic and supercoiled.

(c) Kinetoplasts DNA

The kinetoplast is part of highly specialized mitochondrion found in certain groups of flagellated protozoa such as trypanosome. Its DNA (kDNA) consists of an interlinked series of many thousands of cyclic DNA molecules which vary in size from 0.6 kbp to 2.4 kbp depending from the type of trypanosomes from which they are obtained. These components, which are known as *minicircles*, are further interlinked with a much smaller number of larger circular DNA molecules of above 30 kbp in size known as *maxicircles* in the majority of system analyzed. While maxicircles appear to perform the connectional functions of mitochondrial DNA in trypanosomes, minicircles are microheterogenous in sequence and size and there is no evidence to suggest that they are ever transcribed so that it is possible that they may fulfil some structural rather than coding role.

Satellite DNA

At the other extreme DNA with a $Cot_{1/2}$ value as low as 10^{-3} moles of nucleotide seconds $litre^{-1}$ consists largely of *satellite DNA*. This represents highly repeated sequences of which there may be a million or more copies per haploid genome, which are usually quite short and are arranged in tandem arrays. The origin of the name satellite relates to the method of its isolation on caesium chloride buoyant density gradients of sheared DNA where it will sometimes form a satellite band separate from main DNA band, due to its differing content of adenine and thymine residues.

The simplest known DNA is poly [d(A-T)], which occurs in certain crabs. Other satellites can have any number upto several hundred base pair which are repeated in tandem fashion along the genome. The distribution of satellite DNA among chromosomes varies. Some chromosomes have virtually no satellite sequences while others are largely composed of satellite sequences. In general, satellite DNA appears to be concentrated near the centromere of the chromosomes in the heterochromatin fraction. DNA sequence analysis has shown that the basic repeat unit of satellite DNA is itself made up of sub repeats. For example, the major momse satellite has a repeating structure of 234 base pairs made up of four related 58 and 60 bp segments each in turn made up of 28 and 30 bp sequences. The satellite between and within the related species are themselves related in an evolutionary sense by cyclic rounds of multiplication and divergence of an initial short sequence.

Fold-back DNA

Also known as *palindromic DNA*. It is a special class of DNA sequences comprising 3-6% eukaryotic DNA. The size range is from 300 to 1200 base pairs and the molecules have a $Cot_{1/2}$ value of less than 10^{-5} moles of nucleotide second $litre^{-1}$. This palindromic DNA is represented in all frequency classes and is widely distributed throughout the metaphase chromosomes. Some palindromic DNA arises when two copies of the *Alu* sequences are present close to one another in opposite orientations and it was a result of the ability of such DNA to renative instantaneously.

Tautomeric form of DNA

An important chemical feature of DNA is the position of the hydrogen atoms in the purine and pyrimidine bases. Before 1953, many chemists thought that some of these hydrogen atoms randomly moved from one ring nitrogen or oxygen atoms to another and so could not be assigned a fixed location. Now we realize that although such movements, called *tautomeric shifts* do occur, these hydrogens have preferred atomic locations.

The nitrogen atoms attached to the purine and pyrimidine rings are usually in the *amino* (NH_2) form and only rarely assume the *imino* (NH) configuration. Likewise, the oxygen atoms attached to the C6 atoms of guanine and thymine normally have the *keto* (C-O) from and only rarly take up to *enol* (COH) configuration. It is essential to the biological functioning of DNA that the hydrogen atoms have relatively stable locations. If the hydrogen atoms had no fixed positions, adenine could often pair with cytosine and guanine with thymine, and sequence of the bases on the two intertwined chains would not necessarily be complementary. If that were the case, the DNA could not function as a genetic molecule, for it is the complementary relationship between the opposing chain sequences that gives DNA its capacity for self-replication.

Replication of the double helix involves strand separation followed by formation of complementary DNA chains using the now free single strands as templates to attract the appropriate partner bases dictated by AT and GC base-pairing rules. The amino forms adenine and cytosine and the enol forms of guanine and thymine must indeed occur only very rarely compared to their amino and keto alternatives. Otherwise, the many errors (mutations) incurred during DNA replication would be incompatible with orderly cell growth and division.

Unique DNA

In human cells it is generally classified as that DNA which has a $Cot_{1/2}$ value of about 1000 moles of nucleotide seconds $litre^{-1}$. It comprises about half of the total haploid DNA content and is thought to consist of the sequences coding for most enzyme functions for which there is only one or a small number of genes per haploid genome. The genes coding for the various chains of globin or the enzyme glucose 6-phosphates fall into this category.

Repetitive DNA

In this type of DNA certain bases or nucleotides are repeated many times. Among the repetitive DNA is a fraction which reanneals with a Cot value of between 100 and 1000 moles of nucleotides seconds $litre^{-1}$. The sequences represented in this group are generally thought to be those coding for proteins which form major structural components of the cell such as the histones. The genes for rRNA and tRNA also fall into this category.

Z-DNA or Left Handed DNA

On the basis of the arrangement of pentose sugar forming chains of DNA two forms of DNA have been reported. Usually the sugar chain has the pentose sugar with ascending sequence of carbon atoms. This is right handed DNA or B-DNA. Recently in 1979, another form of DNA have pentose sugar with ascending arrangement of carbon atom has been reported and named Z-DNA or left handed DNA.

(a) Resemblances between Z-DNA and B-DNA

(i) Both are double helical.

(ii) The two strands of double helix are antiparallel in both DNAs.

(iii) Both forms exhibit G≡C pairing.

(b) Differences between Z-DNA and B-DNA

(i) Z-DNA has left handed helical sense as against right handed helical sense of B-DNA.

(ii) Due to a different arrangement of molecules within the Z-DNA polymer, the phosphate backbone follows a zig-zag course, while in B-DNA it is regular.

(iii) In Z-DNA, the sugar residues have alternating orientation so that repeating unit is a dinucleotide as against the B-DNA where repeating unit is a mononucleotide and the orientation of sugar molecules is not alternating.

(iv) In Z-DNA, one complete helix i.e. a twist through 360°, has twelve base pairs or six repeating dinucleotide units (12 base pairs) while in B-DNA, one complete helix has only ten base pairs or ten repeating units.

(v) Because twelve base pairs are accommodated in one helix in Z-DNA, as against ten in B-DNA, the angle of twist per repeating unit (dinucleotide) is 60° as against 36° in B-DNA.

(vi) One complete helix is 45Å in Z-DNA it is 34Å in B-DNA.

(vii) Since bases get more length spread out in Z-DNA and since the angle of tilt is 60°, they are more closer to the axis and hence the diameter of the Z-DNA molecule is 18Å, whereas it is 20Å in B-DNA.

Denaturation and Renaturation of DNA

Single-stranded DNA

Single-stranded DNA molecules rarely occur naturally. The best known example is the DNA molecule from the small spherical bacteriophage ϕX174 first isolated by *Sinsheimer* in 1959 but it is also found in the filamentous bacteriophages such as fd. Among the animal viruses single-stranded DNA is found in the parvo viruses. Some of these naturally occurring single-stranded DNA's are circular and other linear. The former can be distinguished by their insensitivity to exonucleases. In general they are comparatively small molecules with less than 10,000 nucleotides.

The molar proportions of the bases do not show the usual equivalence of A and T, and G and C, required for double helix formation and the bases react readily with formaldehyde since the amino groups are not protected by hydrogen bondings as they are in double stranded DNA. Single-stranded DNA molecules are very much more flexible than double-stranded molecules of the same size since they lack the rigid double helical structure. In general they behave in solution in a manner similar to RNA molecules.

Hypochromic Effect

When double stranded DNA molecules are subjected to extremes of temperature of pH, the hydrogen bonds in the double helix are ruptured and the DNA collapses into two single-stranded molecules. If heat is used as the denaturant, the temperature at which this collapse occurs is known as the T_m or transition temperature. The absorption at 260 nm of any polynucleotide is due to that of its component bases. However, this absorption tends to be suppressed in the double-stranded DNA molecule where the bases are stacked above one another and inhibited from swinging out freely in solution in hydrogen bonds, and consequently it is very much lower than that of equimolar amounts of the component bases or nucleotides free in solution. This inhibition of absorption is to some extent relieved when the DNA molecule goes through the helix-coil transition and the bases are no longer so rigidly stacked although they still interact to some extent.

As a consequence of this absorption of DNA solutions rises by about 20-30 percent as the DNA undergoes the melting process, and the transition process is usually measured by ultraviolet spectroscopy. The increase in absorbance on melting is known as the *hyperchromic effect.*

Such a transition can be characterized by several factors:

1. *The nature of the DNA.* Homogenous DNA, such a viral DNA, melts over a short temperature range but heterogeneous DNA melts over a longer range.
2. *The (G + C) content of the DNA.* The melting temperature of any DNA can be related to its (G+C) content since this base pair confer extra stability on the molecule. In DNA molecules such as those from bacteriophage l, in which some regions are richer in (G+C) than others, the transitions of both regions can be observed.
3. *The nature of the solvent.* In low concentrations of counterion the transition temperature is low and its width broad. At high concentrations of counterion the T_m is raised and with width of the transition becomes sharp. However, if 'denaturing' salts such as sodium perchlorate are present, addition of more salt will lower the T_m since the rupture of apolar bonds caused by the anion will overcome the ionic stabilization of the cation.

The helix-coil transition is also associated with a change in density of the DNA molecule, the single-stranded DNA being more dense than the equivalent double stranded form with the same (G+C) content.

The Renaturation of DNA

When two DNA strands are returned from the extreme conditions which caused them to melt to their original state they may reassociate to form a double helix again. However, whether or not they do so depends on a variety of factors. In principle the process should follow simple second-order kinetics but, because of the high probability of two sequences which were not previously matched forming an imperfect union which is then difficult to dissociate, the degree of renaturation can vary from less than 1 percent according to the nature of the sample and the conditions of renaturation.

1. *The nature of the sample.* Simple sequence DNA molecules such as $d(G)_a$ $d(C)_n$ have no difficulty finding the appropriate sequence with which the reanneal and do so without difficulty. However, if a eukaryotic DNA sample from a large genome is reannealed, clearly some of the sequences will have to encounter many other non-complementary sequences in solution before they find the correct partner. Thus if most of the DNA from eukaryotic cells is quickly cooled to a low temperature, so that the diffusion of the DNA in solution is inhibited, few of the single strands will reassociate.
2. *The temperature of the reassociation process.* At very low temperatures (4°C) not only a diffusion limited but, for a DNA molecule which has become mismatched with a strand having only a few complementary bases, the opportunities to break away and continue its search for the correct complementary strands are reduced. Consequently DNA which is heated beyond the T_m and then quickly cooled to a low temperature will be denatured whereas solutions maintained at high temperatures below the T_m may renature.
3. *The size of the DNA fragments.* Large string-like fragments of single-stranded DNA encounter diffusion problems and frequently cannot reanneal correctly since this effectively reduces their opportunities for finding complementary strands. For this reason DNA is often sheared for reannealing experiments.
4. *The ionic strength of the solution.* Two highly charged DNA molecules are likely to repel one another, and the presence of salt is necessary to mask this repulsion in renaturation.
5. *The concentration of the DNA.* Clearly, at higher concentrations the probability of two complementary strands encountering each other is raised.
6. *The time allowed for reannealing experiments.* If renaturation is allowed under ideal conditions, two DNA samples of identical concentration should take different times to reanneal according to the genome size. For this reason the term *Cot value* has been defined for the study of reannealing of DNA and also for the study of the formations of DNA-RNA hybrids. *Co* represents the DNA concentration and *t* represents time in seconds. This is again monitored by ultraviolet spectroscopy or, more frequently, by hydroxyapatite chromatography since this can be used to separate double- and single-stranded DNA molecules and can be used at higher concentrations and hence for shorter times.

7

Replication of DNA

During cell division the chromosomes which are made up of DNA molecules undergo spletting and the sister chromatids move apart to maintain the usual number of chromosomes. Each chromatid has a single chain of DNA. To make it double and to convert chromatid upto chromosome the duplication of DNA becomes essential.

Replication of the DNA Molecule

Kornberg showed that DNA can replicate in a test tube with no cells present. All that is required is a mixture containing DNA, a specific enzyme (which he called DNA-polymerase), and a mixture of the four precursors: the nucleoside triphosphates deoxy-ATP, deoxy-CTP, deoxy-GTP, and deoxy-TTP. If any one of the four nucleoside triphosphates is omitted from the reaction mixture, DNA does not replicate itself. The intact DNA serves as a template for the reaction–a guide to the exact placement of nucleotides in the new strand. Where there is a T in the template, there must be an A in the new strand, and so forth.

Two other models of how the double helix might replicate were suggested after the original paper by Watson and Crick. In the model of conservative replication, the original double helix would serve somehow as a template but would either be reconstituted or, perhaps, would never unwind at all. Thus the new molecule would contain none of the atoms of the original. According to the model of dispersive replication, fragments of the original molecule would serve as templates, assembling two molecules, each containing old and new parts, perhaps at random. In *semiconservative replication* (the model proposed by *Watson* and *Crick*), the original two strands would separate, and each would function as the template for a new partner. Each molecule produced by semiconservative replication would therefore consist of one old and one new strand. After a short time, experimental work confirmed the model of semiconservative replication.

Synthesis of DNA from monomeric units of deoxyribonucleotides using a pre-existing DNA molecule as a template is DNA replication. Three mechanistic models have been proposed regarding DNA replication. These are:

(i) Dispersive
(ii) Conservative
(iii) Semi-conservative.

(i) Dispersive

According to this model, it was proposed that the original DNA is broken into many fragments, which are more or less uniformly distributed among the progenies. Inside the progeny, these fragments grow to form complete DNA segment. However, according to this model one would expect unequal distribution of genes among progenies leading to great variations among them. It would lead to irregular

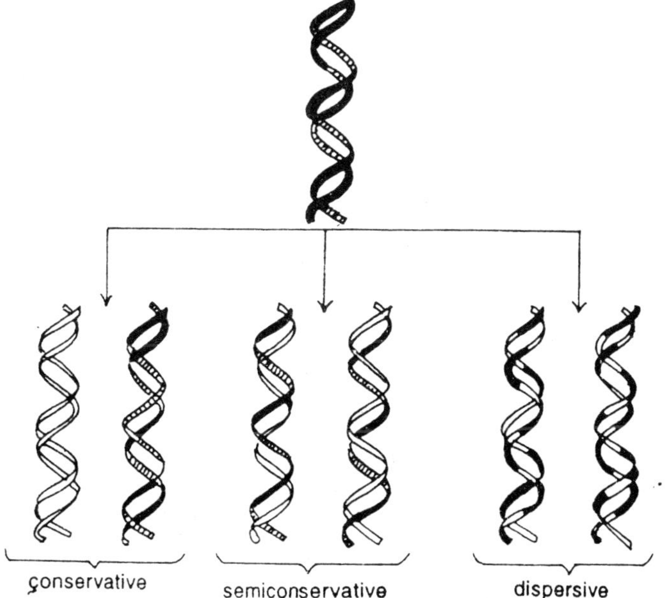

Fig. 7.1. A diagram illustrating three possible modes of DNA replication.

arrangement of nucleotides. However, these expectations are not realized and hence the dispersive mechanism of DNA replication does not find support.

(ii) Conservative

According to this model, the original DNA is copied as such and then one copy of the complete DNA is transmitted to one of the progenies. Thus, according to this model, both primary and secondary structures of the DNA are conserved. There is neither breakage nor unwinding of the parental DNA during replication. The experimental evidence however, do not support this model as well, although such a mechanism of DNA replication is possible structurally and chemically.

(iii) Semi-conservative

This mechanism is based on Watson-Crick's model of DNA structure and is now accepted mechanism of DNA replication. The replication of DNA involves the unwinding of double delical DNA molecule but no breakage of the separated strands. Thus, the primary but not the secondary structures of parental DNA are conserved in progenies.

The idea of semiconservative mechanism of DNA replication came from the careful studies of the double helical model of DNA structure. The complimentary strand structure is an indication of the possible strand separation and formation of complimentary strands on each of the free single strands. During semiconservative replication, the hydrogen bonds between the complimentary base pairs are broken and the two separated strands start replicating as soon as a few bonds are broken. The two strands do not separate completely before the new strands are formed. That is why during the process of DNA replication some 'Y' shaped regions in the DNA molecule are observed. Each strand can act as a template or mould for the formation of new complimentary chains. The nucleotides are put together into short pieces of DNA by one of the DNA polymerases. The short segments of the newly synthesized DNA are known as *Okazaki fragments.* The complimentarity of the strands is maintained because of the selective pairing between A and T and G and C. The small Okazaki fragments are joined together to form complete DNA strands by the enzyme polynucleotide ligase. In experiments, where ligase is selectively inhabited, the accumulation of Okazaki fragments take place.

Experimental Proof

A clever experiment by Matthew Meselson and Franklin Stahl convinced the scientific community that semiconservative replication is the correct model. Working at the California Institute of Technology in 1957, they devised a simple way to distinguish old strands of DNA from new ones. The key was to use a "heavy" isotope of nitrogen. Heavy nitrogen (^{15}N) is a rare, nonradioactive isotope that makes molecules more dense than chemically identical molecules containing the common isotope ^{14}N. To study DNA of different densities (that is, DNA containing ^{15}N versus DNA containing ^{14}N), Meselson, Stahl, and Jerry Vinograd invented a new centrifugation procedure that allowed them to determine the density of DNA from a specific sample. At a certain molarity, a solution of cesium chloride (CsCl) has a density very close to that of DNA; at high gravitational forces produced in an ultracentrifuge, cesium ions sediment to some extent, thus establishing a density gradient. When a DNA sample is dissolved in CsCl and centrifuged at about 100,000 times the force of gravity, the DNA gathers in a layer in the centrifuge tube at a position where the density of the CsCI solution equals that of the DNA. To cluster in this way, DNA that is initially lower in the tube, where the density is greater than its own, must rise; DNA that is in a region of lower density must sink.

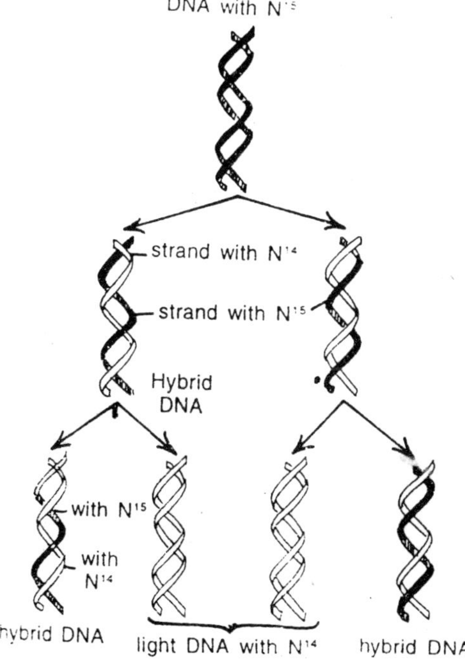

Fig. 7.2. Meselson and Stahl's experiment.

After developing this method of measuring DNA densities, Meselson and Stahl could begin experimenting. They grew a culture of the bacterium *Escherichia coli* for 17 generations on a medium in which the nitrogen source (ammonium chloride, NH_4Cl) was made with ^{15}N instead of ^{14}N. As a result, all the DNA in the bacteria was "heavy." Another culture was grown on medium with ^{14}N. They extracted DNA from both cultures. When these extracts were combined and centrifuged with CsCl, two separate DNA bands formed, showing that this method would work for separating DNA samples of slightly different densities.

Meselson and Stahl then conducted their main experiment. They grew another culture on ^{15}N medium and *transferred* it to normal ^{14}N medium. *E.coli* reproduces every 20 minutes, and Meselson and Stahl collected some of the bacteria from each generation after the transfer. They extracted DNA from the samples. After DNA was duplicated and the cells divided to produce each new generation, the DNA banding in the density gradient was different from the original banding. Initially, the DNA was uniformly labeled with ^{15}N and hence was relatively dense. After one generation, when the DNA had been duplicated once, all the DNA was of an intermediate density. After two generations, there were two equally large DNA bands: one of low density and one of intermediate density. In samples from subsequent generations, the proportion of low-density DNA increased steadily.

These data can be explained by the semiconservative model of DNA replication. The high-density DNA had two ^{15}N strands, the intermediate-density DNA had one ^{15}N and one ^{14}N strand, and the low-density DNA had two ^{14}N strands. In the first round of DNA replication, the strands of the double helix, both heavy with ^{15}N, separated; during the process of separation, each acted as the template for a second strand, which contained only ^{14}N and hence was less dense. Each double helix then consisted of one ^{15}N and one ^{14}N strand and was of intermediate density. In the second replication, the ^{14}N-containing strands directed the synthesis of partners with ^{14}N, creating low-density DNA, and the ^{15}N strands got new ^{14}N partners.

If the DNA had replicated in accord with either of the other models, the results would have been quite different. Under the conservative model, after one generation there would have been two bands–one for heavy DNA (^{15}N–^{15}N) and the other for light (^{14}N–^{14}N); no DNA of intermediate density would have formed at any time. If dispersive replication had taken place, the first round of replication would have produced DNA of intermediate density, but the density of all the DNA would have decreased with each subsequent replication. The crucial observation proving the semiconservative model was that intermediate-density DNA ($^{15}N-^{14}N$) appeared in the first generation and continued to appear in subsequent generations.

Cairn's Autoradiography Experiment

J. Cairns demonstrated the semiconservative type of replication by using a technique of autoradiography as given in the following lines:

In autoradiography technique, the material is first supplied with a suitable radioactive material like tritiated thymidine (H^3-TdR; H^3 is heavy isotope of hydrogen and it replaces normal hydrogen in thymidine to give rise to tritiated thymidine). Tritiated thymidine is used since this will selectively label only DNA and will not label RNA, since thymine base is absent in RNA. The tritiated thymidine gets incorporated into DNA and replaces ordinary thymidine. The material is then sectioned or else the cells may be broken down to release the intact bacterial chromosome on slides. These slides are then covered by photographic emulsion and stored in the dark. During this storage the particles emitted by tritiated thymidine will expose

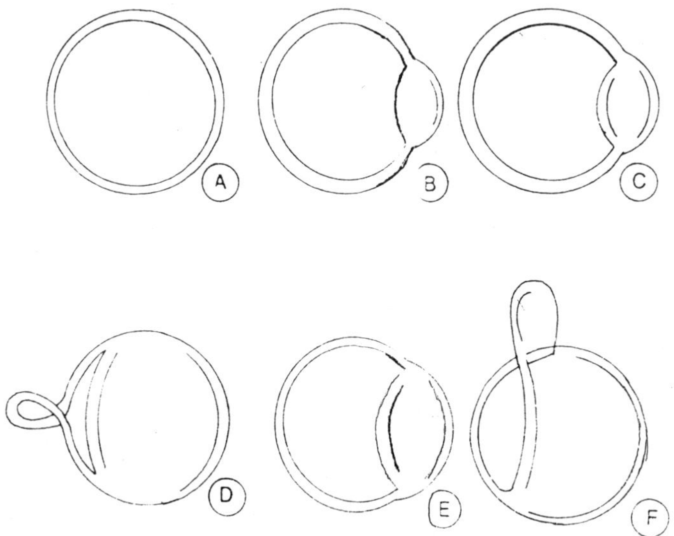

Fig. 7.3. Cairn's model for circular DNA replication.

the film, which can be developed. This photograph will then show the regions of the presence of tritium and thus indirectly the presence of labelled DNA.

Using this above technique, the replication of DNA could be easily followed by *J. Cairns* and the results were reported in 1963. During incorporation of tritiated thymidine, autoradiographs could be prepared at regular known intervals. Autoradiographs from this replicating material prepared at regular known intervals demonstrated semi-conservative mode of replication. It would be expected that in regions, where one cycle of replication of DNA has been completed, the density of dots will be higher than in the region where the replication has not taken place. The lighter density of dots is considered to indicate that only one of the two strands is labelled, while a heavier density of dots will indicate that only one of the two strands is labelled, while a heavier density of dots will indicate that both strands are labelled. Such a situation was actually observed. The rate at which the replication proceeds could also be worked out by measuring the length of DNA undergoing replication in a known interval of time. The generation time was worked out by *Cairns* as 30 minutes in *E.coli.* The length of the chromosome was worked out to be about 1 mm. The rate of replication as obvious would be approximately 30µ to 40µ per minute (1 mm = 1000µ).

That after replication, one of the two strands in the daughter DNA molecules is derived from the parent molecule and the other is synthesized afresh. In θ shaped figure, which is obtained in the second cycle of replication is presence of label, the two areas in the split region would never be equally labelled. For instance, one arc would be twice as heavily labelled as the other arc. This is what was actually observed by *Cairns.* Thus the observation support the semiconservative type of replication of DNA.

Taylor's Experiment on Vicia faba Root Tips

J.H. Taylor (1957) and his coworkers also demonstrated the semiconservative method of DNA duplication in the root tip cells of *Vicia faba* by autoradiography. The roots were grown in a medium containing radioactive thymidine, so that the radioactivity is incorporated in the DNA of these cells. The outline of this labelled chromosome on a photographic film appears in the form of scattered black dots of silver grains. When these root tips with labelled chromosomes were transferred to the unlabelled medium containing colchicine and studied for radioactivity, the following observtions were made:

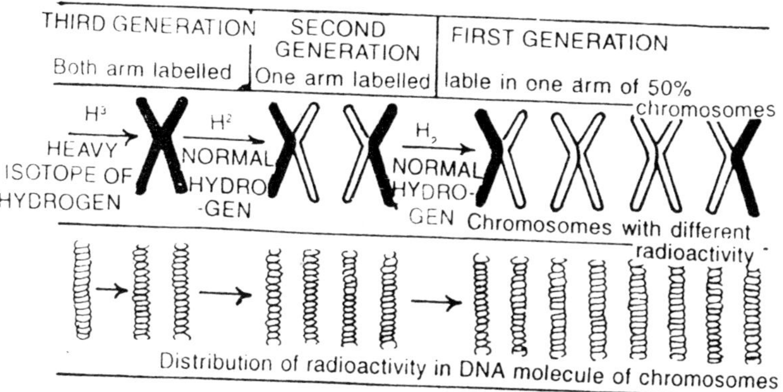

Fig. 7.4. Taylor's experiment on Vicia faba.

1. In the chromosomes of first generation the radioactivity was found to be uniformly distributed in both the chromatids, because in them the original strand of DNA double helix was labelled with radioactivity and the new one was non-labelled.
2. In the chromosomes of second generation only one of the two chromatids in each chromosome was radioactive.

This experiment demonstrates the semiconservative method of chromosome replication. But as it is known that each chromatid is composed of only one double helical molecule of DNA, it is the replication of DNA.

Role of Enzymes and Proteins Replication

The complicated process of DNA replication in *E.coli* requires many different proteins and enzymes, some of which also function in other cellular process such as the repair of DNA damage and genetic recombination. The key enzyme in the synthesis of new DNA is DNA polymerase.

In *E.coli* there are three DNA polymerases. Each catalyzes the DNA template-directed condensation of deoxyribonucleoside 5′-triphosphates. There are significant differences in their activity in DNA synthesis and in the nuclease (DNA or RNA break-down) activities associated with them. All three polymerases catalyze new DNA synthesis in the 5′-3′ direction. The rate of DNA synthesis catalyzed by the three enzymes varies circular prokaryotic DNA molecules replicate bidirectionally (in both directions away from the origin of replication). In these cases there are two replication forks that are mirror images of one another; picture two Ys joined head-to-head by the two pairs of arms. The area of a double-stranded DNA molecule that has denatured for replication is called a *replication bubble.*

In *E.coli* the unwinding of the DNA is catalyzed by an enzyme called helicase, the product of a gene called *rep*. For each 10 base pairs of DNA unwound, one turn of the helix becomes untwisted. As a result, as the helix unwinds a torsional strain is imposed on the DNA. That is, since the DNA molecule is circular, the pulling apart of the helix at one location will cause increased tightening of the molecule elsewhere, much like what happens when the strands of a piece of rope that is fixed at each end are pulled apart. This torsional strain is relieved by the action of enzymes called topoisomerases, which cause single-stranded breaks in DNA away from the replication forks and then allow one strand to rotate relative to the other strand. The unwinding of the DNA produces single-stranded regions, which are stabilized by DNA *single-stranded binding* (SSB) *proteins.* Each SSB protein is a tetramer of a 74,000 dalton subunit, and this tetramer binds to eight bases of single stranded DNA. Over 250 tetramers bind to each replication fork. Once the strands have started to unwind, the internal bases become available for the formation of bonds with bases in the new chain. Initiation of DNA replication next takes place. In *E.coli*, primases bind to the single-stranded DNA of both arms of the replication fork and synthesize short primers. The primers are lengthened by the

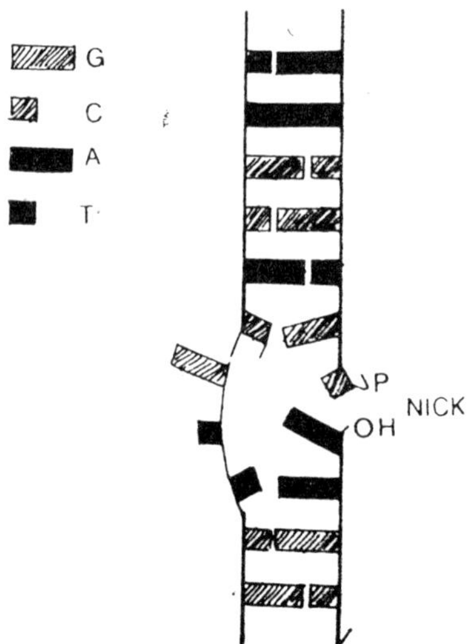

Fig. 7.5. A nicked duplex DNA molecule such as is formed by an endonuclease enzyme.

action of DNA polymerase III, which synthesizes the complementary DNA chains to the template strands. Note that, because of the opposite polarities of the two DNA template strands and the requirement for a 5′- to -3′ direction of new DNA synthesis, the primers are located at different relative positions on the two template strands.

The DNA helix continues to unwind. On the template on the left the new strand continues to be synthesized continuously (in the same direction as the direction of unwinding). However, because DNA polymerases can only synthesize DNA in the 5′-to-3′ direction, the new synthesis on the template on the right has gone as far as it can. Thus, a new initiation of DNA synthesis occurs on the right-hand template on the newly exposed single-stranded template close to the site of helicase activity. As before, a primer is made, which is lengthened by DNA polymerase III action. The overall result is that the new DNA strand being made on the left-hand template is synthesized continuously while the new DNA being made on the right-hand template is synthesizd discontinuously so that Okazaki fragments are produced. Eventually the Okazaki fragments are joined together into a continuous DNA strand. This process requires the activities of the enzymes DNA polymerase I and DNA ligase. If we consider two adjacent Okazaki fragments, the 3′ end of the newer DNA fragment is adjacent to but not joined to the primer 5′ end of the previously synthesized fragment. The DNA polymerase III that has just completed the synthesis of the newer fragment now dissociates from the DNA and DNA polymerase I takes its place. This enzyme continuous the 5′-to-3′ synthesis of the newer DNA fragment, at the same time removing the primer of the older fragment through the action of the 5′-to-3′ exonuclease function of DNA polymerase I. When DNA polymerase I is finished, there is a gap between the two new DNA fragments, and this gap is sealed in a reaction catalyzed by DNA ligase, thereby producing a longer DNA strand. The whole process is completed until all the DNA is replicated.

Replication Fork

How is semiconservative DNA replication accomplished? That is depends upon enzymes should not surprise you. We now know that there are different types of DNA polymerases, with different functions, and that the replication process is intricate. Kornberg's basic observations, including the need for a DNA template and for a mixture of nucleoside triphosphates, still hold. He also showed that nucleotides are always added to the growing chain at the same end: the 3′ end, the

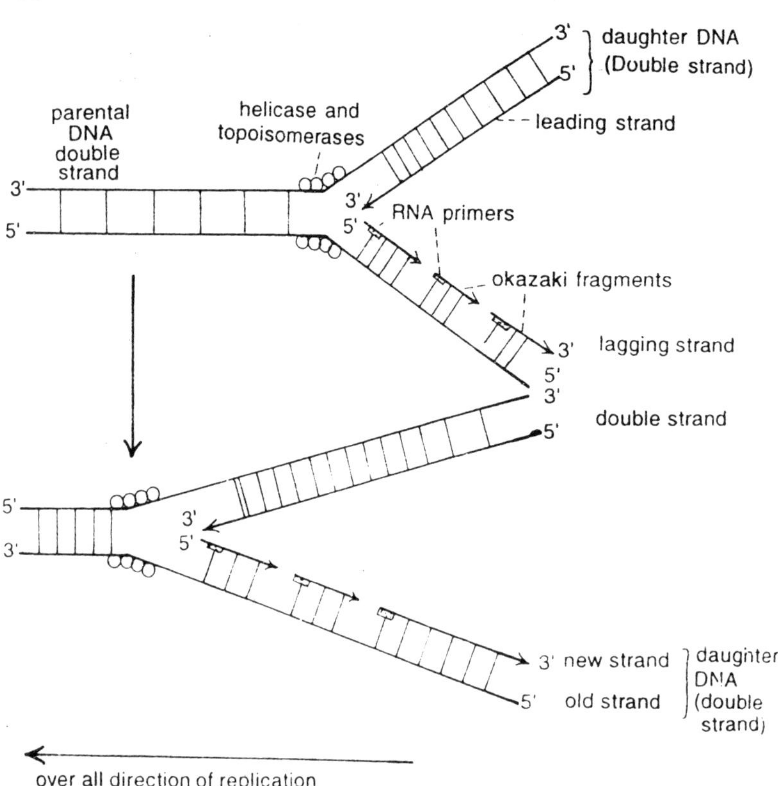

Fig. 7.6. Diagrammatic representation of discontinuous synthesis of DNA.

end at which the DNA strand has a free hydroxyl groups on the 3′ carbon of its terminal deoxyribose. This hydroxyl groups reacts with a phosphate group on the 5′ carbon of the deoxyribose of a deoxyribonucleoside triphosphate, and thus the chain grows. Bonds linking the phosphate groups of the deoxyribonucleoside triphosphate break and thereby release the energy for this reaction. Two of the phosphate groups diffuse away, and one becomes part of the sugar-phosphate backbone of the growing DNA molecule.

For double-stranded DNA to replicate, the strands must unwind and separate from each other. Only then can they function as templates for the synthesis of new, complementary strands. The unwinding results in a *replication fork*–a moving, Y-shaped structure that is the region where new DNA strands are being synthesized. Recall that the two original strands are antiparallel, with the 3′end of one strand paired with the 5′ end of the other. As the replication fork moves along the parent DNA molecule, an enzyme, DNA polymerase III, catalyzes the replication of both strands. How can this be accomplished, given that new nucleotides are added only at the 3′ end of a polynucleotide chain?

One parent strand is being exposed beginning at its 3′ end, which presents no problem–its complementary strand is synthesized continuously as the replication fork proceeds. This daughter strand is called the *leading strand*. The other daughter strand, the *lagging strand*, is produced in discontinuous spurts (100 to 200 nucleotides at a time in eukaryotes; 1,000 to 2,000 at a time in prokaryotes). The discontinuous stretches are synthesized just as the leading strand is, by adding the 5′ end of a nucleotide to the 3′ end of the daughter strand, but the stretches are synthesized in the opposite direction with respect to the replication fork. These stretches of new DNA for the lagging strand are called *Okazaki fragments* after their discoverer, the Japanese biochemist Reiji Okazaki. While the leading strand grows continuously "forward," the lagging strand grows in shorter, "backward," stretches with gaps between them. The gaps between the Okazaki fragments are then filled in by DNA polymerase I, and another enzyme, *DNA ligase*, links the fragments.

Working together, two DNA polymerases, DNA ligase, and several other proteins do the complex job of DNA synthesis with a speed and accuracy that are almost unimaginable. In *E.coli*, the complex makes new DNA at a rate in excess of 1,000 base pairs per second and makes mistakes in fewer than one base in 10^8–10^{12}.

On a bacterial chromosome, DNA synthesis begins at just one point, the *origin of replication*. Each chromosome of a eukaryote, on the other hand, has many origins of replication. DNA synthesis may thus proceed simultaneously in many areas of a single eukaryotic chromosome. Synthesis proceeds in both directions from an origin of replication as two replication forks move away from it.

Enzymes used in Replication

The enzymatic synthesis of DNA is a complex process, primarily because of the need for high fidelity in copying the base sequence and for physical separation of the parental strands. The number of steps that must be completed is far too great to be accomplished by a single enzyme and, in fact, about twenty proteins are known at present to be necessary. Thus, in an effort to provide some understanding with a minimum of confusion, each step in the process will be treated separately. We will consider the basic chemistry of polymerization, the source of the precursors, the problems raised by the chemistry of polymerization, the means of initiating and terminating synthesis, and the mechanisms for eliminating replication errors.

Polymerases

In 1957, Arthur Kornberg showed that in extracts of *E.coli* there exists a DNA polymerase (now called *polymerase I* or *pol I*). This enzyme was able to synthesize DNA from four precursor molecules–namely, the four deoxynucleoside 5′-triphosphates (dNTP), dATP, dGTP, dCTP, and dTTP–as long as a DNA molecule to be copied (a *template* DNA) was provided. Neither 5′-monophosphates nor 5′-diphosphates, nor 3′-(mono-, di-, or tri-) phosphates can be polymerized–only the 5′-triphosphates are substrates for the polymerization reaction; soon we will see why this is the case. Some years later, it was found that pol I, though playing an essential role in the replication process, is not the major polymerase in *E.coli*; instead, the enzyme responsible for advance of the replication fork is polymerase III or pol III. Pol III also exclusively uses 5′-triphosphates as precursors and requires a DNA template before polymerization can occur. Pol I and pol III have many features in common and, in fact, a few types of DNA molecules replicate by using only pol I. The overall chemical reaction catalyzed by both DNA polymerases is:

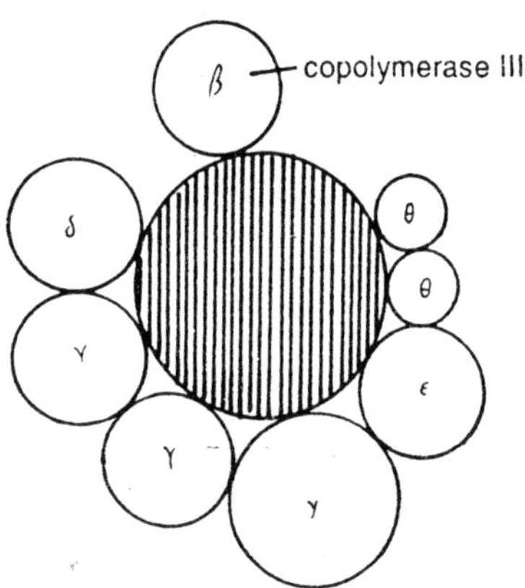

Fig. 7.7. DNA polymerase-III holoenzyme. It consists of subunits α, β, γ, ε, θ, δ and ε.

$$\text{Poly(nucleotide)}_n\text{-3}'\text{-OH} + \text{dNTP} \rightarrow \text{Poly(nucleotide)}_{n+1}\text{-3}'\text{-OH} + \text{PP}$$

in which PP represents pyrophosphate cleaved from the dNTP.

Pol I and pol III have many features in common. Both enzymes only polymerase deoxynucleoside 5′-triphosphates and can do so only while copying a template DNA. Furthermore, polymerization can only occur by addition to a *primer*–that is, an oligonucleotide hydrogen-bonded to the template strand and whose terminal 3′-OH group is available for reaction (that is, a "free" 3′-OH group). The meaning of a primer is already made clear, which depicts six potential template molecules; of these, only three can be said to be active–(c), (e), and (f)–each of which has a free 3′-OH group. The lack of activity with (d) and the direction of synthesis with (e) and (f) indicate that nucleotides do not add to a free 5′-P group. The lack of any synthesis with (a) or (b) indicates that addition to a 3′-OH group cannot occur if there is nothing to copy. Thus we draw two conclusions:

1. Both a primer with a free 3′-OH group and a template are needed.
2. Polymerization consists of a reaction between a 3′-OH group at the end of the growing strand and an incoming nucleoside 5′-triphosphate. When the nucleotide is added, it supplies another free 3′-OH group. Since each DNA strand has a 5′-P terminus and a 3′-OH terminus, strand growth is said to proceed in the 5′→3′ (5′-to-3′) direction.

Occasionally polymerases add a nucleotide terminus that cannot hydrogen-bond to the corresponding base in the template strand. This may be purely a mistake or may result from the tautomerization of adenine and thymine. In any case, it is important that the unpaired base be removable while its incorrectness is recognizable–namely, as an unpaired base at the 3′-OH terminus of a growing strand.

Pol I responds to an unpaired terminal base by terminating polymerizing activity, because the enzyme requires a primer that is correctly hydrogen-bonded. When such an impasse is encountered, a 3′→5′ exonuclease activity, which may be thought of simply as pol I running backwards or in the 3′→5′ direction, is stimulated, and the unpaired base is removed. After removal of this base, the exonuclease activity stops, polymerizing activity is restored, and chain growth begins again. This exonuclease activity is called the *proofreading* or *editing function* of pol I.

Another function of polymerase I is that of a 5′→3′ exonuclease. This activity has the following features:

1. Nucleotides are removed from the 5′-P terminus only, one by one.
2. More than one nucleotide can be removed by successive cutting.
3. The nucleotide removed must have been base-paired.

4. The nucleotide removed can be either or the deoxy-or the ribo-type.
5. Activity can be at a nick as long as there is a 5´-P groups.

The main function of the 5´→3´ exonuclease activity is to remove ribonucleotide primers. The 5´→3´ exonuclease activity at a single-strand break (nick) can occur simultaneously with polymerization. That is, as a 5´-P nucleotide is removed, a replacement can be made by the polymerizing activity. Since pol I cannot form a bond between a 3´-OH group and a 5´-monophosphate, the nick moves along the DNA molecule in the direction of synthesis. This movement is called *nick translation.*

Experimental conditions can be chosen so that polymerization will occur at a single-strand break without concomitant 5´→3´ exonuclease activity. The growing strand then displaces the parental strand. This is thought to be an important step in the mechanism of genetic recombination. Of all *E.coli* polymerases known to date, polymerase I is the only one capable of carrying out an unaided displacement reaction. In other strand displacement reactions, auxilary proteins are required and ATP is cleaved to fuel the unwinding of the helix; this will be discussed when the events at a replication fork are described.

Pol III is a very complex enzyme. In its most active form it is associated with eight other proteins to form the *pol III holoenzyme,* occasionally termed pol III. The term holoenzyme refers to an enzyme that contains several different subunits and retains some activity even when one or more subunits is missing. The smallest aggregate having enzymatic activity is called the *core enzyme.* The activities of the core enzyme and the holoenzyme are usually very different. Genes encoding five of the subunits have been identified; these are called *dnaE, dnaN, dnaQ, dnaX,* and *dnaz.* The *dnaE* protein possesses the major polymerizing activity but each of the subunits, except for the *dnaQ* protein is essential for replication. Pol III shares wtih pol I a requirement for a template and a primer but its substrate specificity is much more limited. For instance, pol III cannot act at a nick nor is it active with single-stranded DNA primed by either a DNA or RNA nucleotide fragment. The principal activity *in vitro* is on gapped DNA in which the gap is less than 100 nucleotides long. Such a gap is akin to the state of the DNA at a replication fork–that is, the parental strands are separated and bear short single-stranded regions ahead of the growing daughter chain.

Pol III cannot carry out strand displacement either, and another system is needed to unwind the helix in order that a replication fork will be able to proceed. The enzyme, like pol I, possesses a 3′→5′ exonuclease activity which performs the major editing function in DNA replication. This function is carried out by the *dnaE* subunit, which is also the major polymerizing subunit, as we have just mentioned. The *dnaQ* subunit plays an important role in editing, also, but the biochemical basis of the role is not yet known; it probably interacts with the *dnaE* subunit. The principal evidence supporting the view that it is involved in providing fidelity to the replication process is that bacteria containing a mutation in the *dnaQ* gene have a somewhat higher mutation frequency. Pol III also possesses a 5′→3 exonuclease activity; however, the enzyme acts only on single-stranded DNA so that it cannot carry out nick translation. The biological role of the 5′→3′ exonuclease activity of pol III is unknown at present. Although pol III holoenzyme is the major replicating enzyme in *E.coli*, much less is known about it than about pol I, because it is a more complex enzyme. Study of pol III is currently an active field of research.

All known polymerases (for both DNA and RNA) are capable of chain growth in only the 5′→3′ direction; that is, the growing end of the polymer must have a free 3′-OH group. It is possible for the following reasons that the enzymes evolved in this way to facilitate editing. If 3′→5′ growth were to occur, the growing strand would also be terminated with a 5′-triphosphate and the 3′-OH group of the incoming nucleotide would react with it. Chemically this is certainly acceptable but since the bonds formed contain only a single phosphate, an editing function would leave a free 5′-monophosphate. In order for chain growth to proceed, an enzymatic system would be needed to enter the replication fork and convert the monophosphate to a triphosphate. There is already a great deal going on in the replication fork, so that it would seem more economical for the cell to require 5′→3′ growth exclusively. However, the observation that chain growth proceeds in only one direction introduces what is probably the greatest complication in the entire replication process; this will be described shortly.

DNA Ligase

Neither replication from a primed circular single strand nor gap filling results in a continuous daughter strand. Discontinuity results because no known polymerase can join a 3′-OH and a 5′-monophosphate group. The joining of these groups is accomplished by the enzyme *DNA ligase*, which functions in replication and other important processes. *E.coli* DNA ligase can join a 3′-OH group as long as both are termini of adjacent base-paired deoxynucleotides–the enzyme cannot bridge a gap.

In the usual polymerization reaction, the activation energy for phosphodiester bond formation comes from cleaving the triphosphate. Since DNA ligase has only a monophosphate to work with, it needs another source of energy. It obtains this energy by hydrolyzing either ATP or nicotine adenine dinucleotide (NAD); the energy source depends upon the organism from which the DNA ligase is obtained. The *E.coli* DNA ligase uses NAD.

Continuous and Discontinuous DNA Replication

Autoradiographic evidence leads us to believe that replication is occurring simultaneously on both strands. *Continuous* replication is,

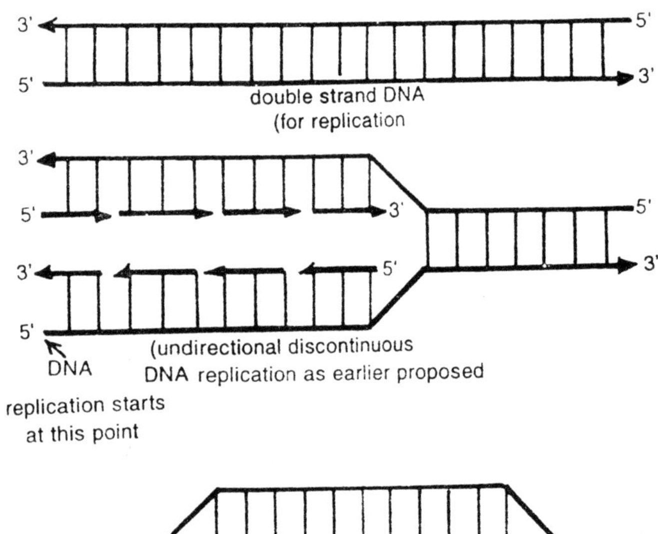

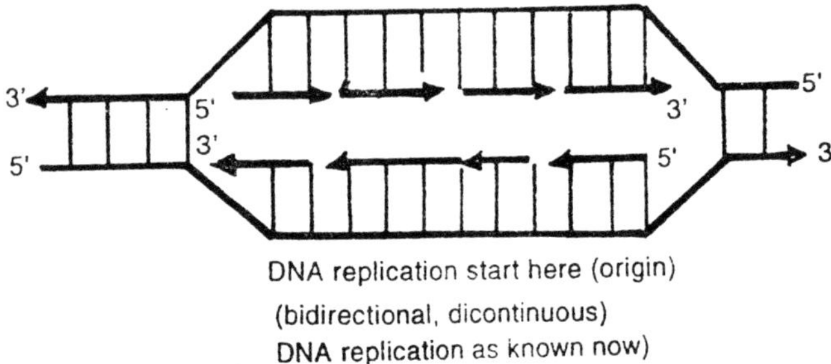

Fig. 7.8. Discontinuous DNA replication.

of course, possible on the 3′→5′ template strand, which begins with the necessary 3′-OH *primer.* (Primer is double-stranded DNA–or, as we shall see, a DNA-RNA hybrid–continuing as single-stranded DNA template. The strand being synthesized has a 3′-OH available). A *discontinuous* form of replication takes place on the complementary strand, where it occurs in short segments, backward, away from the Y-junction. These short segments, called *Okazaki fragments* after *R. Okazaki* who first saw them, average about 1,500 nucleotides in prokaryotes and 150 in eukaroytes. The strand synthesized continuously is referred to as the *leading strand,* and the strand synthesized discontinuously is referred to as the *lagging strand.*

Once initiated, continuous DNA repliction can proceed indefinitely. DNA polymerase III on the leading strand template has what is called high *processivity*: once it attaches, it does not release until the entire strand is replicated. Discontinuous replication, however, requires the repetition of four steps primer synthesis, elongation, primer removal with gap filling, and ligation.

Primer synthesis and elongation

In order for Okazaki fragments to be synthesizd, a primer must be created de novo (Latin, from the beginning). None of the DNA polymerases can create that primer. Instead, one of two enzymes, either *RNA polymerase,* the transcribing enzyme or, more commonly, *primase,* an RNA polymerase coded for by the dnaG gene, creates the primer. It is from two to sixty nucleotides, depending on the species, at the site of Okazaki fragment initiation. The result is a short RNA primer that provides the free 3′-OH group that DNA polymerase III needs in order to synthesize the Okazaki fragment. DNA polymerase III continues until it reaches the primer RNA of the previously synthesized Okazaki fragment. At that point it stops and releases from the DNA.

All three prokaryotic polymerases not only can add new nucleotides to a growing strand in the 5′→3′ direction but also can remove nucleotides in the opposite 3′→5′ direction. This property is referred to as *3′→5′ exonuclease activity.* Enzymes that degrade nucleic acids are classified as *exonucleases* if they remove nucleotides from the end of a nucleotide strand or as *endonuclease* if they can break the sugar-phosphate backbone in the middle of a nucleotide strand. At first glance, exonuclease activity seems like an extremely curious property for a polymerase to have–curious unless we think about its ability to check complementarity. If the complementarity is improper, which is to say that the wrong nucleotide has been inserted, the polymerase can remove the incorrect nucleotide, put in the proper one, and continue on its way. This is known as the *proofreading* function of the DNA polymerase. In addition, the RNA primers of Okazaki fragments can be removed by exonuclease activity.

Role of primer

DNA polymerase I is a polymerase when it adds nucleotides, one at a time, and an exonuclease when it removes nucleotide one at a time. To complete the Okazaki fragment, DNA polymerase I acts in both capacities. (Mutants of DNA polymerase I cannot properly connect Okazaki fragments.) DNA polymerase I completes the Okazaki fragment by removing the previous RNA primer and replacing it with DNA nucleotides. When DNA polymerase I has completed its nuclease and polymerase activity, the two previous Okazaki fragments are almost complete. All that remains is a single phosphodiester bond to be made.

Ligation

DNA polymerase I cannot make the final bond to join the Okazaki fragment to the previously synthesized DNA. An enzyme, *DNA ligase*, completes the task by making the final phosphodiester bond in an energy-requiring reaction.

A question of evolutionary interest is why RNA is used for priming of DNA synthesis. Why not use DNA directly and avoid the exonuclease and resynthesis activity? One possible answer is that because priming is inherently more error-prone then regular DNA synthesis it is best for the cell to have the primer nucleotides removed and replaced by DNA synthesized in a less error-prone fashion before DNA synthesis is completed. If the priming nucleotides are RNA, then DNA polymerase I can recognize and remove them in the final stage of Okazaki fragment synthesis, at which time it replaces them with DNA nucleotides inserted with a low error rate.

Another question of evolutionary interest is why DNA synthesis cannot take place in the $3' \rightarrow 5'$ direction. Perhaps the answer has to do with proofreading and the exonuclease removal of incorrect nucleotides. When an incorrect nucleotide is found and removed, the next nucleotide brought in, in the $5' \rightarrow 3'$ direction, will have a triphosphate end available to provide the energy for its own incorporation. Consider what would happen if the polymerase were capable of adding nucleotides in the opposite direction. The energy for the diester bond would be coming from the triphosphate already attached in the growing $3' \rightarrow 5'$ strand. Then, if an error in complementary were detected and the most recently added nucleotide were removed from the $3' \rightarrow 5'$ strand by the polymerase, the last nucleotide in the double helix would no longer have a triphosphate available to provide energy for the diester bond with the next nucleotide brought in. Continued polymerization would thus require additional enzymatic steps to provide the energy for the process to continue. This could slow the process down considerably. As it is, the process works at a speed of about four hundred nucleotides incorporated per second with an error rate of about one incorrect pairing per one hundred thousand bases, improved to a rate of only one mistake in ten million by exonuclease proofreading. (Other repair system can improve this error rate another thousandfold, to about one error every 10^{10} times an average base is replicated.

The Initiation of DNA Replication

Each replicon (e.g., the *E.coli* chromosome or a segment of eukaryote chromosome) must have a region in which DNA replication is initiated. In *E.coli* this region is referred to as the genetic locus *oriC*. In order for DNA replication to begin, several steps must occur. First, the specific origin site must be recognized by the appropriate protein. Then the site must be opened and stabilized. And, finally, a replication fork must be initiated in both directions, involving continuous and discontinuous DNA replication. Although many of the proteins involved are known, all the steps at the enzymatic level are not, and hence our understanding is a bit sketchy. Following is a description, most of whose steps are known.

OriC, the origin of replication in *E.coli,* is about 245 base pairs long and is recognized by proteins called *initiator proteins* that open up the double helix. The initiator proteins then take part in the attachment of *primosomes*, a complex of two proteins: a primase, which creates RNA primers, and DNA *helicase*, which unwinds DNA at the Y-junction. As the primosomes move along, they create RNA primers used by DNA polymerase III to initiate Okazaki fragments. At some point, leading-strand synthesis begins and Y-junction activity then proceeds as outlined earlier.

In some phages and plasmids, the initiation of replication is not with an RNA primer but with a protein. This protein provides the primer configuration with an OH group from an amino acid. The generality of this type of priming has not been established. Another interesting protein interaction at the origin of replication involves the reverse of initiation of DNA synthesis, the prevention of the initiation of DNA synthesis. This is accomplished by a newly discovered protein, called an "off switch." That is, this protein binds to the DNA at *oriC* and apparently prevents DNA replication from beginning. It does this by its binding activity that prevents the initiator proteins from opening the DNA. Thus this protein may be a very important component in control of the cell cycle stopping the cell cycle from beginning. Presumably, when the appropriate time comes for the cell cycle to begin, the protein is removed.

Events at the Y-Junction

We now have the image of DNA replication proceeding by a primosome, moving along the lagging strand template, opening up the DNA (helicase activity), and creating RNA primers (primase activity). One DNA polymerase III moves along the leading-strand template generating the leading strand by continuous DNA replication, whereas a second DNA polymerase III moves backward, away from the Y-junction, creating Okazaki fragments. *Single-strand binding proteins* (ssb proteins) keep single-stranded DNA stabilized (open) during this process, and DNA polymerase I and ligase are connecting Okazaki fragments.

This simple picture is slightly complicated by the fact that a single DNA polymerase seems to do the entire lagging strand, rather than dropping off the DNA at the completion of an Okazaki fragment and being replaced by a new one at the newest primer near the Y-junction. In addition, there is evidence that the lagging and leading-strand synthesis is coordinated. The *replisome* model has arisen in which both copies of DNA polymerase III are attached to each other and work in concert with the primosome at the Y-junction. According to this model, a single replisome consisting of two copies of the DNA polymerase III *holoenzyme* (each actually made of seven subunits), a helicase, and a primase, move along the DNA. The leading-strand template is immediately fed to a polymerase, whereas the lagging -strand template is not acted on by the polymerase unit an RNA primer has been placed on the strand, meaning that a long (fifteen hundred base) single strand has been opened up.

As the replisome moves along, another single-stranded length of the lagging-strand template is formed. At about the time that the Okazaki fragment is completed, a new RNA primer has been created. The Okazaki fragment is released and a new Okazaki fragment is begun, starting with the latest primer taking the replisome back to the same configuration, but one Okazaki fragment farther along.

Supercoiling

The simplicity and elegance of the DNA molecule masks an inevitable problem of coiling. Since the DNA molecule is made from two strands that wrap about each other, certain operation, such as DNA replication and its termination, meet topological difficulties. Up to this point, we have seen the circular *E.coli* chromosome in its "relaxed" state. However, there are enzymes in the cell that cause DNA to become overcoiled (positively *supercoiled*) or undercoiled (negatively supercoiled). Positive supercoiling comes about either from too many turns of the DNA in a given length or from the molecule wrapping around itself.

Positive supercoiling comes from having the circular duplex wind about itself in the same direction as the helix twists (right handed), whereas negative supercoiling comes about by having the duplex wind about itself in the opposite direction as the helix twists (left handed). The former state increases the number of turns of one helix around the other side (the *linkage number*, L), whereas the latter decreases it. The three forms of DNA, all have the same sequence yet differ in their linkage number. They are referred to as topological isomers (*topoisomers*). The enzymes that create or alleviate these states are called *topoisomerases.*

Topoisomerases affect supercoiling by either of two methods. Type I topoisomerases break one strand of a double helix and, while binding the broken ends, pass the other strand through the break. The break is then sealed. Type II topoisomerases (e.g., *DNA gyrase* in *E.coli*) do the same sort of thing only instead of breaking one strand of a double helix, they break both and pass another double helix through the temporary gap.

As DNA replication proceeds, positive supercoiling builds up ahead of the Y-junction. This is eliminated by the action of topoisomerases that either create negative supercoiling ahead of the Y-junction in preparation for replication or alleviate positive supercoiling after it has been created. The major components of DNA replication in *E.coli* are summarized in table 7.1.

Termination of Replication

The termination of the replication of a circular chromosome presents no major topological problems. The theta-structure replication finishes with both Y-junctions having proceeded around the molecule. The leading strand on one template closes in on the lagging strand begun in the other direction with the same happening on the other template. The process stops with about twenty-five twists remaining at no particular spot on the chromosome (there is no "termination" locus). A topoisomerase then release the two circles and DNA polymerase I and ligase close them up.

Table 7.1. Summary of the Enzymes Involved in DNA Replication in E.coli.

Enzyme (Protein)	*Genetic Locus*	*Function*
DNA polymerase I	*pol A*	Gap filling and primer removal
DNA polymerase II	*pol B*	?
DNA polymerase III		
α subunit	*dnaE (polC)*	DNA replication
β subunit	*dnaN*	DNA replication
γ subunit	*dnaX*	DNA replication
δ subunit	?	DNA replication
ε subunit	*dnaQ*	3′→5′ exonuclease
θ subunit	?	DNA replication
τ subunit	*dnaX*	DNA replication
Initiator protein	*dnaA*	Binds to origin of replication
RNA polymerase subunit	*rpoA, B, C, D*	RNA primer in some system
Primase	*dnaG*	RNA primer in some system
DNA ligase	*lig*	Closes nicked DNA strands
Helicase	*rep*	Unwinds DNA for replication
Ssb proteins	*ssb*	Single-strand stability
DNA topoisomerase I	*topA*	Supercoiling of DNA
DNA topoisomerase II		
α subunit	*gyrA (nalA)*	ATPase
β subunit	*gyrB (cou)*	Cutting, closing of DNA

Several different mechanisms have been explored for the termination of the linear chromosomes of some viruses and all eukaryotic genomes. Linear molecules have the problem of completing the last Okazaki fragment. An RNA primer on the very tip of the 3′→5′ template cannot be replaced by DNA polymerase I, assuming even that a final primer can be put on the very tip of the molecule. In eukaryotes, an enzyme, telomerase, attaches repeats of a short sequence at each chromosome tip.

Replication Models

The model of DNA replication that we have presented here comes primarily from evidence gathered in *E.coli*, which replicates by way of the *theta*-structure intermediate. However, two other modes of replication occur in circular chromosomes: rolling-circle and D-loop.

Rolling-Circle Model

In the *rolling-circle* mode of replication, a nick (a break in one of the phosphodiester bonds) is made in one of the strands of the circular DNA, resulting in replication of a circle and a tail. This form of replication occurs in the Hfr *E.coli* chromosome, or the F plasmid, during conjugation. The F^+ or Hfr cell retains the circular daughter while passing the linear tail into the F^- cell. This method is also used in several phages, which fill their heads (protein coats) with linear DNA replicated from a circular parent molecule.

In the model for rolling-circle replication, the nick made in one strand creates a free 3´-OH end and a free 5´-PO_4 end. Synthesis of a new circular strand occurs by addition of nucleotides to the 3´ end using the complementary intact strand as a template. No primer is needed because the original break produces a primer configuration (3´-OH). As nucleotides are added to one end of the broken strand in a continuous fashion, the other end is displaced as a 5´-PO_4 tail. As replication of the circular templates occurs, the 5´-PO_4 tail is replicated in a discontinuous manner, and the resulting double helix can be severed from the double-helical circle by a nuclease. DNA ligase closes the replicated circular strand and can join the ends of the replicated tail into a circle in the F^- cell, or can package the linear molecule in a phage head, depending upon which type of circular DNA has been replicated.

D-Loop Model

Chloroplasts and mitochondria have their own circular DNA molecules that appear to replicate by a slightly different mechanism than those described. The origin of replication is at different point on each of the two parental template strands. Replication begins on one strand, displacing the other while forming a displacement loop or *D-loop* structure. Replication continues until the process passes the origin of replication on the other strand. Replication is then initiated on the second strand, in the opposite direction. The result is two circles. Some species have chloroplasts and mitochondria with circular DNAs that have multiple D-loops formed.

Eukaryotic DNA Replication

As we saw earlier, linear eukaryotic chromosome usually have multiple origins of replication resulting in figures referred to as "bubbles" or "eyes." Multiple origins allow eukaryotes to replicate their larger quantities of DNA in a relatively short time, even though eukaryotic DNA replication is considerably showed by the presence of histone proteins associated with the DNA to form chromatin. For example, the *E.coli* replication fork moves about twenty-five thousand base pairs per minute, whereas the eukaryotic Y-junction moves only about two thousand base pairs per minute. The number of replications in eukaryotes varies from about five hundred in yeast to as many as sixty thousand in a diploid mammalian cell.

Much less is understood about eukaryotic DNA replication because of the complexity of eukaryotes and the relatively shorter time during which they have been studied effectively. We presume that eukaryotes have solved the same problems faced by prokaryotes in a similar, but not identical, fashion. For example, eukaryotes have five types of DNA polymerases, named DNA polymerase α, β, γ, δ, and ε. DNA polymerases γ, δ, and ε have exonuclease activity. DNA polymerases α and δ are the major replicating enzymes, with polymerase α replicating the lagging strand and polymerase δ replicating the leading strand. The role of polymerase ε is unclear; it seems capable of regular leading- or lagging-strand replication. DNA polymerase β is the major repair polymerase (like polymerase I in prokaryotes). DNA polymerase γ appears to be concerned primarily with mitochondrial DNA replication.

Table 7.2. Eukaryotic DNA Polymerases

Enzyme	*Function*
DNA polymerase α	Replication of nuclear chromosomes (lagging strand)
DNA polymerase β	Repair of nuclear chromosomes
DNA polymerase γ	Replication of mitochondrial chromosomes
DNA polymerase δ	Replication of nuclear chromosomes (leading strand)
DNA polymerase ε	Probably replication of nuclear chromosomes

8

RIBONUCLEIC ACID (RNA)

Although double stranded RNAs are not uncommon, mainly RNA is found as single stranded molecules. The chemical analysis shows the following two main differences in the polynucleotide chains of DNA and RNA:

1. While deoxyribose sugar is found in DNA, in RNA, ribose sugar is found. This difference between DNA and RNA is the most important and would distinguish the two under all conditions.
2. While the DNA, adenine, guanine, thymine and cytosine are the four common bases; in RNA, the four bases are adenine, guanine, uracil and cytosine. Since there is a large number of other unusual bases found both in RNA as well as in DNA and also since thymine may rarely be found in RNA also, this distinction is not as important as that of sugar molecule.

There was yet a third distinction described, i.e, double stranded nature of DNA and single stranded nature of RNA. This distinction is no longer tenable in so far as single stranded DNA and double stranded RNA are now known in number of cases.

While DNA is perhaps always genetic in nature, RNA is only rarely genetic in nature. This would mean that although the primary function of storing and carrying the genetic information is associated with RNA also in some organisms, but mainly RNA found, in all organisms, would perform different important functions during protein synthesis. Therefore, two types of RNA are really known namely genetic RNA and non-genetic RNA.

Time of Transcription

In most eukaryotes most transcription occurs only during interphase, a time when the chromosomes are in their highly diffuse

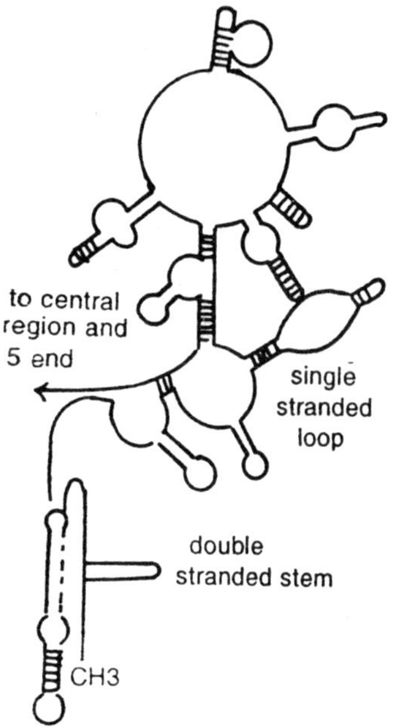

Fig. 8.1. Showing the secondary structure of RNA.

state. However, in case of dipteran polytene chromosomes and lampbrush chromosomes of amphibian oocytes, condensed chromosomes are transcribed, but in them transcription occurs only in that portion of chromosome which becomes diffuse due to puff formation (in polytene chromosomes) or due to loop formations in lampbrush chromosomes.

Table 8.1. Different RNA viruses and the nature of RNA associated with them.

Virus	*Type of RNA*
Plant Virus	
TMV	Single stranded
Wound tumour	Double stranded
Animal Viruses	
Influenza virus	Single stranded
Rous sarcoma	Single stranded
Poliomyelitis	Single stranded
Reovirus	Double stranded
Bacteriophages	
MS2, F2, r17	Single stranded

TYPES OF NON-GENETIC RNA

According to their specific functions during the process of protein synthesis, following kinds of non-genetic RNA molecules have been recognized in prokaryotic and eukaryotic cells:

Ribosomal RNA (rRNA)

Ribosomal RNA (rRNA) or insoluble RNA constitutes the largest part (up to 80%) of the total cellular RNA. It is found primarily in the ribosomes although, since it is synthesized in the nucleus it is also detected in that organelle. It contains the four major RNA bases with a slight degree of methylation, and shows differences in the relative proportions of the bases between species. Its molecules appear to be single polynucleotide strands which are unbranched and flexible. At low ionic strength, rRNA behaves as a random coil, but with increasing ionic strength the molecule shows helical regions produced by base pairing between adenine and uracil and guanine and cytosine.

Types and Synthesis of rRNA

The eukaryotic cells have three kinds of rRNA molecules, namely 28S rRNA (the sedimentation constant varies between 25S and 30S depending up on the species), 18S rRNA and 5S rRNA. The 28S rRNA and 5S rRNA occur in 60S ribosomal subunit, while 18S rRNA occurs in 40S ribosomal subunit of 80S ribosomes of eukaryotes. The prokaryotic cells also contain three kinds of rRNA molecules, namely 23S rRNA, 16s rRNA and 5S rRNA. The 23S rRNA and 5S rRNA occur in 50S ribosomal subunit, while 16S rRNA occur in 30S ribosomal subunit of 70S ribosomes of prokarytes. In eukaryotic cells of plant and animals the 18S and 28S rRNA molecules are transcribed by rDNA of nucleolar organizer (No) region of chromosome and nucleolus acts as a site of synthesis and maturation of these rRNA molecules. 5S rRNA is transcribed by rDNA which occur outside the nucleolar organizer region.

The function of rRNA molecules are still little understood. However, they are supposed to have some definite role in protein synthetic process, besides being the major constituent molecules of ribosomes.

Messenger RNA (mRNA)

The RNA molecules which are transcribed from large number of genes of the total genome (i.e., 99 per cent genes of the total genome of *E. coli*), and have base sequence complementary to DNA, carry DNA's genetic informations for the assembly of amino acids into the polypeptide chains (protein molecules) to the cytoplasmic sites of protein syntheis, the ribosomes, to which they become associated to participate in codon-anticodon interactions with tRNA, are called informational or messenger RNAs (mRNA). The name messenger RNA has been proposed by Jacob and Moned (1961). The molecule of a mRNA is a single-stranded like the rRNA molecule and it is DNA-like in its base composition so that GC contents of mRNA correspond to the GC contents of the genome's total DNA.

Synthesis of mRNA

The origin and fate of mRNA in eukaryotic cells is much more complex than in bacteria. Very recently, it has been found that the formation of a functionally active mRNA is the consequence of a complex series of steps that comprise : (1) the actual transcription of DNA into mRNA precursors, (2) the intranuclear processing (tailoring) of these precursors, and (3) the transport of the mRNAs into the cytoplasm and their association with ribosomes to initiate the process of translation or protein synthesis.

(a) Heterogeneous nuclear RNA

It has been suggested that mRNA is synthesized in the nucleus as part of a heterogeneous population of large RNA molecules which constitute the so-called heterogeneous nuclear RNA (het-RNAs or Hn-RNAs). The Hn-RNas are also known as high molecular weight RNA (HMW RNA) or DNA-like RNA (dRNA). The Hn-RNA molecules range in size from 5 × 105 to 107 daltons and are degraded, for the most part within the nucleus at a relatively rapid rate (Darnell, 1968). Only about 20 percent of the Hn-RNA, in terms to total nucleotide is not degraded and has been postulated to be converted into mRNAs.

The Hn-RNA occurs in nucleoplasm outside the nucleolus and it contains DNA-like base composition and readily hydribizes with DNA.

(b) Polyadenylation and transport of mRNA

Recently it has been disocovered that Hn-RNA contain a sequence of polyadenylic acid (poly A), approximately 200 nucleotides long at their 3' ends. This poly A sequence is added to the Hn-RNA after the transcription is completed and is attached to the 3′ end of the RNA molecule have suggested that once the Hn-RNA molecule is transcribed from DNA, the poly A sequence is added, in a stepwise fashion, by the action of a poly synthetase. Simultaneous with or after polyadenylation, and starting from the 5' end a selective degradation of the Hn-RNA moleules takes place by the action of nucleases. The final product is

PolyA(+) mRNA, which wll finally reach thc cytoplasm. Most of mRNA moleucles of eukaryotes contain a 3' terminal poly A chain. The post-transcriptional attachment of poly A may be a special step in the processing of mRNA and may be related to its transport to the cytoplasm.

Recently, it is also discovered that the mRNAs for the histones called histone mRNAs) are directly transcribed from repetitive sequences in the DNA without the need for a giant precursor moleucle (Hn-RNA) or polyadenylation. It has been demonstrated that the histone-mRNA enter the cytoplasm without delay, whereas the poly A(+)mRNAs have a 15 minute delay before appearing attached to the polyribosomes. This lag period may be attributed to the processing steps that poly A(+)mRNAs probably undergo before passing into the cytoplasm.

Heterogeneity and Types of mRNA

When the total mRNA population of an organism is considered, it is found to be heterogeneous in size, showing a wide range of S values of 6 to 30. This property of mRNA reflects the fact that the size of length of the mRNA molecule is directly related with the size of the codons for different protein molecules, the sizes of which may be quite variable. According to the size, following two types of mRNA molecules can be recognized :

(a) Monocistronic mRNA

Mostly the mRNA carries the codons of single cistron (i.e., codes for one complete protein molecule) of the DNA. Such mRNA molecule is called monocistronic mRNA. For example, for the synthesis of a polypeptide chain of 300 to 500 amino acid residues, a monocistronic mRNA of E. coli contains 900 to 1500 nucleotides in its molecule.

(b) Polygenic or Polycistronic mRNA

Sometimes a mRNA molecule carries the codes from several adjacent DNA cistrons and become much longer in size. This type of mRNA is called polygenic or polycistronic mRNA. For example, for the metabolism of the histidine protein the cell synthesizes about 10 specific enzymatic proteins and a mRNA in this case may carry codons for all the 10 enzymes.

Life-Span of mRNA

In most prokaryotic and eukaryotic cells, mRNA has short life-time. For example, the average life of mRNA of E. coli is about 2 minutes because it is attacked by the cytoplasmic ribonuclease enzyme. So that, at most times, mRNA takes up only 5% of the total cellular

RNA. Likewise, in most eukaryotes the average life span of mRNA is one to four hours. However, in both bacterial and eukaryotes mRNAs are known that are metabolically stable and are apparently resistant to nucleases and survive for long period of time. For example, mRNA with life time of six hours has been detected in the bacterium Bacillus cerus at a time when the cells are induced to become long lived spores. Likewise, in differentiating eukaryotic cells mRNAs with a life time of days have been detected. For example, in the immature red blood cells (reticulocytes) of the mammals the mRNA is synthesised originally by the nucleus in early stages and expelled to the cytoplasm but the mRNA exists upto 2 days for prolong utilization in the synthesis of globin protein of haemoglobin. Further, in extreme cases, such as in the state of dormancy adopted by many animal eggs and plant seeds, mRNA is maintained in a stable form for months or even years.

Informosomes

In the eukaryotic cells, the stability to the mRNA is provided by certain proteins. Several investigators, e.g., Spirin, Beltisina andLerman (1965); Perry and Kelle (1968) and Hanshwaw (1968) have reported that in certain eukaryotic cells the mRNA does not enter in the cytoplsm as a naked RNA strand but often remains ensheathed by certain proteins. Spirin (1965) has coined a new term informosome to his mRNA and protein complex. The informosome is used by the cell when there is a delay in the translation. For instance, in the embryo the genetic expression is manifested late during organogenesis. In such cases mRNA occurs in the form of informosomes. The proteins of informosomes protected the mRNA form the degradaing actions fo the enzyme, ribonuclease. These proteins may also control the synthesis at the level of the the translation and thus may regulate or modulate the protien synthetic process.

Transfer RNA (tRNA) Genes

There are 61 codons in mRNAs that code for amino acids in proteins, and each codon must be read by an anticodon on an appropriate tRNA molecule. Hence, large number of tRNA types is found in the cell, meaning that there is a large number of tRNA genes in the genome.

In the *E. coli* chromosome, the typical case is that there is one gene for a give tRNA type. Some genes are represented two or more times, and this is called *gene redudnacy.* Both nonredundant and redundant tRNA genes may be clustered or scattered in the chromosome.

Eukaryotes have many more tRNA genes than do prokaryotes. For example, in *Xenopus* there are more than 200 copies of each gene

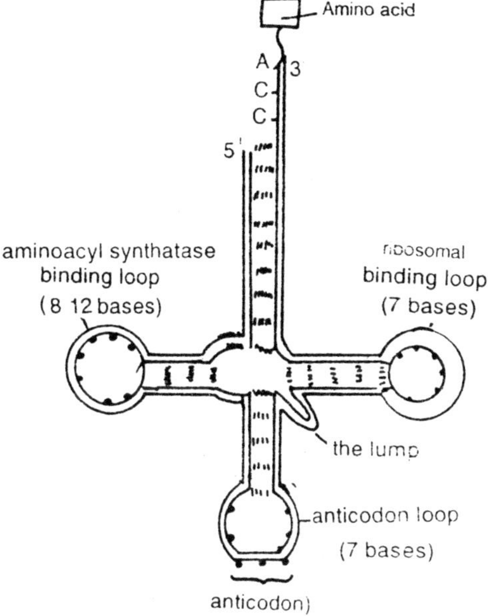

Fig. 8.2. Clover leaf model of tRNA.

per haploid genome. All possible arrangements of tRNA genes can be found; that is, genes may be arranged singly or in clusters, and those genes found in clusters may be redundant or nonredundant tRNA genes.

Structure

Transfer RNAs have an S value of about 4, and a molecular weight of about 25,000–30,000 daltons. The nucleotide chainlength is remarkably uniform for all the tRNAs of prokaryotes and eukaryotes, ranging from 76 to 85.

All the tRNA molecules sequenced to date (more than 40) appear to conform to the cloverleaf model with a stem, three large arms consisting of a stem and a loop, and occasionally an extra arm.

Certain features of the primary sequence of tRNAs appear to be constant. For example, the 3' terminal sequence – CCA.OH (which is added to all tRNAs post-transcriptionally) and the sequence T-ψ-C (where ψ is pseudouridine) in loop IV appear universal. The anticodon is a sequence of three nucleotides that must bond with the codon of mRNA during polypeptide elongation, and hence this region varies according to the tRNA in question. However, the nucleotide to the 5'-

side of the antocodon is always U and that to the 3' side is always a modified purine. The nucleotides present in the stem regions are variable, but the number of base pairs in a particular stem is fairly constant.

In spite of having more than 40 tRNA nucleotide sequence to compare, no general conclusions can be drawn regarding the functions of the various parts of the moecules. One exception to this is the sequenct T-y-C-G, which probably represents the common ribosome binding site of tRNA. Thus, it is likely that the functions of the tRNA are dependent upon the tertiary structure of the molecules.

Tertiary Structure of tRNAs

As was indicated earlier, all of the sequenced tRNA molecules can be arranged in a similar cloverleaf structure. However, this model is derived entirely from analysis of the primary nucleotide sequence and the maximization of hydrogen bonding. The application of x-ray crystallography to stable crystals of tRNA (yeast phenylalanine tRNA was the first) has permitted the elucidation of its tertiary structure at the 0.3 nm resolution level, including the location of the major groups. From the data, it was possible to conclude that all of the double-helical stems predicted by the cloverleaf model do exist. In addition, other hydrogen bonds bend the cloverleaf into a stable tertiary structure

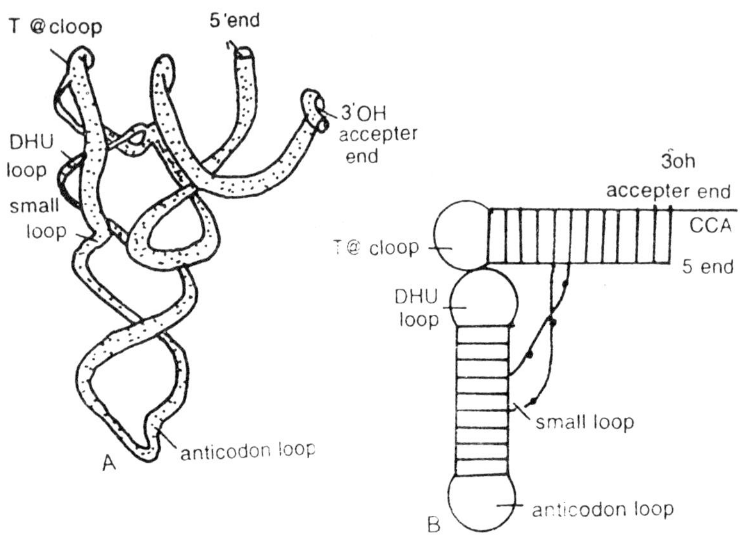

Fig. 8.3. Three dimensional structure of yeast phenylalanine tRNA.

that has a rough L-shaped appearance. In this structure, the amino acid acceptor CCA group at the 3' end of the chain is located at the opposite end from the anticodon loop.

tRNA Biosynthesis

It is apparent that, unlike mRNAs, the tRNA molecules are extensively modified. These modifications occur post-transcriptionally. Indeed, the mature tRNAs not only have modified bases but also are considerably shorter (up to 40 nucleotides or so) than the primary gene transcripts. Thus, in both prokaryotes and eukaryotes there is evidence for *precursor-tRNAs* (*pre-tRNAs*), which must be clipped and trimmed to produce the mature molecules. In prokaryotes, the precursors either contain only one tRNA molecule with extra leader and trailer sequences at the 5' and 3' ends, or they contain several species of tRNA molecules within a long precursor. In the latter case, the tRNA molecules are separated by spacer sequences. These multi-tRNA precursors reflect the tandem arrangement of tRNA genes from which they were transcribed, and they may be described as 5' leader-(tRNA-spacer)-tRNA-trailer-3'.

At least two enzymes are necessary for the processing of these pre-tRNAs. One of these, RNase P, catalyzes the removal of the 5' leader sequence, and the other, RNAse Q, catalyzes the removal of the 3' trailer sequence. Evidence for this has come from studies of mutant strains of E. coli with temperature-sensitive defects in the enzymes. at the nonpermissive temperature, partially processed pre-tRNAs accumulate and can be isolated for sequence comparison with mature tRNAs. Another enzyme (or enzymes) is involved in removing the spacer sequence between tRNAs when they are present in clusters.

In eukaryotes, there is also good evidence for the existence of pre-tRNAs, although studies have generally been done with mixtures of molecules rather than with purified individual species as was done in prokaryotes. Pre-tRNAs from several eukaryotes sediment at about 3.8S compared with 3.8S for mature tRNAs. These molecules are transcribed by RNA polymerase III, and the data obtained from studies with denaturing agents indicate that there are an additional 15–35 nucleotides on the pre-tRNAs. Processing of the pre-tRNA molecules presumably occurs in the nucleus, although, as yet, the enzymes involved and their sites of action have not been defined in any detail.

Interestingly, about 10% of the approximately 400 yeast tRNA genes contain an intron. In every case,the intron in the pre-tRNA is just to the 3' side of the anticodon and is usually les than 15 bp long.

After leader and trailar sequences have been removed from the tRNA, the intron is spliced out of the transcript in a reaction catalyzed by RNA ligase. This reaction is methyl groups are added to the bases, with only a few added to the 2-OH of the ribose moiety. All of the methylations of the 23SrRNA occurs on the p30S RNA component, whereas most if not all of the 16S rRNA methyl groups are added when it is in a mature from. Thus methylation of 23S rRNA is an early event in RNA processing and methylation of 15S rRNA is a late event.

BIOSYNTHESIS OF EUKARYOTIC RIBOSOMES

In outline, eukaryotic ribosome are made as follows: The rRNA genes are transcribed into a high-molecular-with ribosomal precursor RNA (pre-rRNA), which is then modified and cleaved at specific sites to yield a number of discrete intermediate RNAs and finally the mature 18S, 5.8S and 28S rRNAs. Assembly with ribosomal proteins and 5S rRNA (which is transcribed from separate rRNA genes) occurs in the nucleolus during pre-rRNA processing. The resultant ribosomal subunits are then released from the nucleus into the cytoplasm.

Ribosomal RNA maturation has been studied in a number of eukaryotes, including mammalian cells, amphibians, insects, higher plants, and fungi. The results indicate that all eukaryotes have similar pathways for the biosynthesis of ribosomes. In the following section, ribosome production in higher eukaryotes and fungi (yeast) will be described for comparison purposes and to point out the generalities involved.

Ribosomal DNA

In most eukaryotes studied to date, the genes for 18, 5.8, and 28S rRNA are multiple and clustered at the sites of the nucleolar organizer regions. Saturation hybridization experiments between purified rRNA and nuclear DNA have shown the extent of this gene multiplicity, and this varies from about 100 to 1000, depending on the organism being studied. In general there are more copies the more complex the eukaryotes in an evolutoinary sense. In most organisms, the 5S rRNA is coded by extranucleolar genes whose multiplicity is usually higher than that for the other rRNA genes. The 5S rRNA genes may be clustered in the genome as in humans or scattered throughout the genome as in Xenopus laevis, the South African clawed toad. At least in yeast and slime moulds, the 5S rRNA genes are intersperesed with the other rRNA genes. In a few eukaryotes, an intron is found in the 28S.

9

PROTEIN SYNTHESIS

Proteins are the molecules responsible for catalyzing most intracellular chemical reactions (enzymes), for regulating gene expression (regulatory proteins), and for determining many features of the structues of cells, tissues, and viruses (structural proteins). A protein is composed of one or more chains of amino acids that are covalently joined. The chains of amion acids are called *polypeptides.* The 20 different amino acids commonly found in natural polypeptides can be in any number and any order. Because the number of amino acids in a polypeptide molecule usually ranges from 100 to 1000, the number of different protein molecules that is possible is enormous.

Each amino acid contains a carbon atom (the α carbon) to which is attached one carboxyl group (–COOH), one amino group ($-NH_2$), and a side chain commonly called an R group. The R group are generally chains or rings of carbon atoms bearing various chemical groups. The simplest side chains are those of glycine (–H) and of alanine (CH_3). Polypeptide chains are formed when the carboxyl group of one amino acid joins with the amion grioup of a second amino acid; the resulting chemical bond is an ordinary covalent bond called a *peptide bond.* Thus, the basic unit of a protein is a polypeptide chain in which α-carbon atoms alternate with peptide units to for a backbone having an ordered array of side chains.

The two ends of every polypeptide molecule are distinct. One end has a free $-NH_2$ group and is called the *amino terminus*; the other end has a free -COOH group and is the *carboxyl terminus.* Polypeptides are synthesized by adding individual amion acids to the carboxyl end of the growing chain. Conventionally the amion acids of a polypeptide chain are numbered starting with the amino acid at the amion end.

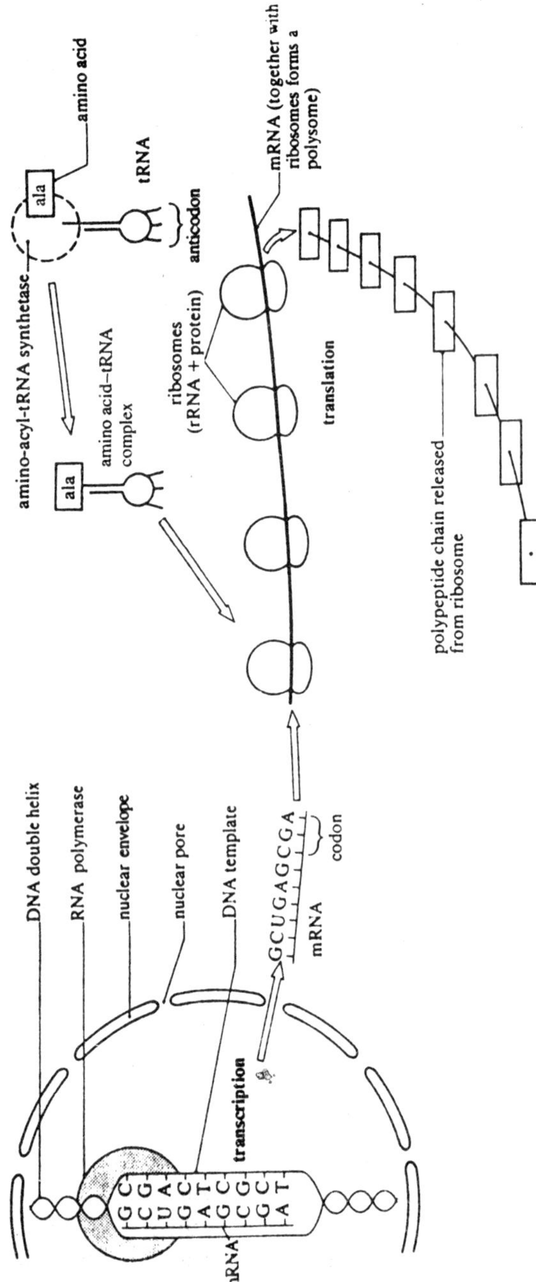

Fig. 9.1. Simplified summary diagram of the major structures and processes involved in protein synthesis in the cell.

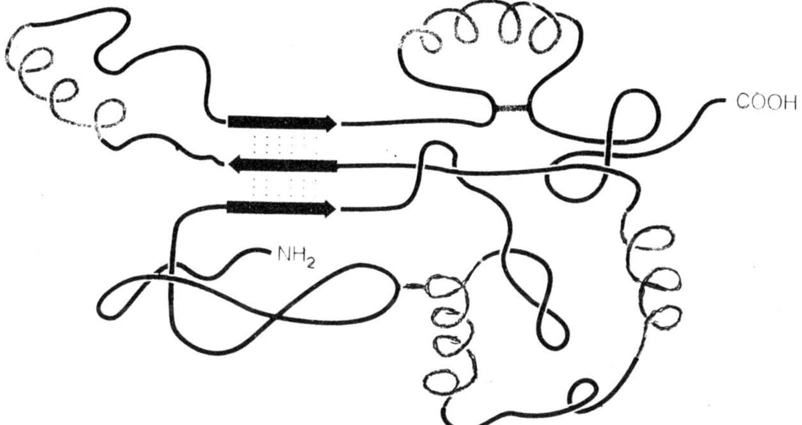

Fig. 9.2. A schematic diagram of the path of the backbone of a polypeptide, showing possible ways in which the polypeptide may be folded. Heavy black arrows represent β sheets; the dots joining the arrows represent hydrogen bonds.

Most polypeptide chains are highly folded, and a variety of three-dimensional shapes have been observed. The manner of folding is determined primarily by the sequence of amino acids –in particular, by noncovalent interactions between the side chains–and so each polypeptide chain tends to fold into a unique three-dimensional shape as it is being synthesized. In some cases, protein folding is interactions with other proteins in the cell. The rules of folding are complex and shape cannot usually be predicted from the amino acid sequence except for the simplest proteins. On the average. the molecules fold so that amino acids with charged side chains tend to be on the surface of the protein (in contact with water) and those with uncharged side chains tend to be internal. Specific folded configurations also result from hydrogen-bonding between prepaid groups. Two fundamental polypeptide structures are the α helix and the β sheet. Covalent bonds may also form between the sulfur atoms of some pairs of cysteines.

Many protein molecules consist of more than one polypeptide chain. When this is the case, the protein is said to contain *subunits.* The subunits may be identical or different. For example, hemoglobin, the oxygen carrier of blood, consists of four subunits, two each of two different types.

Relations Between Genes and Polypeptides

Most genes contain the information for the synthesis of only one polypeptide chain, Furthermore, the *sequence* of nucleotides in a gene determines the *sequence* of amino acids in polypeptide. This point was first proved by studies of the tryptophan synthetase gene *trpA* in *E. coli*, a gene in which many mutations had been obtained and accurately mapped. The effects of numerous mutations on the amino acid sequence of the enzyme were determined by directly analyzing the amino acid sequences of the wildtype and mutant enzymes. Each mutation was found to result in a single amino acid substituting for the wildtype amino acid in the enzyme; more importantly, *the order of the mutations in the genetic map was the same as the order of the affected amino acids in the polypeptide chain.* This attribure of genes and polypeptdes is called *colinearity*, which means that the sequnce of base pairs in DNA determines the sequence of amino in the polypeptide in a colinear or point-to-point manner. Colinearity is universally found in prokaryotes. However, we will see later that in eukaryotes noninformational DNA sequences interrupt the continuity of most genes, the order but not the spacing between the mutations correlates with amino acid substitution.

TRANSCRIPTION

The first step in gene expression is the synthesis of an RNA molecule copied from the segment of DNA that constitutes the gene. The basic features of the production of RNA are described in this section.

General Features of RNA Synthesis

The essential chemical characteristics of the enzymatic synthesis of RNA resemble those of DNA synthesis.

1. The precursors is the synthesis of RNA are the four ribonucleoside 5'-triphosphates– namely, adenosine triphosphate (ATP), guanosine triphosphate (GTP), cytidine triphosphate (CTP. and uridine triphosphate (UTP). They differ from the DNA precursors only in that the sugar is ribose rather than deoxyribose and the base uracil (U) replaces thymine(T).
2. In the formation of RNA, a sugar-phosphate bond is formed between the 3'-hydroxyl group of one nucleotide and the 5'-triphosphate of a second nucleotide. This is the same chemical reaction as that which occurs in the synthesis of DNA, but the enzyme is different.

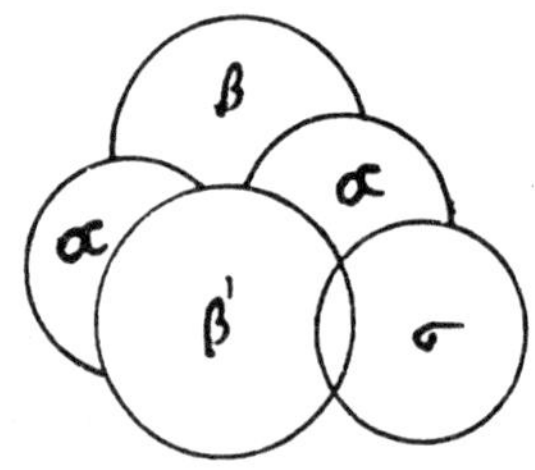

Fig. 9.3. A model of the structure of prokaryotic RNA polymerase showing association of five polypeptides ($\alpha_2\beta\beta$ σ)

3. The sequence of bases in an RNA molecule is determined by the base sequence of the DNA template. Each base added to the growing end of the RNA chain is chosen for its ability to base-pair with the DNA template strand: thus, the bases C, T, G, and A in a DNA strand cause G, A, C, and U, respectively, to be added to the growing end of an RNA molecule.
4. Nucleotides are added only to the 3'-OH end of the growing chain: as a result, the 5' end of a growing RNA molecule bears a triphosphate group. (The 5' → 3' direction of chain growth is the same as that in DNA synthesis.)

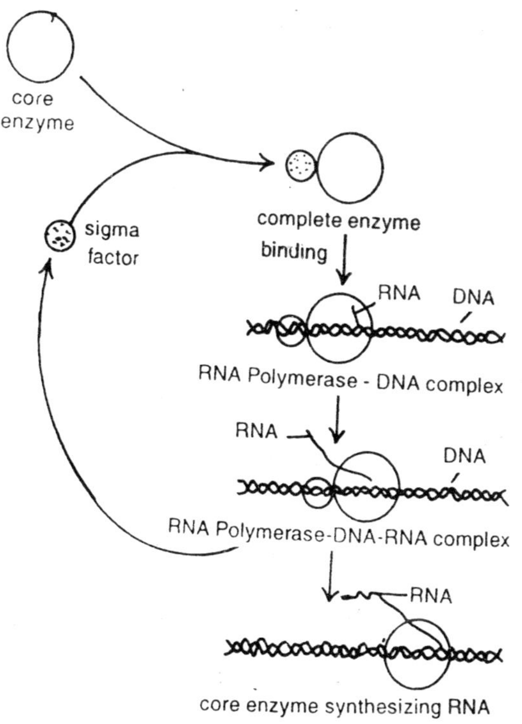

Fig. 9.4. Role of sigma factor and core enzyme of RNA polymerase during transcription.

A significant difference between DNA polymerase and RNA polymerase is that *RNA polymerase is able to initiate chain growth without a primer.*

An important feature of RNA synthesis is the following:

Introns and Exons

In 1977 biologists were surprised to discover that the DNA of a eukaryotic gene is longer than its corresponding mRNA. It should be the same length because the mRNA is a direct copy. It was discovered that immediately after the mRNA is made, certain sections of the molecule are cut out, before it is used in translation. The sections of the gene that code for these unused pieces of RNA are called *introns*. The remaining sections of the gene are the code for the protein and are called *exons*. The size and arrangement of introns is very variable and characteristic for a particular gene. In prokaryotes there are no introns.

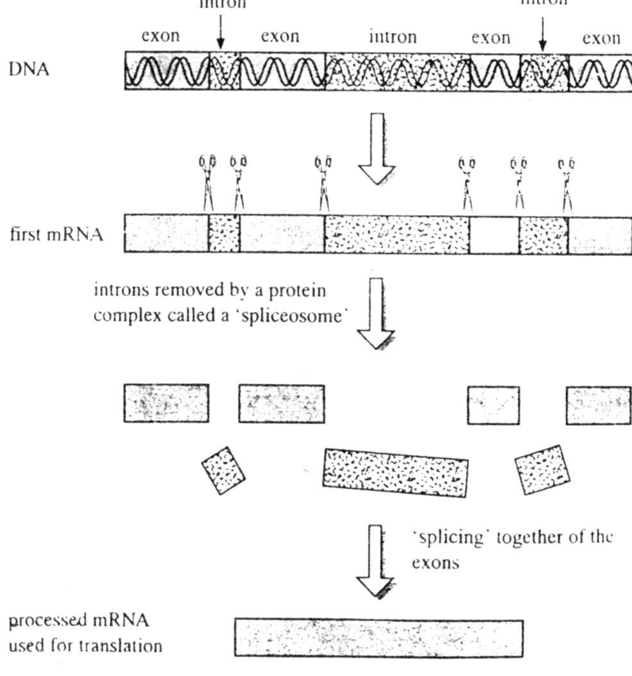

Fig. 9.5. Exons, introns and intron splicing.

One possible function for introns has come with the discovery that the same mRNA may have different introns removed in different cells. The gene therefore has alternative introns and can code for different, though similar, proteins. This increases its potential use.

An example is the calcitonin gene. Two different forms of mRNA can be produced by this gene, depending on which introns are removed.

One is produced in the thyroid gland and codes for the protein calcitonin, which had 32 amino acids. Calcitonin is a hormone which acts to lower calcium levels in the blood. The other is produced in the hypothalamus and codes for a protein with 37 amino acids which is similar to calcitonin and is called CGRP (calcitonin gene-related peptide). This is a powerful vasodilator agent. It is also released from nerve endings in some parts of the peripheral nervous system.

Protein Synthesis in Prokaryotes

There are three major steps in protein synthesis: initiaion, elongation, and termination. These will be discussed in turn.

Initiation

Initiator tRNA. The first amino acid in the syntheis of all bacterial polypeptides is N-formyl-methionine (fmet), which is a modified methionine amino acid in which the α-amino group is "blocked" and therefore cannot participate in peptide bond formation. The formyl group is added to the methionine after the amino acid has become attached to a specific tRNA, called tRNA. fmet. This reaciton is catalyzed by the enzyme *transformylase.* In many cases the fmet that starts a polypeptide chain is subsequently removed by enzymatic action.

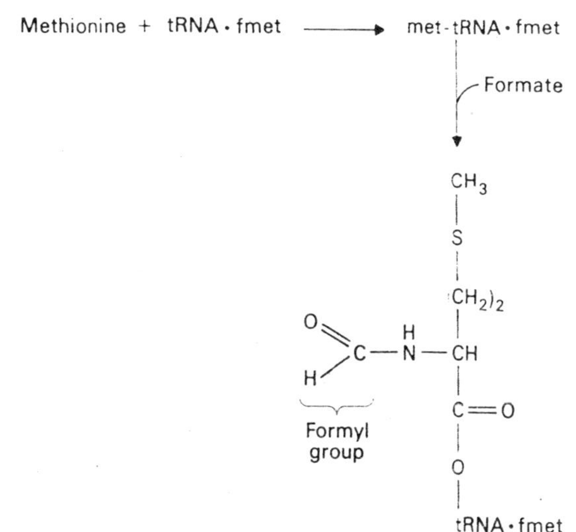

Fig. 9.6. Synthesis of N-formylmethionyl-tRNA.

Biochemical analysis has shown that at least two species of tRNA that can be charged with methionine are present in all prokaryotic organisms: one is that tRNA involved with initiaton, and the other species is responsible for the insertion of methionine elsewhere in the polypeptide chain. The latter rRNA is designed rRNA. met. Both of these tRNAs in bacteria are aminoacylated by the same enzyme, but only tRNA, fmet is a substrate for the transformylase-catalyzed reaction. Both tRNAs read AUG (the only methionine codon), but in addition tRNA.fmet can recognize GUG and UUG codons. RNA-sequencing studies have shown that both molecules havae an anticodon that is complementary to AUG. The two tRNA molecules do differ in some other properties. For example, the binding of fmet-TRNA.fmet to ribosomes is catalyzed by an initiator factor whereas the binding of met-tRNA.met is catalyzed by elongation factors. The two tRNAs apparantly bind to the ribosome at different sites. Thus it is clear that fmet.tRNA.fmet must have a strucutre that is specific for its role in initiation.

Ribosome binding sites

In bacteria, the first step in initiation is the formation of a complex between the 30S ribosomal subunit, fmet-TrNA, and an mRNA molecule. The 50S subunit is added later to form the active 70S ribosome (monosome). The mRNA may contain information for one to several distinct polypeptide chains. For each of the segments coding for a polypeptide, there is a specific nucleotide sequence for orienting the mRNA correctly and in the right reading frame on the ribosome. These sequences are called the *ribosome binding sites.*

Message origin	Ribosome binding site sequence
E. coli lac Z	UUC ACA CAG GAA ACA GCU AUG ACC AUG AUU
E. coli trp B	AUA UUA AGG AAA GGA ACA AUG ACA ACA UUA
E. coli RNA polymerase β	AGC GAG CUG AGG AAC CCU AUG GUU UAC UCC
Phage λ cro	AUG UAC UAA GGA GGU UGU AUG GAA CAA CGC

Fig. 9.7. Some prokaryotic ribosome binding sites. The initiation codon, AUG, is boxed. The larger boxed regions indicate the regions of contiguous complementarity (including allowable G-U base pairs) to the 3' end of 16S rRNA.

Most of the ribosome binding sites have apurine-rich sequence about 8 to 12 bases upstream from the AUG start codon. this equence and other bases in this region are complementary to a pyrimidine-rich region, including at least CCUCC at the 3' end of 16S rRNA. The mRNA region that binds in this way is called the *Shine-Dalgarno sequence* after the discoverers of this relationship. Thus it appears that the formation of complementary base pairs between mRNA and 15 S rRNA in the 30S ribosomal subunit allows the ribosomes to locate and bind to the initator regions in the mRNA.

Initiation factors and initiation

In addition to mRNa, fmet-tRNA, and ribosomal subunits, three protein initiaion factors (IF-1, IF-2, and IF-3) and GTP are required for the initiation process to occur. First the properties of the initiation factors are discussed and then the scheme proposed for the initiation processes in protein synthesis is presented.

1. *IF-3.* The IF-3 factor weighs 23,000 daltons and functions in binding mRNA to the 30S subunit. It also acts as a dissociation factor for separating the 30S and 50S subunits after polypeptide synthesis is complete. Like all the IFs, IF-3 is found bound to free 30S subunits and can be released by washing the subunits in 0.5 M ammonium chloride.

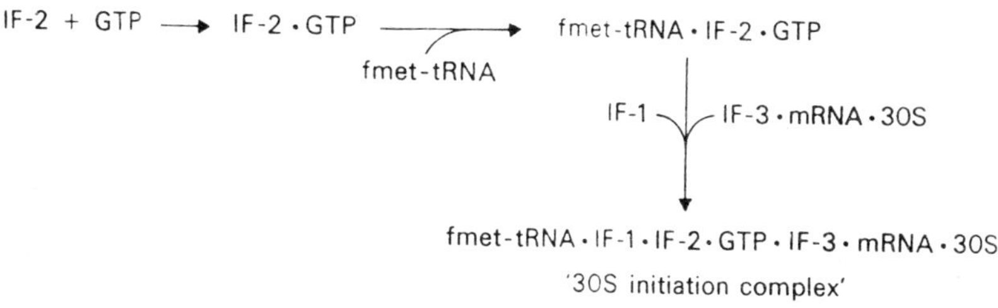

Fig. 9.8. Initiation of protein synthesis: steps in the formation of the 30S initiation complex.

Experiments with radioacative If-3 have shown that it is capable of binding to both 30S subunits and to mRNA molecules. In an in vitro protein-synthesizing system, IF-3 enhances the binding of fmet-tRNA to mRNA.30S subunit complexes. It is attractive to suppose that IF-3 recognizes mRNAs by the AUG or GUG initiation codons, but there is no solid evidence on this point.

In summary, the initiation reaction in which IF-3 is involved is:

IF-3 + mRNA + 30S subunit → (IF-3.mRNA.30S) complex

2. *IF-2.* The 80,000 dalton IF-2 protein in involved with the binding of the initiator tRNA to the IF-3.mRNA.30S complex. The high-energy molecule GTP is used in this reaction. In vitro experiments have shown that IF-2 and GTP will bind to form a complex that is stabilized when it is turn forms a complex with fmet-tRNA. This latter complex then binds with the IF-3.mRNA.30S complex and

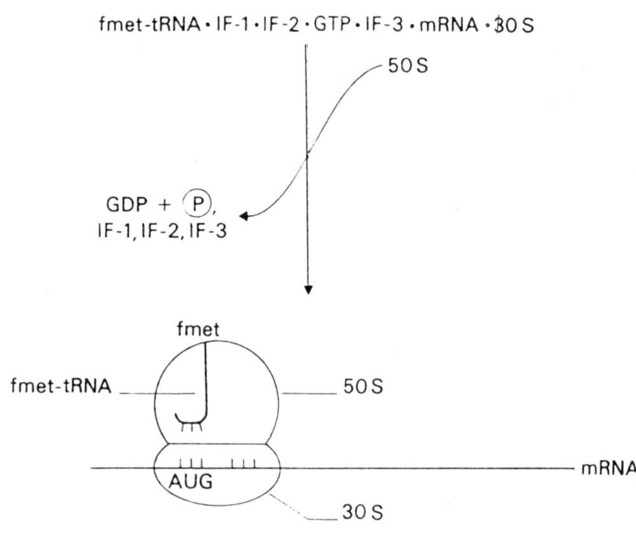

Fig. 9.9. Initiation of protein synthesis: addition of 50S ribosomal subunit to 30S initiation complex leads to formation of 70S ribosome in frame on the mRNA.

the IF-1 protein factor (9000 daltons) to forms the 30S initiation complex.

3. *Dissociation of initiation factors from the initiation complex.* The initiation factors function to bring fmet-tRNA, mRNA and 30S subunits into a stable association. The next step is the addition of a 50S subunit to form a 70S initiation complex. This leads to the hydrolysis of GTP to GDP + P and the release of three intiation factors. The factors can then be used for further initiaton reactions on the same or different mRNA.

Elongation

The 70S ribosome has two sites for binding aminoacyl-tRNA. In protein synthesis, charged tRNA binds first to a site called the A (aminoacyl) site. Then the amino acid it carries becomes joined to the growing polypeptide chain carried by the tRNA at the site called the P (peptidyl) site by the formation of a peptide bond.

It is not known whether the fmet-tRNA enters the A site and then moves to the P site or whether it enters the P site directly. Before further protein synthesis can occur, however, the fmet-tRNA must become located in the P site hydrogen-bonded to the start codon on the mRNA. Once this has occurred a cyclic sequence of events commences in which one amino acid at a time is added to the growing polypeptide chain. This is called ***elongation.***

Binding of aminoacyl-tRNA

The charged tRNA with the complementary anticodon to the codon in the reading frame of the A site becomes bound to that site of the

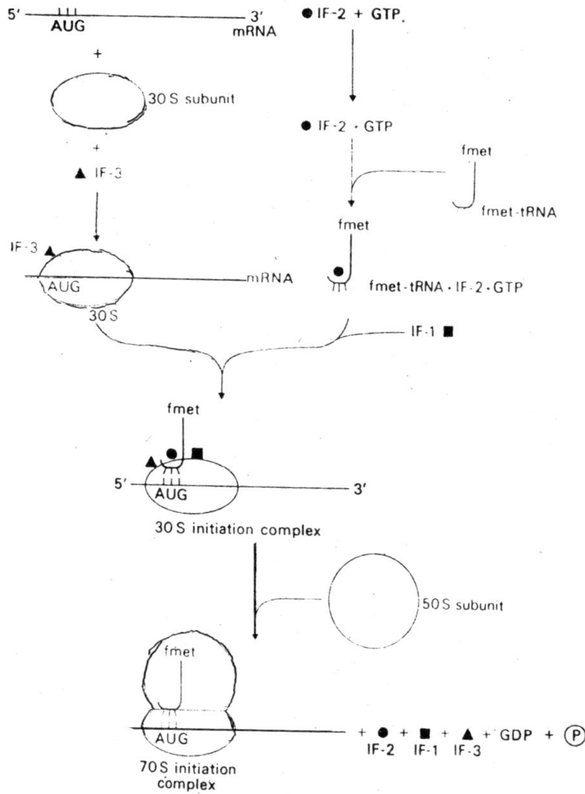

Fig. 9.10. Summary of the steps in the initiation of protein synthesis in prokaryotes.

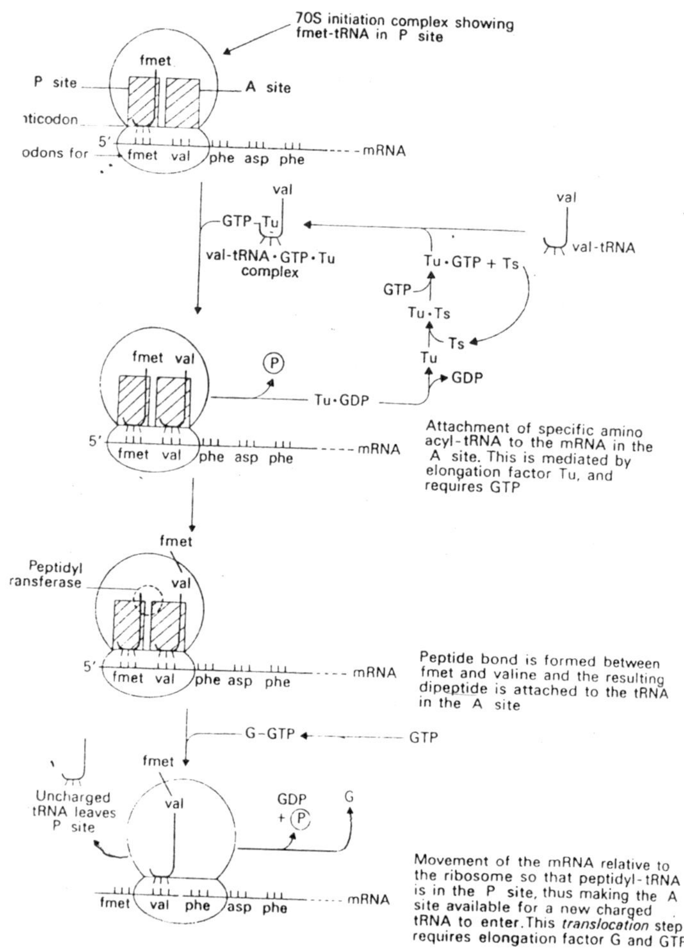

Fig. 9.11. Summary of the elongation (peptide bond formation) and translocation steps in protein synthesis.

ribosome in a reaction requiring *elongation factor* T (EE-T) and GTP. This factor can be isolated from the soluble proteins of *E. coli*, and by column chromatography it can be separated into two polypeptides: Ts, which is stable and weighs about 30,000 daltons, and Tu, which is unstable and weights 42,000 daltons.

EF-T has been shown to bind with GTP and this is postulated to bring about the dissociation of the factor into the two polypeptides, resulting in the formation of an EF-Tu.GTP complex and releasing free EF-Ts. The next step in the elongation process is the binding of aminoacyl-tRNA to the complex to produce an aminoacyl-tRNA.Tu.GTP complex. There is evidence that this complex is an intermediate in aminoacyl-tRNA binding to ribosomes. Once the charged tRNA is bound in the A site, GTP is hydrolyzed as a result of the enzymatic action of one or more 50S ribosomal proteins. This hydrolysis cause the release of EF-Tu in a complex with GTP. the latter is released and the elongation factor can reassociate with EF-Ts. The process can then be repeated with another aminoacyl-tRNA.

(a) Tu + Ts (Elongation factor T) + GTP → Tu + GTP + Ts

(b) Tu + GTP + aa-tRNA → aa-tRNA · Tu · GTP
(amino acyl-tRNA) complex

(c) aa-tRNA · Tu · GTP + active 70S ribosome → aa-tRNA·70S (charged tRNA) enters A site)
+
Tu·GDP + P_i (released from ribosome)

(d) Tu · GDP + Ts → Tu · Ts

Experiments with an analog of GTP that cannot be hydrolyzed have shown that GTP hydrolysis is required for release of EF-Tu from the ribosome but it is not needed for aminoacyl-tRNA binding to the ribosome. Other experiments have shown that binding of fmet-tRNA to the ribosome does not require EF-Tu.

Peptide bond formation

At the beginning of the this stage, a tRNA carrying the growing polypeptide chain is located in the P site, and an aminoacyl-tRNA is located in the A site. Thee tRNAs are maintained in positions conducive for peptide bond formation of the hydrogen bonds between the

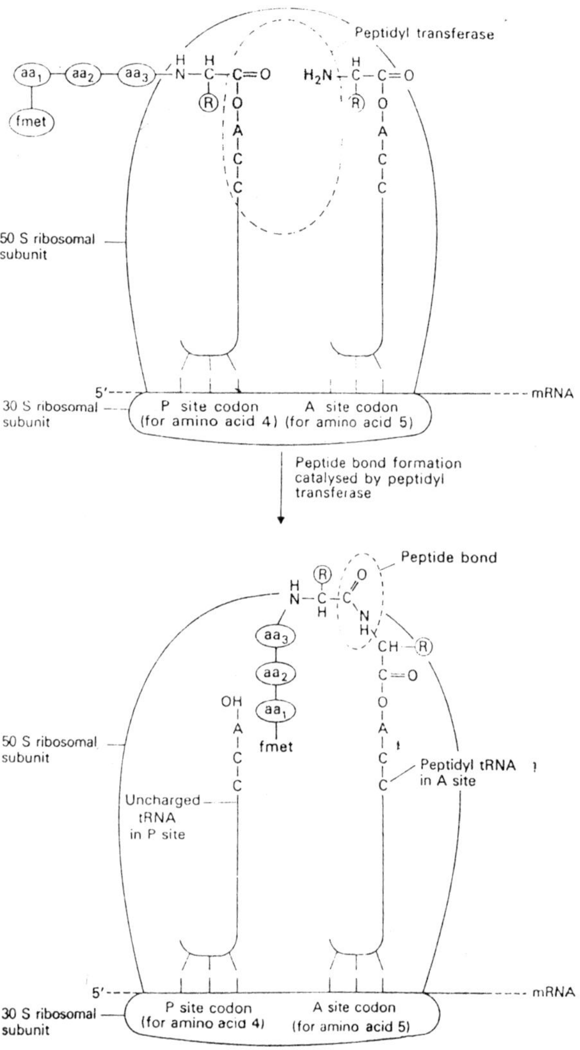

Fig. 9.12. Diagrammatic representation of peptide bond formation on ribosomes catalyzed by peptidyl transferase.

respective codons and anticodons and by the tertiary structure of the ribosome. The peptide bond is formed with the aid of the enzyme, *peptidyl transferase*, which is a ribosomal protein of the 50S subunit. The end result of the reaction is that the polypeptide chain is one amino acid longer, and the growing polypeptide chain has been transferred from the tRNA in the P site to the tRNA in the A site. The tRNA in the P site, which now has no amino acid bound to it, is called an *uncharged tRNA.*

Translocation

Once the peptide bond has been formed and the polypeptide chain is on the tRNA in the A site, the next step is advancement of the ribosome precisely one codon (three nucleotides) down the mRNA, a process called *translocation.* During this translocation event, the peptidyl-tRNA remains attached to the mRNA by codon-anticodon–pairing properties and thus becomes located in the P site. The A site is then vacant, and the aminoacyl-tRNA specified by the new codon there becomes bound by the process already described. The uncharged tRNA left in the P site after peptide bond formation is also released from the ribosome during translocation.

Elongation factor G (EF-F), a 72,000–84,000 dalton protein, and GTP hydrolysis are needed for translocation to occur, but it is not yet known how the translocation mechanism works. One GTP molecule is hydrolyzed for each translocation event. It appears that EF-G molecule is hydrolyzed for each translocation event. It appears that EF-G leaves the ribosome after translocation, since EF-Tu and EF-G cannot interact with the ribosome at the same time.

Termination

The end of the polypeptide chain is indicated on a mRNA molecule by a specific *chain-terminationg* (stop) *codon.* Three such codons are known: UAA, UAG, and UGA. No naturally occurring tRNA has an anticodon for any of these stop codons, and therefore no amino acid can be put into the polypeptide. Three specific termination factors have been shown to be involved in regarding the stop codon. they differ in their codon specificity and GTP requirement (Table 9.1).

RF1 and RF2 have overlapping specificities for the stop codons. They have been shown to interact with the termination codons by interaction at the A site. The RF3 factor apparently plays a stimulatory role in RF1 and RF2 activity. There is some evidence for a GTP requirement in the RF3 factor's activty. In any event, chain termination, as mediated by these factors, involves the cleavage of the carboxyl group of the C-terminal end of the polypeptide chain from the tRNa in the P site. This results in the release of the polypeptide and the now uncharged tRNA. The ribosome will then move along the mRNA until a new initiation sequence is encountered (as it may be in polycistronic mRNAs), or it will dissociate from the mRNA. If none is found, when the ribosome is released from the mRNA, IF-3 functions to keep the two subunits apart. Thus, when a new 70S initiation complex is formed, the two subunits are drawn randomly from the free pools of 30S and 50S subunits.

Table 9.1. Properties of prokaryotic termination factors.

Termination factor	*Molecular weight (daltons)*	*Stop codons recognized*	*GTP requirement*
RF1	44,000	UAA and UAG	No
RF2	47,000	UAA and UGA	No
RF3	46,000	None	Yes

While the polypeptide chain is being synthesized, the primary sequence of amino acids directs the three-dimensional shape. In other words, the elongating chain begins to assume its final shape as it is being made. Indeed, some enzyme activity can be detected on ribosomes that have not yet completed the synthesis of an enzymatic polypeptide.

Polysomes

Efficient translation of an mRNA molecule cannot be achieved by a single ribosome moving along it. The amount of space a ribosome

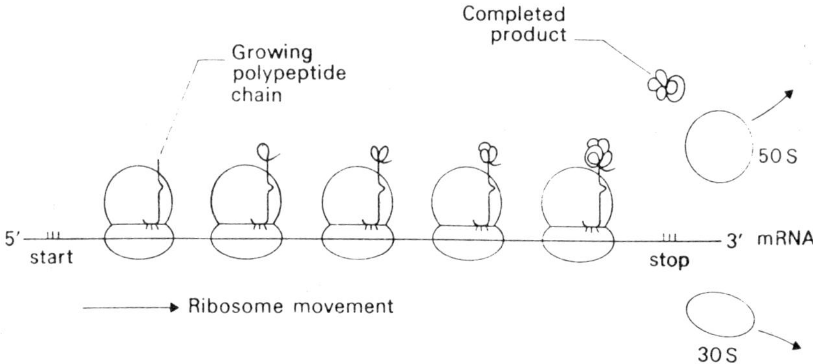

Fig. 9.13. Diagrammatic representation of a polysome engaged in protein synthesis.

takes up on a mRNA is relatively small, and thus several ribosomes can work on the mRNA at once. The association of a number of ribosomes on a single mRNA chain is called a *polyribosome* or *polysome*, and this allows several polypeptide chains to be made from each mRNA. The length of the polypeptide chain on a given ribosome will be directly proportional to how far the ribosome has moved along the mRNA from the 5' end of the molecule. The existence of polysomes explains why a cell needs so little mRNA, while at the same time it contains so much more protein.

Relationship of Transcription and Translation

In bacteria the mRNA typically becomes associated wth ribosomes while synthesis of the mRNA molecule is continuing . This is possible owing to the lack of a nuclear membranes so that as the 5' end of the growing mRNA molecule is displaced form the DNA as the double helix reforms, the ribosome binding site becomes available. Ribosomes then load on to the mRNA in rapid sequence, the first being close behind the RNA polymerase.

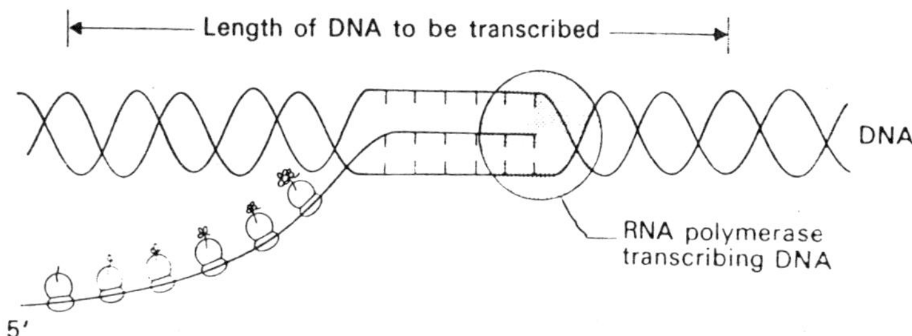

Fig. 9.14. Schematic of the possible translation of an mRNA while it is still being transcribed in prokaryotes.

To give some idea of the rates of these processe, the mRNA for the tryptophan biosynthetic operon is transcribed at a rate of about 1000 nucleotides per minute and the translation process proceeds at about the same rate. Thus approximately 350 amino acids can be polymerized into polypeptide chains each minute.

Another important aspect of translation is the lability of the mRNA. As mentioned in an earlier chapter, prokaryotic mRNAs are considered to be short-lived in that degradation of the molecule by 5'-exonuclease action competes wth the ribosome-mediated initiation of protein synthesis. Continued mRNA synthesis, then, is necessary for continued polypeptide synthesis.

Protein Synthesis in Eukaryotes

The steps and mechanisms of protein synthesis are similar in both eukaryotes and prokaryotes. As previously discussed, the ribosomes are different, and this is also the case with the soluble protein factors.

Initiation

Initiation of protein synthesis involves the binding of mRNA to the ribosomes. In eukaryotes it has now been established that the 5'-cap structure is necessary to get efficient binding but the 3' poly(A) sequence apparently is not needed. There is good evidence, however, that the poly(A) sequence stablizes the mRNA during the translation process. The precise mechanism whereby the eukaryotic message binds to the ribosome is not known, although it is most likely that RNA–RNA and RNA–protein interactions are involved. To date the nucleotide sequence has been determined to the 5'-side of the AUG start codon for a number of eukaryotic mRNAs and there are very few common features. Strikingly, while the last 50 nucleotides of *E. coli* 16S rRNA and eukaryotic 18S rRNA are highly homologous and similiar secondary structure models can be built, the eukaryotic rRNA does *not* have the CCUCC sequence characteristic of prokaryotic rRNA. Also, there is no Shine-Dalgarno sequence in the eukaryotic mRNa. Thus it appears that, unlike the case in prokaryotes, there is not a fixed nucleotide sequence that plays a role in ribosome binding to the message.

As in prokaryotes the initiation codon is AUG, and a special initiator methionyl-tRNa recognizes that signal in the message. Unlike its prokaryotic counterpart the methionine carried by the tRNA does not become formylated, since the appropriate enzyme system does not exist in eukaryotic cells. The initiator tRNAcan be distinguished from the met-tRNA that reads AUG codons elsewhere in the message, however, by the fact that it can be formylated in vitro in the presence of an *E. coli* extract. Hence in eukaryotes it is also appropriate to define the two methionine-accepting tRNAs as tRNA.fmet and tRNA.met.

At least in mammals, there are many more initiation factors (labeled eIFs for eukaryotic initiation factors) than in prokaryotes. In many cases the proteins have not been purified to homogeneity, and thus their absolute roles in protein synthesis are uncertain. Between them all, they carry out the initiation events performed by the three prokaryotic IFs. It remains to be seen to what extent the situation in mammals is generalizable throughout the eukaryotes.

Elongation

As in prokaryotes, there are two elongation factors: eEF-1 (equivalent to prokaryotic EF-T) and eEF-2 (equivalent to prokaryotic

EF-G). eEF-1 has been studied in a number of systems, and in general it exists in multiple forms. Purified eEF-1 from rabbit reticulocytes, for example, has a molecular weight of 186,000 and consists of three subunits weighing 62,000 daltons each. In some systems the subunits aggregate to produce molecules of greater than 1 million daltons. Regardng the function of eEF-1, much less is known than in the prokaryotes. The eEF-1 from rabbit reticulocytes, for example, has been shown to bind to aminoacyl-tRNA and to GTP, and thus to facilitate bindng of the aminoactyl-tRNA to the A site in ribosomes. During this step, GTP is hydrolyzed to DGP as a result of GTPase activity of the elongation factor, and an eEF-1.DGP complex is released from the ribosome.

The eukaryotic eEF-2 is similar to prokaryotic EE-G, although the two are not interchangeable in *in vitro* systems. The factor from rabbit reticulocytes has a molecular weight of 96,500–110,000 daltons and, after binding with GTP, binds to the ribosome. This event results in hydrolysis of GTP to GDP, translocation of the ribosome one codon down the message, and release of an eFF-2.GDP complex. All these steps are similar to the events that take place in prokaryotes, one exception being that the eukaryotic factor forms a stable complex with GTP whereas the prokarotic factor does not.

Termination

The same chain termination codons are functional in eukaryotes as in prokaryotes. One release factor has been identified in rabbit reticulocytes, and this has a molecular weight of 115,000 daltons and may be a dimer. This factor recognizes all three chain termination codons and it requires GTP to carry out its function. No stimulatory factor analogous to the porkaryotic RF-3 has been found in eukaryotes.

Protein Synthesis and Cellular Compartmentation in Eukaryotes

In prokaryotes there is no nuclear membrane to separate the transcription process from the translation process. The process of a nuclear membrane and the various modification process peculiar to mRNAs in eukaryotes present many levels at which the regulation of gene expression can be affected. These include transcription itself, the processing of the primary transcript to produce mature mRNA, RNA-RNA splicing, and the movement of mRNA from the nucleus to the cytoplasm.

In eukaryotic cells, the cytoplasm contains a network of interconnecting channels bounded by membranes called the *endoplasmic reticulum* (ER). The membranes involved are continuous and may in fact connect to the nuclear membrane and the cell membrane. Close examination of the ER reveals that it is differentiated into two types, *smooth* (SER) and *rough* (RER), which are distinguished by the fact that the latter has ribosomes bound to it (hence the rough appearance) whereas the former does not. Thus, ribosomes in the cytoplasm are either membrane bound or free. The membrane-bound ribosomes synthesize proteins that are either secreted from the cell or that are packaged in lysosomes (where the proteins degrade other proteins). Thus, for example, pancreatic cells that secrete enzymes into the intestine are extremely rich in RER. The free ribosomes synthesize all other proteins found in the cell, that is, those in the cytoplasm, nucleus, mitochondria and chloroplasts (if present).

The proteins is to be secreted are made on the ribosomes of the RER and then transferred across the membrane into the channel system. The mRNAs for the secreted proteins must somewhow become associated specifically with the ribosomes of the RER. In 1975, G. Blobel and B. Dobberstein proposed a *signal hypothesis* to explain this. They suggested that there is a unique sequence of codons located to the 3' side of the initiator AUG codon which is present only in mRNAs for proteins that must be transferred across membranes. Translation of these codons results in a specific amino acid sequence at the N-terminal end of the protein. Then they postulated that the special end of the protein facilitates the attachment of the ribosome to the membrane so that the protein can be transferred across it. Once the protein has been completed, they proposed that the ribosome dissociates from the ER.

10

THE LAC OPERON

Jacob and Monod proposed the operon model in 1961 for the co-ordinate regulation of transcription of genes involved in specific metabolic pathways. The operon is a unit of gene expression and regulation which typically includes.

- The *structural genes* (any gene other than a regulator) for enzymes involved in a specific biosynthetic pathway whose expression is co-ordinately controlled.
- Control elements such as an *operator sequence*, which is a DNA sequence that regulates transcription of the structural genes.
- *Regulator gene(s)* whose products recognize the control elements, for example a repressor which binds to and regulates an operator sequence.

The Lactose Operon

Escherichia coli can use lactose as a source of carbon. The enzymes required for the use of lactose as a carbon source are only synthesized when lactose is available as the sole carbon source. The lactose operon (or lac operon) consists of three structural genes: lacZ, which codes for b-balactosidase, an enzyme responsible for hydrolysis of lactose to balactose and glucose; lacY which encodes a galactoside permease which is responsible for lactose transport across the bacterial cell wall; and lacA, which encodes a thiogalactoside transacetylase. The three structural genes are encoded in a single transcription unit, lacZYA, which has a single promoter P_{lac}. This organization means that the three lactose operon structural proteins are expressed together as a polycistronic mRNA containing more than one coding region under

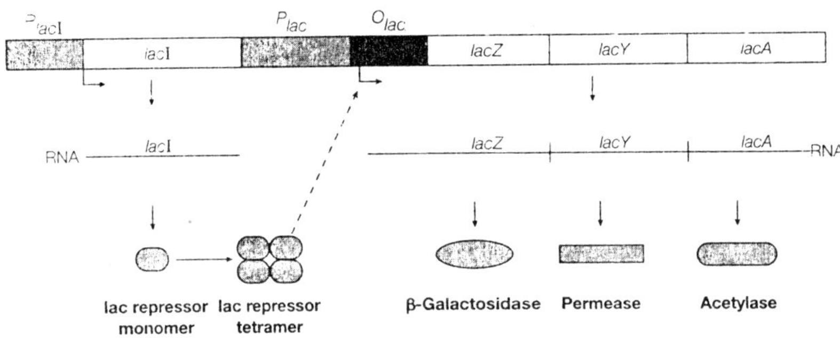

Fig. 10.1. Structure of the lactose operon.

the same regulatory control The lacZYA transcription unit contains an operator site o_{lac} which is positioned between bases –5 and +21 and the 5'-end of the P_{lac} promoter region. This site binds a protein called the lac repressor which is a potent inhibitor of transcription when it is bound to the operator. The lac repressor is encoded by a separate regulatory gene lacI which is also a part of the lactose operon lacI is situated just upstream from P_{lac}.

The Lac Repressor

The lacI gene encodes the lac repressor which is active as a tetramer of identical subunits. It has a very strong affinity for the lac operator-binding site, O_{lac}, and also has a generally high affinity for DNA. The lac operator site consists of 28 bp which is palindromic. (A palindrome has the same DNA sequence when one strand is read left to right in a

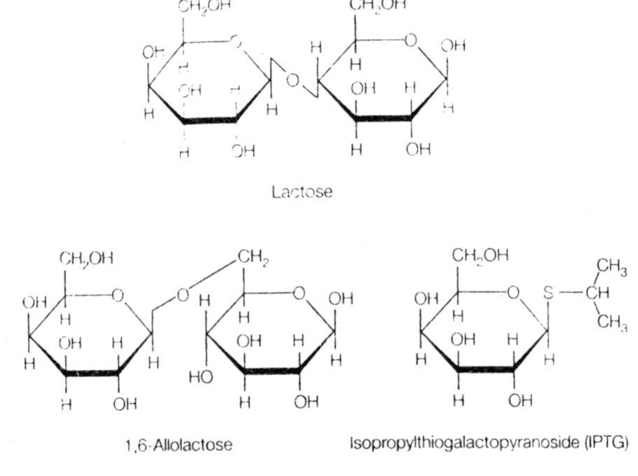

Fig. 10.2. Structure of lactose, allolactose and IPTG.

5' to 3' direction and the complementary strand is read right to left in a 5' to 3' direction. This inverted repeat symmetry of the operator matches the inherent symmetry of the lac repressor which is made up of four identical subunits. In the absence of lactose, the repressor occupies the operator-binding site. It seems that both the lac repressor and the RNA polymerase can bind simultaneously to the lac promoter and operator sites. The lac repressor actually increases the binding of the polymerase to the lac promoter by two orders of magnitude. This means that when lac repressor is bound to the O_{lac} operator DNA sequence, polymerase is also likely to be bound to the adjacent P_{lac} promoter sequence.

Induction

In the absence of an inducer, the lac repressor blocks all but a very low level of transcription of lacZYA. When lactose is added to cells, the low basal level of the permease allows its uptake, and b-galactosidase catalyzes the conversion of some lactose to allolactose.

Allolactose acts as an inducer and binds to the lac repressor This causes a change in the conformation of the repressor tetramer, reducing its affinity for the lac operator. The removal of the lac repressor from the operator site allows the polymerase (which is already sited at the adjacent promoter) to rapidly begin transcription of the lazZYA genes. Thus, the addition of lactose, or a synthetic inducer such as isopropyl-β-D-thiogalactopyranoside (IPTG) very rapidly stimulates transcription of the lactose operon structural genes. The subsequent removal of the inducer leads to an almost immediate inhibition of this induced

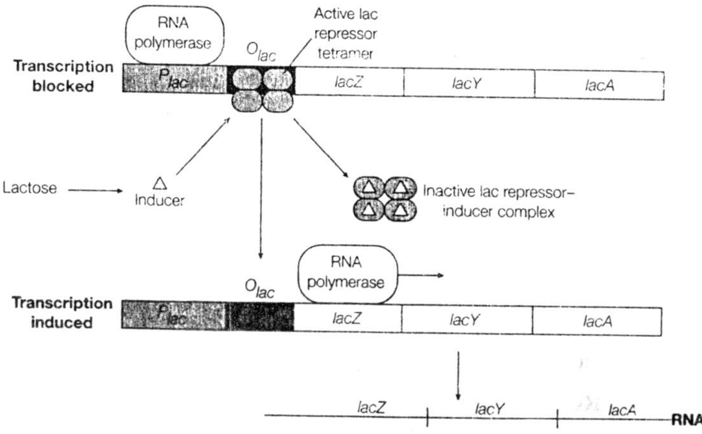

Fig. 10.3. Binding of inducer inactivates the lac repressor.

transcription, since the free lac repressor rapidly re-occupies the operator site and the lacZYA RNA transcript is extremely unstable.

cAMP Receptor Protein

The P_{lac} promoter is not a strong promoter. P_{lac} and related promoters do not have strong-35 sequences and some even have weak–10 consensus sequences. For high level transcription, they require the activity of a specific activator protein called *cAMP receptor protein* (CRP). CRP may also be called catabolite activator protein or CAP. When glucose is present, E. coli does not require alternative carbon sources such as lactose. Therefore, catabolic operons, such as the lactose operon, are not normally activated. This regulation is mediated by CRP which exists as a dimer which can not bind to DNA on its own, nor regulate transcription. Glucose reduces the level of cAMP in the cell. When glucose is absent, the levels of cAMP in E. coli increase and CRP binds to cAMP. The CARP-cAMP complex binds to the lactose operon promoter P_{lac} just upstream from the site for RNA polymerase. CRP binding induces a 90° bend in DNA, and this is believed to enhance RNA polymerase binding to the promoter, enhancing transcription by 50-fold.

The CRP-binding site is an inverted repeat and may be adjacent to the promoter (As in the lactose operon), may lie within the promoter itself, or may be much further upstream from the promoter. Differences in the he CRP-binding sites of the promoters of different catabolic operons may mediate different levels of response of these operons to cAMP in vivo.

THE TRP OPERON

The Trypotophan Operon

The trp operon encodes five structural genes whose activity is required for trypotophan synthesis. The operon encodes a single transcription unit which produces a 7 kb transcript which is synthesized downstream from the trp promoter and trp operator sites P_{trp} and O_{trp} Like many of the operons involved in amino acid biosynthesis, the trp operon has evolved systems for co-ordinated expression of these genes when the product of the biosynthetic pathway, trytophan, is in short supply in the cell. As with the lac operon, the RNA product of this transcription unit is very unstable, enabling bacteria to respond to changing needs for tryptophan.

The Trp Repressor

A gene product of the separate trpR operon, the trp repressor, specifically interacts with the operator site of the trp operon. The

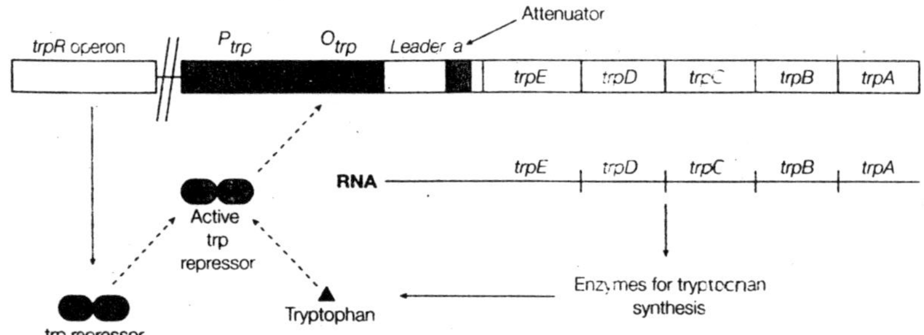

Fig. 10.4. Structure of the trp operon and function of the trp repressor.

symmetrical operator sequence, which forms the trp repressor-binding site, overlaps with the trp promoter sequence between bases -21 and +3 and +3. The core binding site is a palindrome of 18 bp. The trp repressor binds tryptophan and can only bind to the operator when it is complexed with tryptophan. The repressor is a dimer of two subunits which have structural similarity to the CRP protein and lac repressor. The repressor dimer has a structural with a central core and two flexible DNA-reading heads each formed from the carboxyl-terminal half of one subunit. Only when tryptophan is bound to the repressor are the reading heads the correct distance apart, and the side chains in the correct conformation, to interact with successive major grooves of the DNA at the trp operator sequence. Tryptophan, the end-product of the enzymes encoded by the trp operon, therefore acts as a co-repressor and inhibits its own synthesis through end-product inhibition. The repressor reduces transcription initiation by around 70-fold. This is a much smaller transcriptional effect than that mediated by the binding of the lac repressor.

The Attenuator

At first, it was thought that the repressor was responsible for all of the transcriptional regulation of the trp operon. However, it was observed that the deletion of a sequence between the operator and the trpE gene coding region resulted in an increase in both the basal and the activated (depressed) levels of transcription. This site is termed the attenuator and it lies towards the end of the transcribed leader sequence of 162 not that precedes the trpE initiator codon. The attenuator is a rho-independent terminator site which has a short GC-rich palindrome followed by eight successive U residues. If this sequence is able to form a hairpin structure in the RNA transcript, then it acts as a highly efficient transcription terminator and only a 140 bp transcript is synthesized.

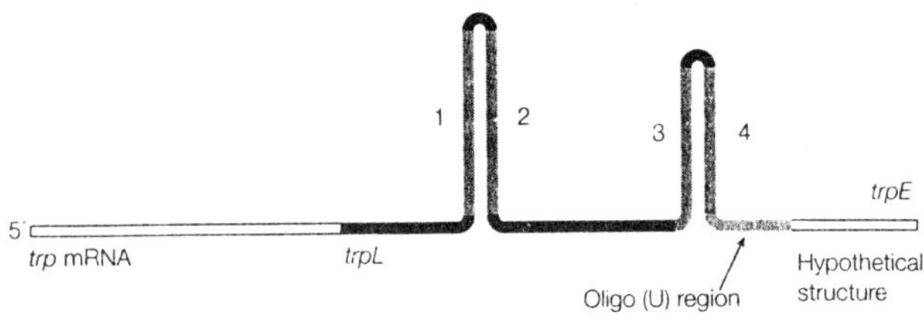

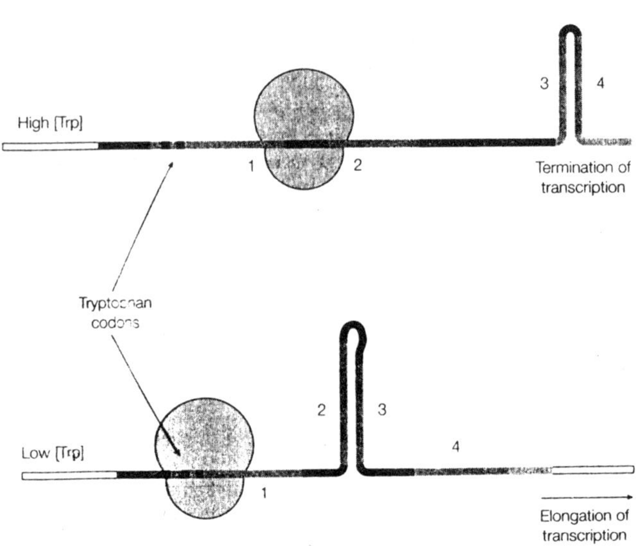

Fig. 10.5. Transcriptional attenuation in the trp operon.

Leader RNA Structure

The leader sequence of trp operon RNA contains for regions of complementary sequence which can form different base-paired RNA structures. These are termed sequences 1, 2, 3, and 4. The attenuator hairpin is the product of the base pairing of sequences 3 and 4. The attenuator hairpin is the product of the base pairing of sequences 3 and 4 (3:4 structure). Sequences 1 and 2 are also complementary and can form a second 1:2 hairpin. However, sequence 2 is also complementary and can form a second 1:2 hairpin. However, sequence 2 is alsc complementary to sequence 3. If sequences 2 and 3 form a

2:3 hairpin structure, the 3:$ attenuator hairpin cannot be formed and transcription termination will not occur. Under normal conditions, the formation of the 1:2 and 3:4 hairpins is energetically favourable.

The Leader Peptide

The leader RNA sequence contains an efficient ribosome-binding site and can form a 14-amino-acid leader peptide encoded by bases 27–68 of the leader RNA. The 10th and 11th codons of this leader peptide encode successive tryptophan residues, the end-product of the synthetic enzymes of the trp operon. This leader has no obvious function as a polypeptide, and tryptophan is a rare amino acid; therefore, the chances of two tryptophan codons is succession is low and, under conditions of low tryptophan availability, the ribosome would be expected to pause at this site. The function of this leader peptide is to determine tryptophan availability and to regulate transcription termination.

Attenuation

Attenuation depends on the fact that transcription and translation are tightly coupled in E. coli; translation can occur as an mRNA is being transcribed. The 3'-end of the trp leader peptide coding sequence overlaps complementary sequence the two trp codons are within sequence overlaps complementary sequences the two trp codons are within sequence 1 and the stop codon is between sequences 1 and 2. The availability of tryptophan (the ultimate product of the enzymes synthesized by the trp operon) is sensed through its being required in translation, and determines whether or not the terminator (3:4) hairpin forms in the mRNA.

As transcription of the trp operon proceeds, the RNA polymerase pauses at the end of sequence 2 until a ribosome begins to translate the leader peptide. Under conditions of high tryptophan availability, the ribosome rapidly incorporates tryptophan at the two trp codons and thus translates to the end of the leader message. The ribosome is then occluding sequence 2 and, as the RNA polymerase reaches the terminator sequence, the 3:4 hairpin can form, and transcription may be terminated. This is the process of attenuation.

Alternatively, if tryptophan is in scarce supply, it will not be available as an aminoacyl tRNA for translation, and the ribosome will tend to pause at the two trp codons, occluding sequence 1. This leaves sequence 2 free to form a hairpin with sequence 3, known as the anti-terminator. The terminator (3:4) hairpin cannot form, and transcription continues into trpE and beyond. Thus the level of the end product, tryptophan, determines the probability that transcription will terminate early (attenuation), rather than proceeding through the whole operon.

Importance of Attenuation

The presence of tryptophan gives rise to a 10-fold repression of trp operon transcription through the process of attenuation alone. Combined with control by the trp repressor (70fold), this means that tryptophan levels exert a 700-fold regulatory effect on expression from the trp operon. Attenuation occurs in at least six operons that encode enzymes concerned with amino acid biosynthesis. For example, the His operon has a leader which encodes a peptide with seven successive histidine codons. Not all of these other operons have the same combination of regulatory controls that are found in the trp operon. The His operon has no repressor-operator regulation, and attenuation forms the only mechanism of feedback control.

The Arabinose Operon of E. coli

The arabinose operon is another example of glucose-sensitive operon. As with lactose, when arabinose is absent, only a few molecules of the enzymes needed for arabinose catabolism are present in the cell. When arabinose is added (provided glucose is absent), there is a vary rapid increase in the number of arabinose catabolic enzymes. This is controlled by a different mechanism from that described for the lactose operon.

The gene governing the metabolism of arabinose comprise what is called a *regulon*, which is composed of at least three operons. The *araBAD* operon contains the gene for the enzymes involved with the conversion of L-arabinose to D-xylulose 5-phosphate. The controlling sites for this operon are located adjacent to it. Two operons control the transport of arabinose into the cell: *areE*, which is the structural gene for the L-arabinose binding protein, and *araF*. The regulator gene for the system, *araC*, is located between the *araBAD* controlling site region and the *leu* operon. The *araC* gene controls the expression of *araBAD* by its positive and negative action in the controlling site region. Since *araBAD*, *araE*, and *araF* are inducible by L-arabinose and controlled coordinately by *araC*, it is assumed that the structures of the three controlling site regions are similar. The following discussion will focus on the *araBAD* operon.

The operon is thought to be controlled as follows. The *araC* gene codes for a P1 protein, which has repressor function and exerts its

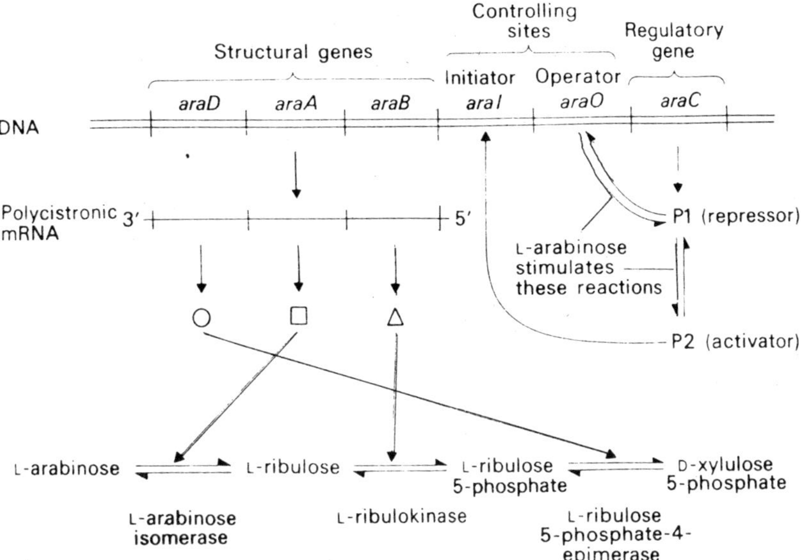

Fig. 10.6. The araBAD (arabinose) operon E. coli and the associated controlling sites and regulatory gene.

effect by binding to that adjacent araO (operator) controlling site and preventing the RNA polymerase from binding. Thus the operon is under negative control by P1. When L-arabinose is present, it stimulates the release of P1 from the DNA and the conversion of P1 to P2, which is an activator of the operon. The P2 binds to rhe araI (initiator) site, facilitating RNA polymerase binding and the initiation of transcription. (All this, of course, requires the prior binding of cAMP-CGA complex, and this is thought to occur in the same vicinity.) The three structural genes are transcribed on a single polycistronic mRNA. The operon, then, is under positive control by P2.

Much of this is hypothesis, but there is good evidence for some parts of it. There is genetic evidence for the existence of the araI site, and it is thought that part of this has promoter activity and another part is involved with cAMP-CGA binding. The function of the *araC* gene has been demonstrated by the study of genetic mutants. Mapping experiments with these mutants have shown that *araC* consists of only one cistron. Moreover, *araC* nonsense mutants have been shown to have no cis effect on the *araBAD* operon, thus indicating that *araC* is not in the BAD operon.

Three classes of *araC* alleles are known:

1. *araC*$^+$. The wild type allele renders the operon inducible; the three enzymes are induced by L-arabinose.
2. *araD*$^+$. These occur with high frequency and result in a pleiotropoically L-arabinose-negative phenotype. In other words, the enzymes are not inducible by L-arabinose.
3. *araC*^{c}. These are quite rare and give a pleiotropically constitutive phenotype in that the enzymes are produced even in the absence of the inducer.

As with the regulatory mutants of the lactose operon, diploid studies have been used to obtain an understanding of the function of the *araC* gene. From these studies it was sown there *araC*$^-$ is recessive to C$^+$ in either the cis or trans arrangement. The C$^-$ alleles are complemented by A$^-$, B$^-$, and D$^-$ alleles, and thus the pleiotropic negative phenotype is not the result of a polarity effect on the araBAD operon. The conclusion from studies of C$^-$ and C$^+$C$^-$ strains was the C$^+$ produces a protein that, in the presence of L-arabinose, is necessary for the expression of the region. This suggested some positive control in the system and contrasts with the lactose operon C$^-$ mutants, which are constitutive because of the loss of negative control

The Cc alleles are cis- and trans-dominant to C$^-$, suggesting that they produce the activator P2 in the absence of L-arabinose; this activator is able to turn on an operon that is either cis or trans to the Cc allele. On the other hand, *C*$^+$ is dominant to *C*^{c}. This suggests that there is some negative control of the operon. Based on the model for regulation of the operon that was presented (which was, of course, proposed on the basis of the data now being discussed), in the absence of arabinose, P1 acts as a repressor, preventing expression of the operon by araCc activator. That is, when P1 is on the operator, transcription ceases even if P2 is present.

There is some biochemical evidence to support the regulatory model. Studies of heat-sensitive *araC*– mutants have shown that both the repressor and activator functions are heat-labile, thereby indicating that arac produces a protein that can serve both functions. The *araC* protein has been purified and it has been shown to have both repressor and activator activity. Indeed there is some evidence that the P2 form of the protein is a dimer of the P1 form. There is also evidence that L-arabinose interacts directly with the *araC* protein to bring about the necessary conversion. Also, there is direct evidence that the activator form of *araC* protein, P2, is required for transcription of the operon. In an in vitro system, synthesis of ara mRNA shows an absolute requirement for *araC* protein.

In conclusion, the L-arabinose regulation is under both positive and negative control, with the *araC* protein playing a pivotal role in the regulatory process. The exact nature of the controlling sites remains to be worked out. In contrast to the lactose operon where inhibition must be removed for the genes to be expressed, the arabinose operon requires activation for transcription to begin.

11

GENETIC CODE

The Genetic Code is a Sequence of Bases

When Watson and Crick proposed the double helical structure for DNA in 1953 they also suggested that the genetic information which passed from generation to generation, and which controlled the activities of the cell, might be stored in the form of the sequence of bases in the DNA molecule. Once it had been shown that DNA is a code for the production of protein molecules it became clear that the sequence of bases in the DNA must be a code for the sequence of amino acids in prtein molecules. This relationship between bases and amino acids is known as the *genetic code.* The problems remaining in 1953 were to demonstrate that a base code existed, to break the code and to determine how the code is translated into the amion acid sequence of a protein molecule.

The Code is a Triplet Code

There are four bases in the DNA molecule, *adenine* (A), *guanine* (G), *thymine* (T) and *cytosine* (C). Each base is part of a nucleotide and the necleotides are arranged as a polynucleotide strand. The sequence of bases in the strand can be indicated by the initial letters of the bases. This 'alphabet' of four letters is responsible for carrying the code that results in the synthesis of a potentially infinite number of different protein molecules. There are 20 common amino acids used to make proteins and that the bases in the DNA must code for. If one base determined the position of a single amino acid in the primary structure of protein, the protein could only contain four different amino acids. If a combination of pairs of bases coded for each amino acid then 16 amino acids could be specified into the protein molecule.

Only a code composed of three bases could incorporate all 20 amino acids into the structure of protein molecules. Such a code would produce 64 combinations of bases, more than enough. Watson and crick therefore predicted that the code would be a triplet code.

It was later proved that the code is indeed a triplet code, meaning that three bases is the code for one amino acid.

Evidence for a Triplet Code

Evidence that the code is a triplet code was provided by Francis Crick in 1961. He produced mutations involving the addition or deletion of bases in T4 phages. Adding or deleting a base changes the way in which the code is read after the point of additon or deletion. The mutton is said to produce a 'frame-shift'. These frame-shifts produced base triplet sequences which failed to result in the synthesis of protein molecules with the original amino acid sequence. Only by adding a base and deleting a base at specific points could the original base sequence be restored.Restoring the original base sequence

A
G
C
T

Single code

AA	AG	AC	AT
GA	GG	GC	GT
CA	CG	CC	CT
TA	TG	TC	TT

Double code

AAA	CAA	CAA	TAA
AAG	GAG	CAG	TAG
AAC	GAC	CAC	TAG
AAT	GAT	CAT	TAT
AGA	GQA	CGA	TGA
AGG	GGG	CGG	TGG
AGC	GGC	CGC	TGC
AGT	GGT	CGT	TGT
ACA	GCA	CCA	TCA
ACG	GCG	CCG	TCG
ACC	GCC	CCA	TCC
ACT	GCT	CCT	TCT
ATA	GTA	CTA	TTA
ATG	GTG	CTG	TTG
ATC	GTC	CTC	TTC
ATT	GTT	CTT	TTT

Fig. 11.1. Singlet code, a double code and a triplet code.

prevented the appearance of mutants in the experimental T4 phages. Adding a single base is referred to as a (+) type mutation and deletin a base a (-) type. (+)(–) restores the correct reading. The double mutants (++) or (– –) also produced frame-shifts which resulted in mutants which produced faulty proteins. However (+++) (– – –) mutations usually had no effect on protein function. Crick argued that this is because such mutations do not cause frame-shifts, only the addition or deletion of one amino acid which often does not affect the performance of a protein. This implies that the code is read three bases at a time, that is in triplets.

These experiments also demonstrated that the code is *non-overlapping*, that is to say no base of a given triplet contributes to part of the code of the adjacent triplet.

Features of the Genetic Code

The Code is a Triplet Code

As already stated, the genetic code is a triplet code, meaning that three bases in DNA code for one amino acid in a protein. The DNA code for a protein is first copies into messenger RNA (mRNA) before a protein is made. mRNA is complementary to the DNA. The complementary triplets in themRNA are referred to as *codons*. Each codon is therefore three bases long and is the code for one amino acid. The DNA code for each amino acid can be obtained by converting the RNA codons back into their complementary DNA triplets of bases.

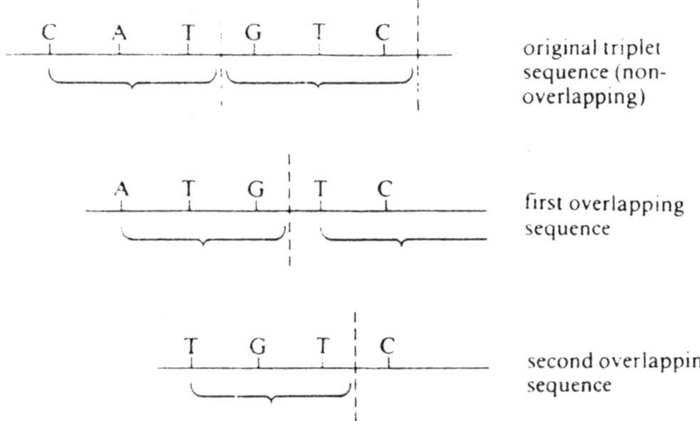

Fig. 11.2. Base sequences indicating non-overlapping and overlapping codes.

Table 11.1. The RNA bases which are complementary to those of DNA.

DNA bases	*Complementary RNA bases*
A (adenine)	U (uracil)
G (guanine)	C (cytosine)
T (thymine)	A (adenine)
C (cytosine)	G (guanine)

The Code is Degenerate

Table shows the genetic code in the form of codons. As can be seen from the table some amino acids are coded for by several codons. This type of code where the number of amino acids is less than the number of codons is termed *degenerate.* Analysis of the code also shows that for many amino acids only the first two letters appear to be significant.

The Code is Punctuated

Three of the codons shown in table act as 'full stops' in determining the end of the code message. An example is UAA. They are sometimes described as 'nonsense codons' and do not code for amino acids. They presumably mark the end-point of a gene. They act as 'stop signals' for the termination of polypeptide chains during translation.

Certain codons act as 'start signals' for the initiation of polypeptide chains, such as AUG (methionine).

The Code is Universal

One of the remarkable features of the genetic code is that it is thought to be universal. All living organisms contain the same 20 common amino acids and the same five bases, A, G, T, C and U.

Advances in molecular biology have reached the point now where it is possible to determine the base sequences for whole genes and for whole organisms. The first organism whose complete genetic code was established was a virus, the phage ϕX174. The phage has only ten genes and its complete genetic code is 5386 bases long. The sequence was discovered by Fred Sanger, the man who first discovered the sequence of amino acids in a protein. He was awarded Nobel prizes for both these sequencing milestones. Whole genes can now be synthesised artificially, a practice which is of use in genetic engineering. Soon after the beginning of the twenty-first century it is anticipated that it will be possible to write out the entire genetic code of a human, an estimated 3000 million base pairs long as a result of the Human Genome Project. (The genome is the total DNA in an organism.) Other organisms whose genomes are being sequenced are *E. coli*, yeast, the fruit fly *Drosophila*, a nematode worm and the laboratory mouse.

Deciphering the Code

The deciphering of the genetic code–that is, determining (1) which codons specify which amino acids, (2) how many of the 64 possible codons are used, (3) how the code is punctuated, and (4) whether different species use the same or different codons–took place during the early 1960s and was one of the most exciting periods in the history of science.

The first major breakthrough came in 1961 when M. W. Nirenberg (1968 Nobel Prize recipient) and J. H. Matthaei and then S. Ochoa (1959 Nobel Prize recipient) and co-workers demonstrated that synthetic RNA molecules could be used as artificial mRNAs to direct *in vitro* protein synthesis. That is, when ribosomes, aminoacyl-tRNAs, and the soluble protein factors required for translation are purified free of natural mRNAs, these components can be combined in vitro and stimulated to synthesize polypeptides by the addition of chemically synthesized RNA molecules. If these synthetic mRNA molecule are of known composition, the composition of the polypeptides synthesized can bae used to deduce which codons specify which amino acids.

The first codon assignment (UUU for phenylalanine) was made when Nirenberg and Matthaei demonstrated that *polyuridylic acid* [poly U = (U)n)] directed the synthesis of polyphenylalanine [(phenylalanine)n]. Ochoa and others continued this approach using synthetic RNAs with random sequences of known nucleotide compositon, such as 50 percent U and 50 percent G. The frequencies of the different triplets in such a random copolymer can be easily calculated. For example the 50 percent U/50 percent G copolymer will contain 12.5 percent (½.½.½ = 1/8) of each of the eight possible codons: UUU, UUG, UGU, GUU, UGG, GUG, GGU, and GGG. These can then be compared with the amino acids incorporated (phenylalanine, leucine,cysteine, valine, tryptophan, and glycine) when this random copolymer is used in an in vitro protein synthesizing system. By varying the composition, for example, to 75 percent U and 25 percent G, one can vary the relative frequencies of the eight codons and correlate them with the relative frequencies of the amino acids in the polypeptides synthesized. Such amino acids in the polypeptides synthesized. Such experiments provided a great deal of information about the nature of the code.

First Letter	Second Letter: U	C	A	G	Third Letter
U	UUU(Phe) UUC (Phe) UUA (Leu) UUG (Leu)	UCU (Ser) UCC (Ser) UCA (Ser) UCG (Ser)	UAU (Tyr) UAC (Tyr) UAA (Ochre) UAG (Amber)	UGU (Cys) UGC (CYS) UGA ? UGG (Tryp)	U C A G
C	CUU (Leu) CUC (Leu) CUA (Leu) CUG (Leu)	CCU (Pro) CCC (Pro) CCA (Pro) CCG (Pro)	CAU (His) CAS (His) CAA (Glun) CAG (Glun)	CGU (Arq) CGC (Arq) CGA (Arq) CGG (Arq)	C A G
A	AUU (Heu) AUC (Heu) AUA (Aeu) AUG (Met)	ACU (Thr) ACC (Thr) ACA (Thr) ACG (Thr)	AAU (ASpN) AAC (ASPN) AAA (LYS) AAG (Lys)	AGU (Ser) AGC (Ser) AGA (Arg) AGG (Arg)	U C A G
	GUU (Aal) GUC (Val) GUA (VaL) GUG (Val)	GCU (Ala) GCC (Ala) GCA (Ala) GCG (Ala)	GAU (Asp) GAC (Asp) GAA (Glu) GAG (Glu)	GGU (Glu) GGC (Gly) GGA (Gly) GGG (Gly)	U C A G

Fig. 11.3. The triplet genetic code.

More definitive data were later obtained by H. G. Khorana using *in vitro* systems that were activated by synthetic mRNAs of known nucleotide sequence. Khorana's experiments permitted direct comparisons between nucleotide sequences and the amino acids incorporated in response to these sequences. The ultimate "cracking" of the code occurred when trinucleotides were found to function as "mini-mRNAs," directing the specific binding of aminoacyl-tRNAs to ribosomes. By using all of the 64 possible trinucleotide sequences in such amino-acyl-tRNA binding experiments, it was possible to verify the codon assignments made from data of earlier experiments.

On the basis of extensive data accumulated over several years, the codon assignments shown in table became firmly established. Two important questions remained to be answered. (1) Are the assignments based on in vitro experiments valid *in vivo*? (2) Is the code *universal*; that is, do the codons specify the same amino acids in all organisms? Several lines of evidence now indicate that these codon assignments are correct for protein synthesis in vivo for most, if not all, species. When the amino acid substitutions that result from mutations induced with chemical mutagens with specific mutagenic effects are determined by amino acid sequencing, the substitutions are almost always consistent with the codon assignments and the known effect of the mutagen. More convincingly, when the nucleotide sequences of genes or of mRNAs are determined and compared with the amino acid sequences of the polypeptides coded for by those genes or mRNAs, the observed correlations are always found to be those predicted from the accepted codon assignments. This can be illustrated by comparing the nucleotide sequence of the gene coding for the protein coat or capsid of bacteriophage MS2 with the amino acid sequence of the capsid polypeptide. Phage MS2 stores its genetic information in RNA (like TMV virus). Its chromosome is the equivalent to an mRNA molecule in organisms with DNA genomes.

Degeneracy and Wobble

All of the amino-acids except methionine and tryptophan are specified by more than one codon. Three amino acids, leucine, serine,and arginine, are each specified by six different codons. Isoleucine has three codons. The other amino acids each have either two or four codons. The occurrence of more than one codon per amino acid is called *degeneracy* (through the usual connotations of the term are hardly appropriate). The degeneracy in the genetic code is not at random; instead, it is highly ordered. Usually, the multiple codons specifying an amino acid differ by only one base, the third or 3' base of the codon. The degeneracy is primarily of two types. (1) Partial degeneracy occurs when the third base may be either one of the two pyrimidines (U and C) or, alternatively, either one of the two purines (A and G). With partial degeneracy, changing the third base from a purine to a pyrimidine, or vice versa, will change the amino acids specified by the codon. (2) In the case of complete degeneracy, any of the four bases may be present at the third position in the codon, and the codon will still specify the same amino acid. For example, valine is specified by GUU, GUC, GUA, and GUG.

It has been speculated that the order in the genetic code has evolved as a way of minimizing mutational lethality. Many base substitutions at the third position of codons do not change the amino acid specified by the codon. Moreover, amino acids with similar chemical properties (such as leucine, isoleucine, and valine) have codons that differ from each other by only one base. Thus, many single base-pair substitutions will result in the substitution of one amino acid for another amino acid with very similar chemical properties (e.g., valine for isoleucine). In most cases, such substitutions will not result in inactive gene products; again, this minimizes the effects of mutations.

Because of the degeneracy of the genetic code, there must either be several different tRNAs that recognize the different codons specifying a given amino acid, or the anticodon of a given tRNA must be able to base-pair with several different codons. Actually, both of the above occur. Several tRNAs exist for certain amino acids, and some tRNAs recognize more than one acids, and some tRNAs reccognize more than one codon. The hydrogen bonding between the bases in the anticodon of tRNAs and the codon of mRNA appears to follow strict base-pairing rules (i.e., be "tight") only for the first two bases of the codon. The base-pairing involving the third base of the codon is apparently less stringent, allowing what Crick had called *wobble* at this site.

On the basis of molecular distances and steric (three-dimensional structure) considerations, Crick proposed that wobble would allow several types, but not all types, of base-pairing at the third codon base in the codon–anticodon interaction. His proposal has since been strongly supported by experimental data. Table shows the base-pairing predicted by the wobble hypothesis. It necessitates that there be at least two tRNAs for each amino acid whose condons exhibit complete degeneracy at the third position. This has indeed been found to be true.The wobble hypothesis predicted the occurrence of three tRNAs for the six serine codons. Three serine tRNAs have now been characterized: (1) tRNAser1 (antiodon AGG) binds to codons UCU and UCC, (2) tRNAser2 (anticodon AGU) binds to codons UCA and UCG, and (3) tRNAser3 (anticodon UCG) binds to codons AGU and

AGC. These specificities were verified by the trinucleotide-stimulated binding of purified aminoacyl-tRNAs to ribosomes *in vitro.*

Table 11.2. Base-pairing between the 5' Base of the Anticodon of tRNAs and the 3' Base of Codons of mRNA according to the Wobble Hypothesis.

Base in Anticodon	*Base in Codon*
G	U or C
C	G
A	U
U	A or G
I	A, U, or C

Finally several tRNAs contain the base inosine (produced by post-transcriptional enzymatic modification). Crick's wobble hypothesis predicted that insoine could pair (at the whobble position) with adenine, uracil, or cytosine (in the codon). In fact, purified alanyl-tRNA containing inosine (I) at the 5' position of the anticodon binds to ribosomes activated with GCU, GCC, or GCA trinucleotides. The same result has been obtained with other purified tRNAs with inosine at the 5' position of the anticodon. The wobble inosine at the 5' position of the anticodon. The wobble hypothesis thus fits several observations; whether it is entirely accurate remains unknown.

Initiation and Termination Codons

The genetic code also provides for punctuation of genetic information at the level of translation. Three codons, UAA, UAG, and UGA, specify polypeptide chain termination. These codons are recognized by protein release factors, rather than by tRNAs. One of these proteins, designated RF-1, is apparently specific for UAA and UAG. The other, RF-2 causes termination at UAA and UGA codons. Two codons, AUG and GUC, are recognized by the initiator tRNA, $tRNA^{Met}_{f}$, but apparently only when they follow an appropriate nucleotide sequence in the leader segment of an mRNA molecule. At internal positions, AUG is recognized by $tRNA^{Met}$, and GUG is recognized by a valine tRNA. In the case of the initiation codons AUG and GUG and $tRNA^{Met}_{f}$, the wobble base appears to be the first of 5' base of the codon. Since wobble at the first base is unique to initiation, it may be related to base-pairing at the *P* site rather than at the A site on the ribosome.

Universality of the Code

A vast amount of data is now available from *in vitro* studies,from amino acid replacements due to mutations and from correlated nucleic acid and polypeptide sequencing–all suggesting that the genetic code is the same or very nearly the same in all organissms. These data (e.g., see the human haemoglobin substitutions. The correlated nucleotide and amino acid sequences in the overlapping genes of the DNA bacteriophage ϕX174, and the RNA bacteriophage MS2, all indicate that the genetic code is largely *universal.*

The major exception to the universality of the code occurs in mitochondria of humans, yeast, and several other species, where UgA is a tryptophan codon. UGA is a termination codon in non-mitochondrial systems. Also, in yeast mitochondria, CUA specifies threonine instead of the usual leucine, and, in mammalian mitochondria, AUA specifies methionine instead of the usual isoleucine. Excluding these exceptions, the code appears to be universal.

12

LINKAGE

Mendel's law of independent assortment is applied only to those genes which are located on separate chromosomes, because, the linked genes of a linkage group (chromosome) inherit together. A dihybrid contains either linked genes or independently assorted genes, can be determined by test crossing it with a double recessive parent. The independently assorted genes give the test cross ratio of 1:1:1:1 and linked genes give the test cross ratio of 1 : 1 as have been illustrated by following examples:

Example I

If genes occur on different chromosomes, they assort independently and give a test ratio of 1:1:1:1 as follows:

P_1	AA BB	×	aa bb
P_1 Gametes :	(AB)		ab
F_1:		AA Bb	
Test cross :	AA Bb	×	aa bb
		↓	↓
Gametes :		(AB) (Ab) (aB) (ab)	(ab)
F_2:		1/4 Aa Bb : 1/4 Aa bb : 1/ 4aa Bb : 1/4 aa	
	or	1:1:1:1 (Test cross ratio).	

Example II

The linked genes do not assort independently, but tend to stay together in the same combinations as they were in the parents. In the following figure, the genes on the left of the slash line (/) are on one chromosome and those on the right are on the homologous chromosome. The linked genes give the test cross ratio of 1:1 as follows:

Parents : AA/ab × ab/ab

Gametes : (AB) (ab) (ab) (ab)

F_1 : 1/2 AB/ab : 1/2 ab ab or 1 : 1 (test cross ratio)

Views of Classical Geneticists on Linkage

Mendel could not notice the phenomenon of linkage because fortunately the seven pairs of factors or alleles studied by him in pea were located on seven different pairs of chromosomes. It was noticed and discovered by some other post-Mendelian geneticists who during their genetic investigations came across to linked genes. The evolution of linkage concept took place by the views of following classical geneticists :

1. Sutton's Views on Linkage

Sutton (1903) was the first classical geneticist who made certain predictions about the linkage merely by performing certain cytological investigations. He suggested that each chromosome must bear more than a single gene and that the genes represented by one chromosome must be inherited together. However, he could not prove his predictions by genetic experiments.

Coupling and Repulsion Hypothesis

W. Bateson, in 1905, described a cross in sweet pea, where a deviation from independent assortment was exhibited. The plats of a variety of sweet pea having blue flowers (B) and long pollen (L) were crossed with those of another variety having red flowers (b) and round pollen (1). The F_1 individuals (BbL1) had blue flowers and long pollen. These were crossed with plants having red flowers and round pollen (bbll). (In this case character for blue colour of flowers is dominant over red colour and long pollen character is dominant over round pollen.)

Normally, if independent assortment takes place, we should expect 1:1:1:1 ratio in a testcross. Instead 7:1:1:7 ratio was actually obtained, indicating that there was a tendency in the dominant alleles to remain together. Similar was the case with recessive alleles. This deviation was, therefore, explained as *"gametic coupling"*by Bateson. Similarly, it was observed that when two such dominant alleles or two recessive alleles come from different parents, they tend to remain separate. This was called "repulsion". In Bateson's experiment in repulsion phase,

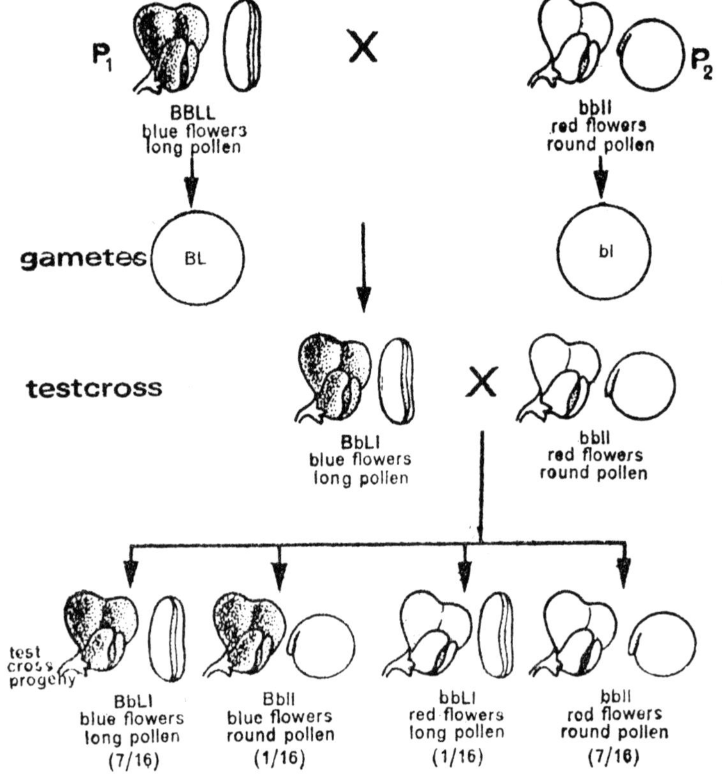

Fig. 12.1. Results obtained by Bateson in sweet pea when two characters, blue flowers and long pollens, were present in coupling phase.

one parent would have blue flowers and round pollen (BBII) and the other would have red flower and long pollen (bbLL).

Bateson explained the lack of independent assortment in the above experiments by means of a hypothesis known as "coupling and repulsion hypothesis". Although coupling and repulsion as explained above were later discovered to be the two aspects of the same phenomenon called "linkage", the terms "coupling phase" and "repulsion phase" are still considered to be useful terms in scientific literature.

Morgan, while performing experiments with *Drosophila*, in 1910 found that coupling or repulsion was not complete. He proposed that the two genes are found in coupling phase or in repulsion phase, because these are present on the same chromosome (*coupling*) or on two different homologous chromosomes (*repulsion*). Such genes are then called as linked and the phenomenon is called linkage. He further suggested

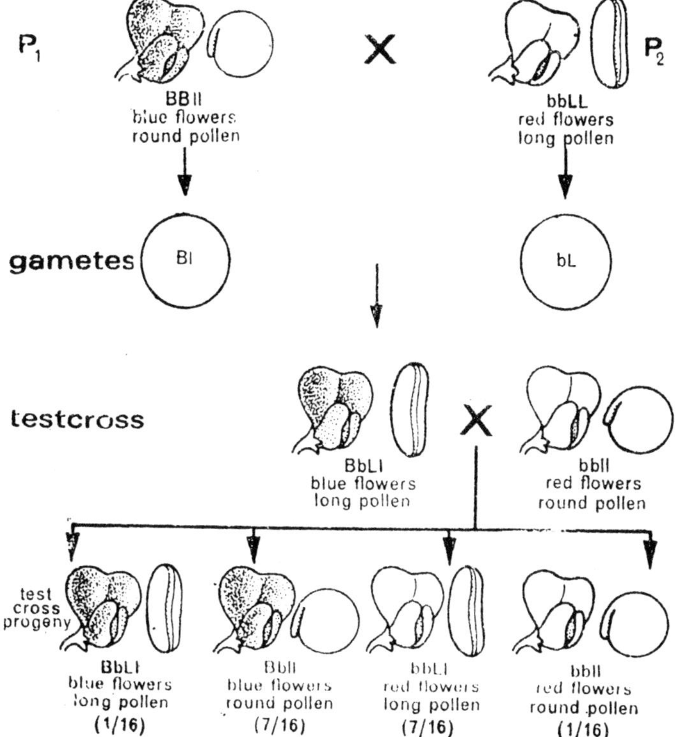

Fig. 12.2. Results obtained by Bateson in sweet pea when two characters, blue flowers and long pollens, were present in repulsion phase.

that the strength of linkage will be determined by the distance between the two genes in question. The greater this distance, lower will be the linkage strength. The linkage is broken down due to the phenomenon of crossing over occurring during meiosis. Crossing over will be relative more frequent, if the distance between the two genes is more than in case where the distance between two genes is lesser. The phenomenon of crossing over involves exchange of chromosome segments.

In other to make the phenomenon of linkage and crossing over easily understandable, let us take another hypothetical example, where genes *A* and *B* are involved, *a* and *b* being their recessive alleles respectively. A cross *AB/AB* × *ab/ab* would give rise to a F_1 dihybrid *AB/ab.* This dihybrid will then be crossed with double recessive parent *ab/ab* to get the testcross progeny.

Depending upon the distance between any two genes which is inversely proportional to the strength of linkage, the non-crossovers will vary from 50%–100% (100% non-crossovers is a state where no crossingover takes place as in male *Drosophila*. The crossovers will similarly vary from 0–50% and will never exceed 50%. The easiest way of finding out the proportion of non-crossovers and crossovers is to make a testcross (*AB*/ *ab* × *ab*/ *ab*). In such a situation, since the phenotypic ratio will depend upon the ratio of different kinds of gametes coming from F_1 (*AB*/ *ab*), the relative proportion of non-crossovers and crossovers can be easily determined from the phenotypic ratio. This is illustrated in the next section using an example from maize.

KINDS OF LINKAGE

The phenomenon of linkage is of following two kinds:

1. Complete Linkage

When the linked genes are so closely located in chromosomes that they inherit in same linkage groups for two or more generations in a continuous and regular fashion, then, they are called completely linked genes and the phenomenon of inheritance of completely linked genes is called completed linkage.

Example

According to Bridge all the gene of male *Drosophila* remain completely linked. Further, in a mutant strain of *Drosophila*, the genes for bent wings (b^+) and shaven bristles (svn) of the fourth chromosome exhibit complete linkage.

2. Incomplete Linkage

The linked genes do not always stay together because homologous non-sister chromatids may exchange segments of varying length (which bearing many linked genes) with one another during meiotic prophase, by the process of crossing over. The linked genes which are widely located in chromosomes and have chances of separation by crossing over are called incompletely linked genes and the phenomenon of their inheritance is called incomplete linkage.

Examples

Incomplete linkage has been observed in pea, Zea mays (maize), tomato, female *Drosophila*, mice poultry and man. Here, the examples of linkage have been considered only for *Drosophila* and Zea mays (maize).

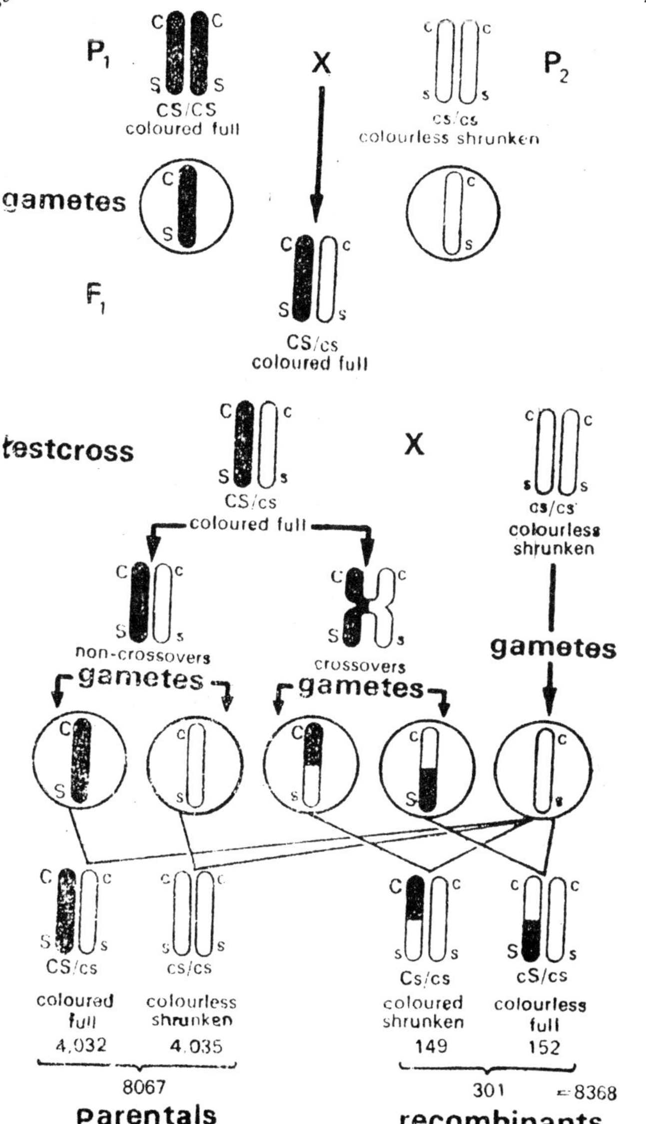

Fig. 12.3. Results obtained in maize by Hutchinson, when characters coloured seeds and full seeds were taken in coupling phase.

(i) Incomplete Linkage in Drosophila

The wild type Drosophila has gray body and long wings (b^+v^+/b^+v^+), where alleles for gray b^+ and long v^+ dominate over the mutant alleles for black b and vestigial v. When, a gray long fly (b^+v^+/b^+v^+) is crossed with a black bodied and vestigial winged fly (bv/bv), the F_1 heterozygote is found with gray long phenotype and b^+v^+/bv genotype. The F_1 heterozygote (b^+v^+/bv) when test crossed with double recessive parent (bv/bv), instead of occurring of two class of phenotypes in the ratio of 1:1, occur four classes of phenotypes.

The test cross results are clearly showing that parental combinations (gray long and black vestigial) are those expected from complete linkage and appeared in 83% cases. The other two (gray vestigial and black, long) are new combinations and appeared in 17% cases. Thus, in 17% cases crossing over has occurred.

(ii) Incomplete Linkage in Maize

In *Zea mays* (maize) a case of incomplete linkage between the alleles for colour and shape of the seed has been observed by Hutchison. When a maize plant with seeds having colour and full endosperms (CS/Cs) is crossed with another plants having recessive alleles for colourless, shrunken seeds (cs/cs) the F_1 heterozygotes are found with the phenotype of coloured full and genotypes of Cs/cs. When F_1 hybrid is test crossed with double recessive parent (cs/cs) four classes of descendants are obtained instead of two as shown in following figure :

Parent : Coloured full × Colourless shrunken
CS/Cs cs/cs

F_1 : Coloured full
(CS/cs)

Test cross : F_1 Coloured full × Colourless shrunken
Cs/cs s/cs

Test cross results :	Coloured,	Coloured,	Colourless,	Colourless
	full	shrunken	full	shrunken
	CS/CS	Cs/cs	cS/cs	cs/cs
	48%	2%	2%	48%

The test cross results are clearly showing that parental combination of alleles (e.g., CS/CS and cs/cs) are those expected from complete linkage and appear in 96% cases, the other two are new combinations (e.g., Cs/cs and cS/cs) and appear in 4% cases. Thus, in 4% cases crossing over have occurred between linked genes.

Linkage Groups

All the linked genes of a chromosome form a linkage group. Because, all the genes of a chromosome have their identical genes (allelomorphs) on the homologous chromosome, therefore, linkage groups of a homologous pair of chromosome is considered as one. The number of linkage groups of a species, thus corresponds with haploid chromosome number of that species.

Example

1. *Drosophila* has 4 pairs of chromosomes and 4 linkage groups.
2. Man has 23 pairs of chromosomes and 23 linkage groups.

13

CROSSING OVER

The process of Crossing over can be defined as a process which produces new combinations (recombinations) of genes by inter-changing of corresponding segments between non-sister chromatids of homologous chromosome." The chromatids in which crossing over has occurred, have new combination of genes and are called cross over. According to its occurrence in the germinal or somatic cells following two types of crossing over have been recognised.

Germinal or Meiotic Crossing Over

Commonly crossing over occurs only in the germinal cells of reproductive organs during the process of gametogenesis which includes meiosis. This type of crossing over is called germinal or meiotic crossing over. It is universal in its occurrence and has great genetic significance.

Somatic or Mitotic Crossing Over

Sometimes crossing over may occur during mitosis of somatic cells. This type of crossing over occurs in rare case, has no genetic significance and is called somatic or mitotic crossing over. It has been observed in body cells of *Drosophila* by Stern and in the fungus *Aspergillus nidulans* by Pontecorvo.

By somatic crossing over two chromosomes with unequal chromatids are formed. In the following division, in 50% of case two different alleles go to the same daughter nucleus and the original situation is restored. In the other 50%, identical alleles go to the same pole, a situation comparable to anaphase II segregation with double reduction. Thus, in one cell two recessive alleles are present and recognizable, marking the event of crossing over.

Theories about the Mechanism of Crossing Over

The exact physical mechanism of crossing over is still not clear to geneticists, however, following theories have been forwarded by different geneticists to explain the exact genetical mechanism of crossing over.

1. Classical Theory

In the classical views proposed by Karl Sax in 1932, chiasmata result from alternate opening out between sister and homologous (non sister) chromosomal strands with kinetochore loops being always reductional. On this hypothesis, chiasmata do not represent the result of crossing over, but crossing over may result from breakage and rejoining at points of overlap.

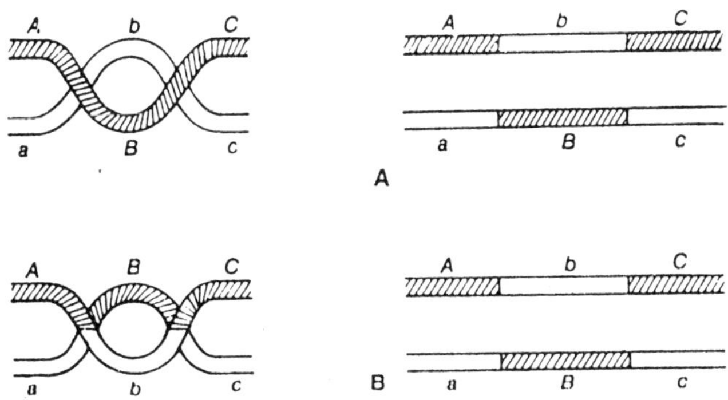

Fig. 13.1. Crossing over and chiasma formation based on classical theory (A) and chiasmatype theory (B).

2. Duplication Theory

The duplication theory has been proposed by Belling in 1927 and 1933. According to its, duplication of chromosomes followed by longitudinal joining of the chromomeres, which, if overlapping of homologous occurs, may result in the joining of chromomeres from homologous (non-sister chromatids) rather than from the same chromosome (sister-chromatids).

3. Precocity Theory

The precocity theory of C.D. Darlington is ordinarily discussed in context with meiotic chromosome pairing. However, Darlington extended this theory to explain recombination also. The precocity theory presumes that prophase is precocious in meiotic cell division

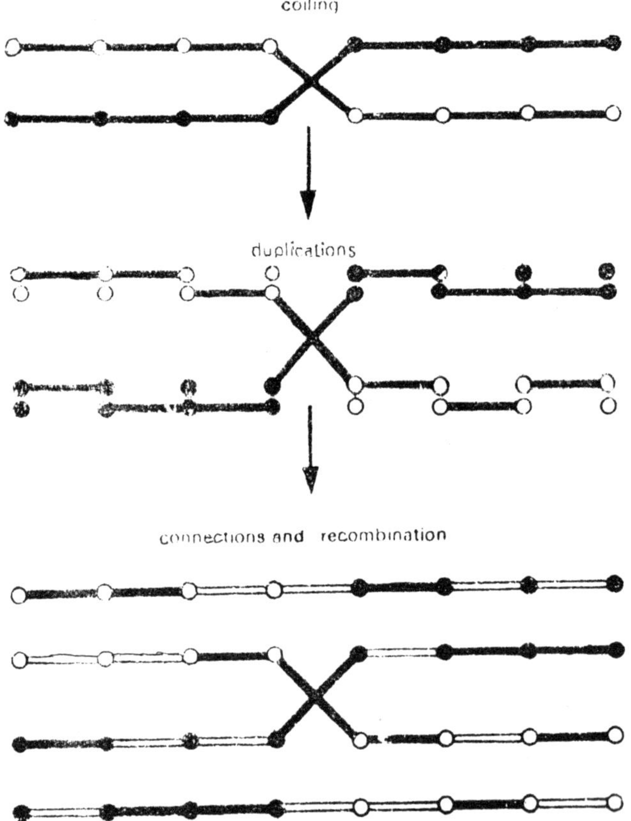

Fig. 13.2. The mechanism of genetic recombination as proposed by Belling.

and therefore involves homologous pairing to satisfy the pairing need, which is achieved in mitosis due to replication of the chromosomes. This theory is now untenable in the light of recent information that DNA synthesis really takes place before the onset of prophase I in meiosis. although synthesis of very small fraction of DNA (0.3%) is delayed till zygotene. Precocity theory presumes that the DNA synthesis or chromosome duplication takes place later in pachytene or diplotene and then result in the separation of homologous chromosomes (terminalization).

On the basis of precocity theory, Darlington explained crossing over to be the result of strain of torsion produced due to coiling of homologous chromosomes and the sister chromatids.

4. Belling's Hypothesis and Copy Choice

It was earlier proposed by J. Belling, that crossing over is really brought about due to novel attachments formed between newly synthesized genes. According to this theory, chromosome duplication consists of two stages (i) the formation of new genes, and (ii) the

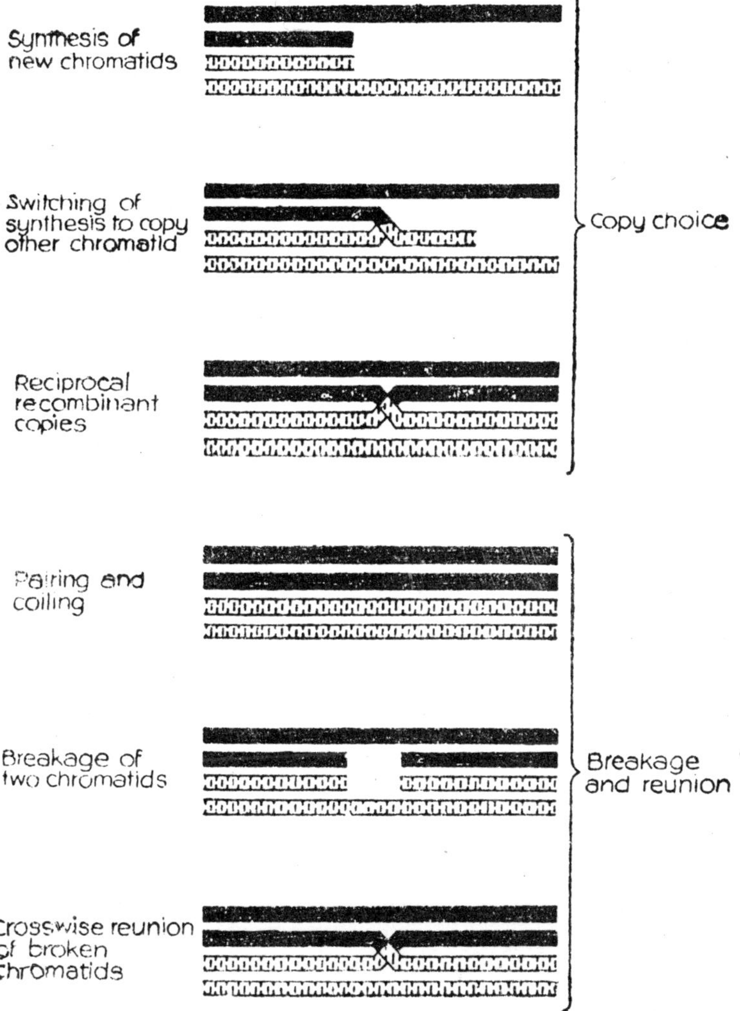

Fig. 13.3. Diagrammatic representation of copy choice theory and break and exchange theory of crossing over.

formation of new connections between these genes. This results in the formation of recombinants.

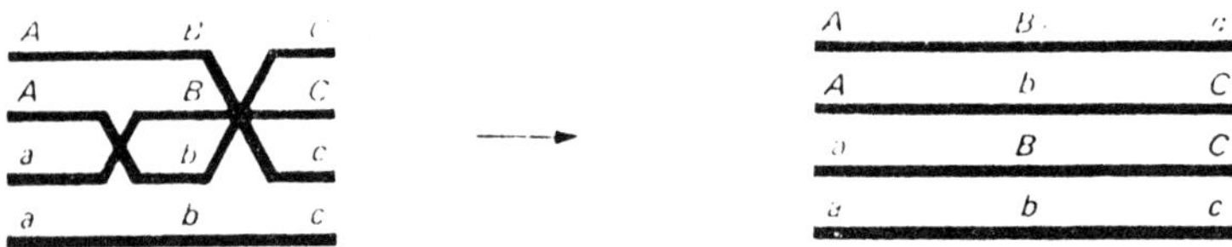

Fig. 13.4. A possible three strands double crossover, making Belling's copy-choice (switch) model very unlikely.

In order to explain some unusual results of recombination in microbial systems, a modified version of Belling's theory was proposed by J. Lederberg in 1955. This is known as copy choice mechanism of recombination where it is assumed that a newly synthesized daughter chromatid is derived due to copying of one chromosome up to a certain distance and then switching on to the other homologous chromosome for copying the remaining distance or region of chromosome. Copy choice theory was also inadequate, because it assumed a conservative rather than a semi-conservative replication of DNA duplexes. In these copy choice theories (several such theories are known), a chromosome with DNA duplex is synthesized *de novo* by copying the parent chromosome. This is hardly the case, since we know that the chromosomes replicate in a semi-conservative manner (refer Taylor's experiment). Although recent modifications of copy choice could overcome the above difficult situation, it still remains a complex and unlikely mechanism of recombination.

Cytological Detection of Crossing Over

The example which have been cited in the section of incomplete linkage (*Drosophila* and maize) have also furnished good evidence for genetical detection of crossing over, but they did not give any cytologically demonstrable evidence in support of genetical crossing over, because, the homologous chromosomes appeared, on microscopic examination, to be exactly alike. It was impossible to observe whether chromosome blocks had changed placed until visible markers of some kind could be incorporated on the chromosomes. The first cytological demonstration of genetic crossing over has been given by Stern (working with *Drosophila*) and *H.B. Creighton* and *B. McClintock* (working with maize) in 1931.

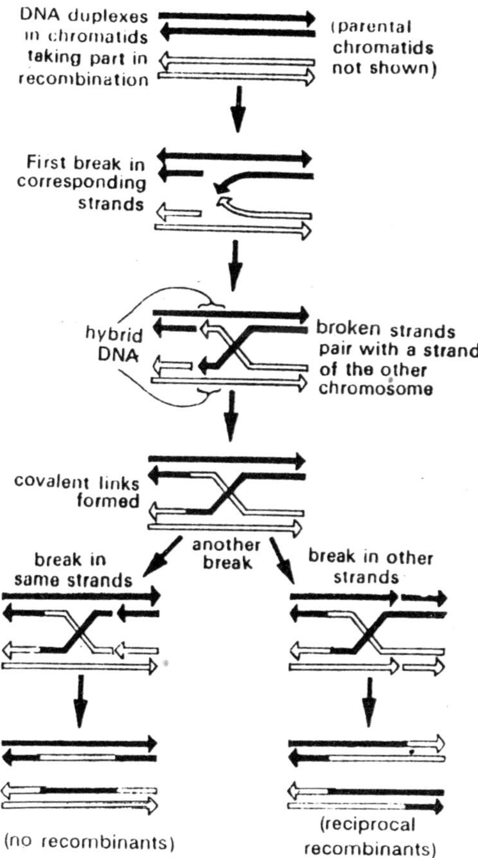

Fig. 13.5. Mechanism of recombination as explained on the basis of hybrid DNA model of R. Holliday.

1. Stern's Experiment

A wild type female *Drosophila* has one recessive gene for round eyes and one dominant gene for red eyes on each of its rod-shaped X chromosome, while, a mutant strain of it, has one mutant dominant gene Bar (B) for narrow eyes and one mutant recessive gene carnation (car) for light red eyes on its X chromosome. By crossing these two strains, Stern obtained a dihybrid having car and B genes on one X chromosome and normal genes (++) on other X chromosome. He made both of the X chromosomes of this heterozygote female aberrant by treating such flies with X-rays. The X chromosome having the

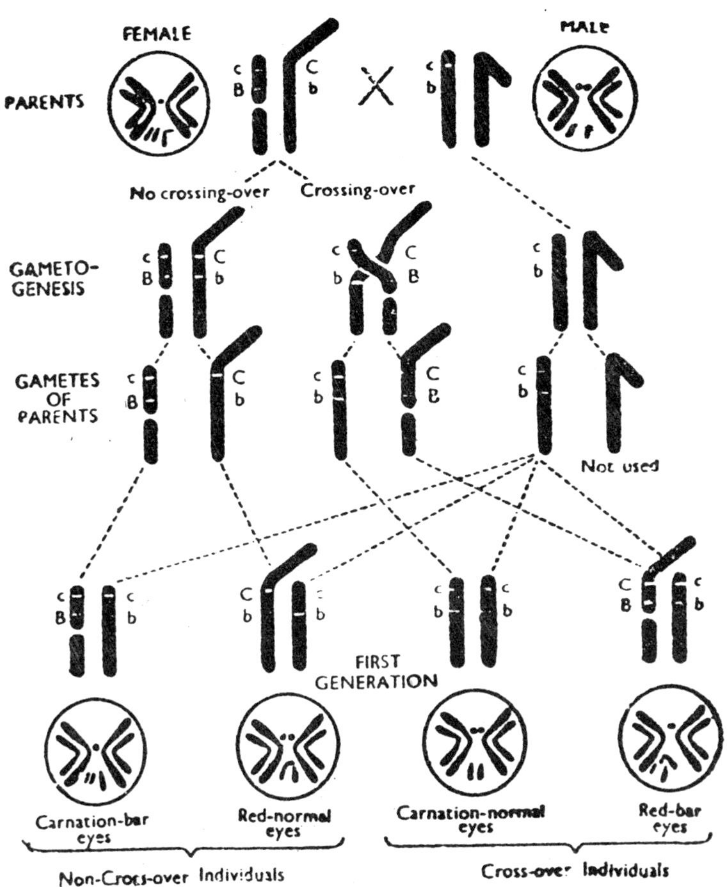

Fig. 13.6. Stern's experiment.

genes car and B was broken into two segments, one fragment having both of the genes, while the other X chromosome having a gragment of the Y chromosome attached to it and contained normal alleles (++). Thus, both aberrant X chromosomes of heterozygote were cytologically detectable. A female heterozygote with the aberrant X chromosomes was mated with a normal male having X chromosome with car + alleles, four classes of eggs (e.g., two types of eggs are cross overs and two types of eggs are non-crossovers) were produced which by fertilization produced following four kinds of females :

1. Carnation Bar females with broken X without any fragment of Y.
2. The red round females with unbroken X with attached Y fragment.

3. The caranation round have the unbroken X without an attached Y fragment.
4. The red Bar have the broken X with the attached Y fragment.

Thus, flies, in which over was indicated phenotypically showed microscopic evidence of exchanges between homologous chromosomes. The physical or cytological basis of crossing over was thus established.

2. Creighton and McClintock's Experiment in Corn

Using corn as the material, Creighton and McClintock (1931) utilized the same principle which Stern utilized in case of fruitfly (*Drosophila*). They obtained a plant which had a knob on the 9th chromosome. This 9th chromosome was also involved in a reciprocal translocation with 8th chromosome. The plant was heterozygous for coloured aleurone and the waxy endosperm characters and carried these genes in the repulsion phase i.e. Cwx/cWx. (c = recessive for colourless seed; wx = recessive for waxy endosperm.) Such a plant was testcrossed with plants homozygous recessive for both characters i.e., colourless and waxy (cwx/cwx).

If the chromosomes region between the knob and c gene is represented as I region and that between c and Wx as II region, then one would expect two types of non-crossover gametes (cWx and Cwx) and six types of crossover gametes including single and double crossovers. The progeny can be classified into eight types based on the phenotype and the cytological observations.

The following observations in the phenotype and cytology of the progeny suggested that actual exchange of chromosome segments was involved in genetic crossing over : (i) Association of knob in the chromosome with the phenotypes, colourless seed (c) and non-waxy endosperm (Wx) indicated crossing over in I region, because in the parent these were located on a knobless chromosome. (ii) At meiotic metaphase I, the presence of a ring of four chromosomes without a knob suggested cytological exchange of chromosome segments, because in the parent the knob was associated with the 9th chromosome carrying translocation and note that whenever a translocation is present in heterozygous condition, a ring of four chromosomes will be formed during meiosis). The classes 4th and 6th shown in the checkborad. (iii) Similarly, if there were no quadrivalents (ring of four) and only 10 bivalents were observed, the presence of the knob in one of these bivalents could be treated as an evidence for cytological crossing over, since the knob was originally associated with the translocation.

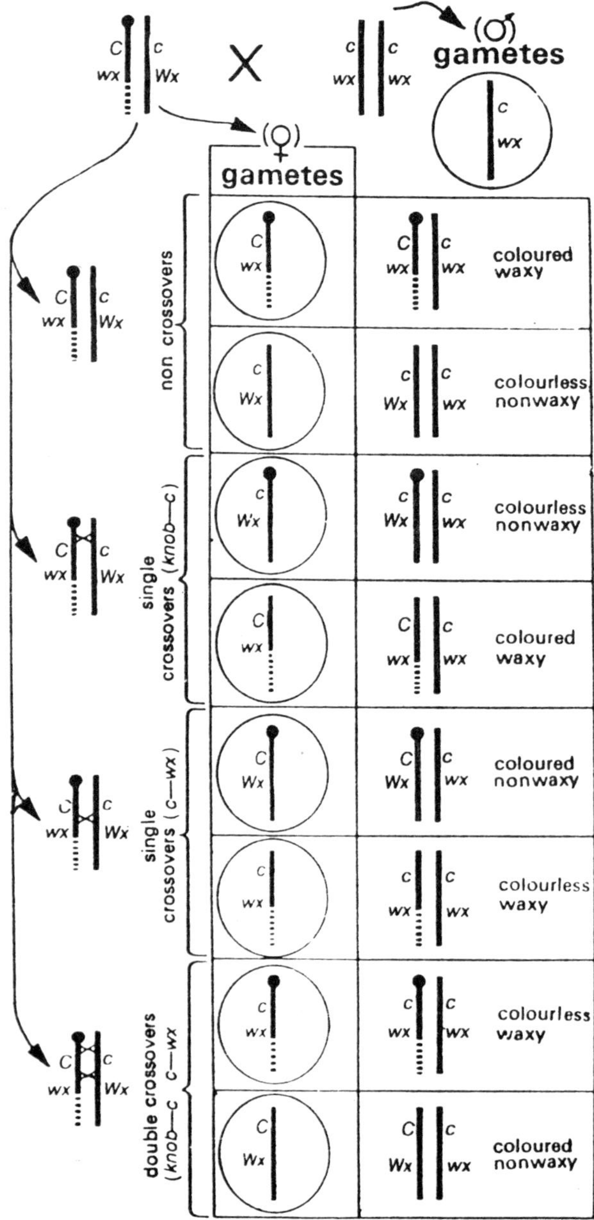

Fig. 13.7. Creighton and McClintock's experiment in corn to give proof for cytological crossing over.

KINDS OF CROSSING OVER

According to the number of chiasma the crossing over may be as following kinds :

1. Single Crossing Over

When only one chiasma occurs only at one point of the chromosome pair. It is called single crossing over. It produces two non-crossover chromatids and two cross over chromatids.

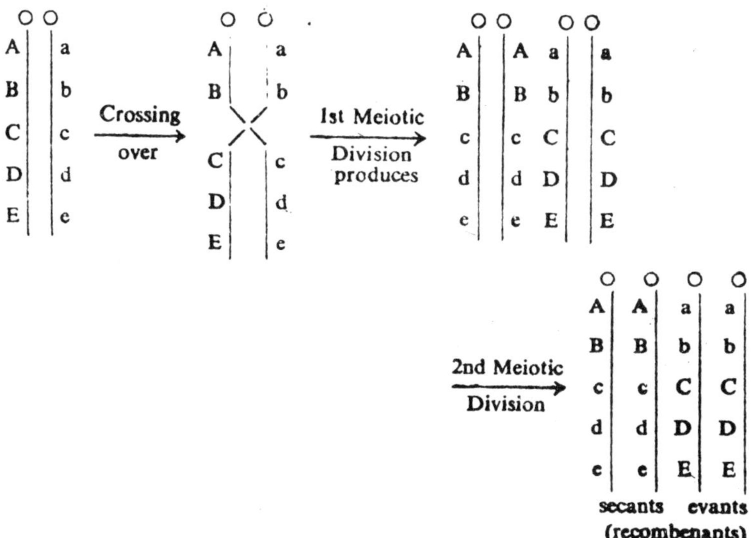

Fig. 13.8. Showing two stranded crossing over.

2. Double Crossing Over

When the crossing over occurs at two points between two given points in the same chromosome pair, it called double crossing over. It produces four crossovers. In double crossing over following two types of chiasma may be formed.

(a) Reciprocal chiasma

In reciprocal chiasma same two chromatids are involved in the second chiasma as in the first. It produces only two non-crossover chromatids, because, it restores the order which was changed by the first chiasma. In this chiasma out of four chromatids only two are involved.

(b) Complimentary chiasma

When both the chromatids taking part in the second chiasma are different from those chromatids involved in the first chiasma, the chiasma is called complimentary chiasma. It produces four single crossover but no non-crossover. In complimentary chiasma all the four chromatids of a tetrad are involved.

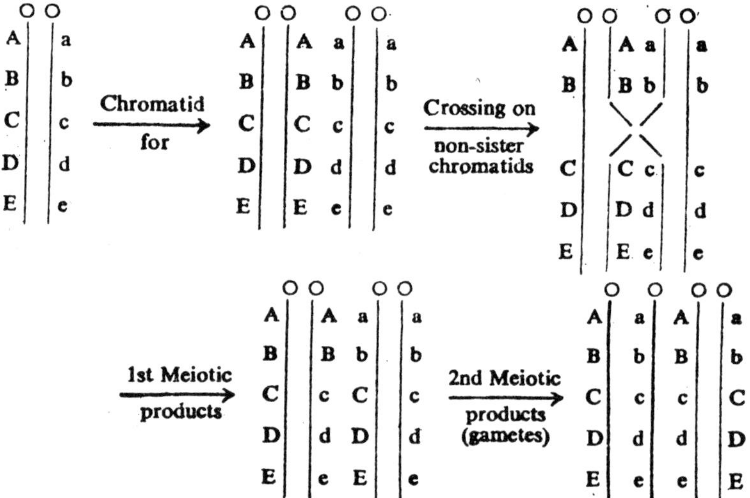

Fig. 13.9. Showing four stranded crossing over.

3. Multiple Crossing Over

When crossing over occurs at three, four, or more points between any two given points in the same chromosome pair these are called triple, quadruple or multiple crossing over.

Two-Factor Crosses

Recombinant combinations of the alleles of two linked genes are produced by crossing over in the interval between the two segregating loci. The rationale behind genetic mapping is that the probability of a crossover occurring between two loci is a function of the length of the interval separating the loci. Intuitively, this seems very reasonable. Consider, for example, the following two crosses:

All three genes, *A, B,* and *C*, are on the same chromosome. The *A* and *B* loci are close together, whereas *A* and *C* are quite far apart. A crossover occurring anywhere within the long interval between the *A* locus and the *C* locus will produce recombinant combinations (*A c* and

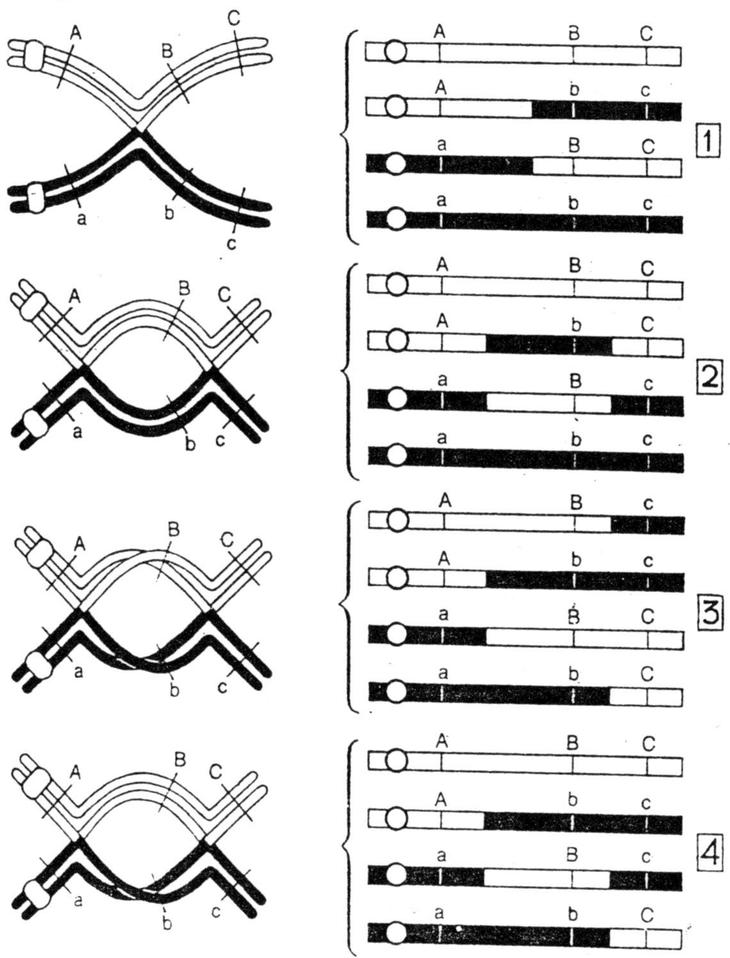

Fig. 13.10. Diagram showing the results of single and double crossing over. 1-Single chiasma. 2-Two strand double cross over. 3-Four strand double cross over. 4-Three strand double cross over.

a C) of the two pairs of alleles segregation in Cross 2. Recombinants (*A b* and *a B*) will be produced in Cross 1 only when a crossover occurs within the short interval between the *A* locus and the *B* locus. It seems very reasonable, therefore, to expect more recombinants to be produced in Cross 2 than in Cross 1.

On the basis of the above type of logic, one of Morgan's students. A.H. Sturtevant (while still an undergraduate), suggested that the frequency of recombinant gametes produced by used as an index of the distance between two loci on a chromosome. *Linkage maps* are made *quantitative* by *defining one map unit as the distance that yields* 1 *percent recombinant chromosomes or gametes.* Thus, the genes *b* (black body) and *vg* (short or vestigial wings) that yielded 18 *percent recombinant* testcross progeny–whether the F_1 dihybrid (double heterozygote) carried the markers in the *cis* (coupling)-configuration or in the *trans* (repulsion)-configuration–are about 18 *map units* apart on linkage Group II of the genetic map of *D. melanogaster.*

It is very important not to confuse the frequency of *crossing over, the event occurring in meiotic tetrads,* with the frequency of *crossover* or *recombinant chromosomes, the products of crossing over. Linkage map distances are determined by the frequencies of crossover* or *recombinant chromosomes. Each meiotic crossing over event yields two crossover chromosomes* (two recombinant chromosomes if the interval within which crossing over occurred is flanked by heterozygous loci). Thus, if a single crossover occurs between two loci in 100 percent of the tetrads, only recombinant (i.e., the recombination frequency will be 50 percent).

If one *assumes* that the probability of a crossover occurring between two loci is directly proportional to the distance between the two loci, that is,

$$\text{Probability of crossover} = K\,(\text{distance})$$

where *K* is a proportionality constant, then one would *predict* that *map distances would be additive.* This property of *additivity* can be illustrated by the following example. If loci *P* and *Q are linked* and are 8 *map units apart* (yield 8 percent recombinant gametes), and loci *P* and *R are linked* and are 3 *map units apart, then loci Q and R are also linked* and are *either (1) 5 map units apart or (2) 11 map units apart.* That is, *additivity* can be achieved only by the following two linkage arrangements:

The assumption that the probability of crossing over is directly proportional to distance is an over-simplification. The assumption is reasonably accurate, however, at least as a first approximation, for linkage distances up to 10-20 map units. As distances become large, significant deviations from additivity are observed. These deviations are of the following nature. Consider, for example, that X and Y are both 15; that is, two-factor crosses, in which alleles of either genes *A* and *B* or genes *B* and *C* are segregating, would produce 15 percent recombinant gametes. Additivity would predict that 30 percent recombinant gametes would be produced in two-factor crosses involving genes *A* and *C.* In actual crosses analogous to the gene *A*-gene *C* cross, recombinant gametes are produced at frequencies significantly lower than 30 percent.

The reason for this systemic deviation from additivity as distances become large is that, when loci are far apart, double crossovers occur at significant frequencies. Consider any two homologous, nonsister chromatids that have recombinant combinations of genetic markers as a result of a crossover occurring between the two marked loci. A second crossover involving these two chromatids in the interval between the two loci will, in effect, cancel out the first crossover. Although two crossovers have occurred, they will not be counted when the data for such a two-factor cross are recorded. For this reason, two-factor crosses involving loci that are far apart (10-20 map units or more) will underestimate linkage distances.

Maximum Frequency of Recombination is 50 Percent

The *maximum frequency of recombination* that can result from crossing over between linked genes is *50 percent.* Fifty percent is also, of course, the frequency of recombination observed for genes on different chromosomes that are assorting independently. Thus, recombination frequencies never exceed 50 percent. For two-factor crosses, recombination frequencies are observed to asymptotically approach 50 percent as the *additive map distance* between the loci involved increases. The additive map distance is the summation of the map distances between genetic markers spanning the shortest marked intervals. The additive map distance between genes *A* and *Z* is 130 map units, the sum of the distances between *A* and *B*, *B* and *C*, *C* and *D*, ..., *X* and *Y*, and *Y* and *Z*. This deviation from additivity results from the increasing number of multiple crossovers as distances become large. All multiple crossovers with an *even number of exchanges* between two segregating loci will yield parental combinations of the segregating loci will yield parental combinations of the genetic markers and thus go undetected. When intervals between loci are small, such multiple crossovers will be rare and have a negligible effect on additivity relationships.

The maximum frequency of recombination for genes located on the same chromosome is 50 percent–the same frequency as observed from genes located on different chromosomes and thus assorting independently. The reason for the 50 percent maximum frequency can be illustrted as follows. Assume that a single crossover occurs between two loci in every meiotic event. In reality, that will not happen, since crossovers are random events such that no crossovers will occur in some meiotic events, double crossovers will occur in others, and so on. But assume that it does happen. The resulting recombination frequency will be 50 percent because each crossover involves only two of the four chromatids, and the gametes carrying the two noncross-over chromatids will have parental combinations of the genetic markers.

Now consider an analogous hypothetical situation. Assume that two crossovers always occur between two loci. Again, the recombination frequency will be 50 percent. Double crossovers can occur in three different ways. (1) *Two-strand double crossovers* occur when both crossovers involve the same two chromatids. (2) *Three-strand double crossovers* are those in which the second crossover involves one of the same two chromatids as the first crossover plus one different chromatid. Three-strand double crossovers can occur in which as many ways as two-strand or four strand double crossovers. (3) *Four strand double crossovers* occur when the second crossover involves the two chromatids not involved in the first crossover. If the second crossover in a retrad involves two chromatids at random, irrespective of which chromatids were involved in the first crossover, then the ratio of two-strand to three-strand to four-strand double crossovers will be 1:2:1. Deviations from this 1:2:1 ratio are said to result from *chromatid interference*: the effect of the involvement of two chromatids in a crossover on which chromatids will be involved in additional crossovers in nearly regions. Ratios of two-to three- to four-strand double crossovers usually approximate 1:2:1, indicating that little, if any, chromatid interference is occurring.

All of the gametes produced by meiotic events in which two-strand double crossovers have occurred between two loci will contain parental combinations of the genetic markers. Half of the gametes produced by meiotic events in which three-strand double crossovers have occurred between two loci will contain recombinant combinations of the genetic markers; the other half will contain parental combinations. All of the gametes resulting from four-strand double crossovers between the loci will be recombinant. Given the 1:2:1 ratio of two-strand to three-strand to four-strand double crossovers, the summation of the gamete types produced from meiotic events in which a double crossover always occurs between two loci shows that 50 percent are parental in genotpe and 50 percent are recombinant in genotype. Similarly, the frequency of recombination can be shown to equal 50 percent if three crossovers, and so on, as long as there is no chromatid interference.

This may be easier to visualize by considering any two nonsister chromatids. As the distance between two loci becomes large, the frequency of multiple crossovers with an *even number of exchanges* (*yielding parental combinations*) will approximately equal the frequency of multiple crossovers with an *odd number of exchanges* (*yielding recombinant combinations*). If chromosomes were infinitely long, the two (even number of exchanges and odd number of exchanges) would ocur with exactly equal frequency, yielding 50 percent recombinant and 50 percent parental gamete types. Only as a result of very unlikely events, such as a preponderance of four-strand double crossovers, would the observed recombination frequency exceed 50 percent.

THREE-FACTOR CROSSES

How can one detect and take into account the double crossovers that go unrecognized in a two-factor cross? This can be accomplished (and then only in part) only by identifying a third gene with distinguishable alleles that maps between the loci involved in the two-factor cross:

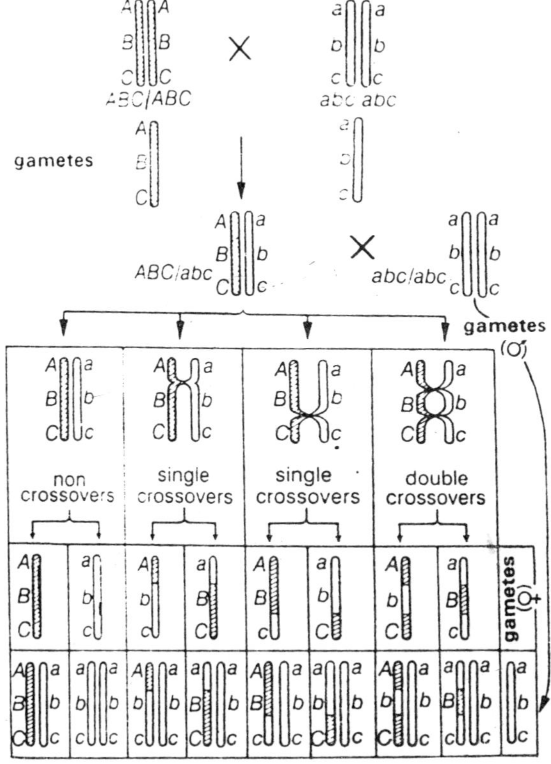

Fig. 13.11. A three-point testcross involving three hypothetical genes A, B and C.

Given a third genetic marker on a chromosome, one can carry out a *three-factor cross*, a cross in which three pairs of alleles are segregating. In addition to allowing the detection of some of the double crossovers not recognizable in two-factor crosses, *three-factor crosses* allow one to *order the markers* involved. The three-factor cross is undoubtedly the most important tool used in chromosome mapping. In diploids, the three-factor cross is usually used in a manner analogous to the use of the two-factor cross in detecting linkage. That is, homozygotes are crossed to produce triple heterozygotes or trihybrids (such as *A B C/a b c*), and the trihybrids are then testcrossed (to *a b c/ a b c* homozygotes), so that the kinds and frequencies of the F_1 gametes can be determined directly from the phenotypes of the testcross progeny.

Chromosome Mapping

Linkage studies using *Drosophila* and other organisms indicated that recombination values ranged between zero and 50% for any two genes linked on the same chromosome. Although a range of values was found for different pairs of linked genes, the *same* values were found for the same pair of genes in repeated experiments. This general observation can be interpreted to mean that each gene occupies a particular location, or *locus* (pl. loci), on a chromosome.

If genes occupy fixed positions, then the different recombination frequencies found for different pairs of genes could be a reflection of the distance separating any two genes along the length of the chromosome. The greater the distance between two genes, the greater the probability that crossing over will occur between them because there is more space for such an event to take place anywhere in that region. If two genes are situated closer together, there is less space and, therefore, less chance for an exchange to take place between them. Thus the second generalization derived from linkage analysis is that the frequency of recombination between linked genes is a consequence of the amount of space between them. These two generalizations provided the basis for mapping genes on a chromosome in fixed order and at specified distances from one another on the chromosome map.

Mapping by Two-Factor Testcrosses

Strutevant summarized testcross data from a number of experiments with X-linked genes in *Drosophila* in the form of a chromosome map. The consistency of recombination values for any two linked genes showed that the gene loci were fixed and that all the genes in the same *linkage group* occurred in a linear order along the same chromosome.

Their relative distances apart, or their relative locations on the chromosome, were expressed in a *chromosome map* based on recombination data from genetic analysis.

We still use Sturtevant's original proposal that the *percentage of recombinants arising from crossing over can be converted into a measurement of distance between two linked genes.* Specifically, *one map unit of distance equals the space between genes in which 1% recombinants arise by crossing over.* A *map unit* is a relative measurement in arbitrary units and not an absolute or actual measurement of chromosome length in micrometers or other physical units.

Suppose we found 6% recombinants in test cross progeny produced in both the coupling and repulsion phases of linkage.

coupling (cis)

P $\quad \frac{AB}{AB} \times \frac{ab}{ab}$

F_1 $\quad \frac{AB}{ab}$

Testcross $\quad \frac{AB}{ab} \times \frac{ab}{ab}$

progeny class	genotype	number of individuals
parental	$\frac{Ab}{ab}$	390
parental	$\frac{ab}{ab}$	410
recombinant	$\frac{Ab}{ab}$	26
recombinant	$\frac{aB}{ab}$	24
	total progeny	850

$$\text{percent recombinants} = \frac{50}{850} = 0.058$$

$0.058 \times 100 = 5.8\%$

repulsion (trans)

P $\quad \frac{Ab}{Ab} \times \frac{aB}{aB}$

F_1 $\frac{Ab}{aB}$

Testcross $\frac{Ab}{aB} \times \frac{ab}{ab}$

progeny class	genotype	number of individuals
parental	$\frac{Ab}{ab}$	305
parental	$\frac{aB}{ab}$	295
recombinant	$\frac{AB}{ab}$	18
recombinant	$\frac{ab}{ab}$	22
	total progeny	640

$$\text{percent recombinants} = \frac{40}{640} = 0.062$$

$0.062 \times 100 = 6.2\%$

We would place these linked genes 6 map units apart on the chromosome map, as follows:

In further studies we may find that gene *C* is linked to *A* and shows 10% recombination, or crossovers. Gene *C* is 10 map units distant from gene *A* in the same linkage group, but is the order of the three genes *ABC* or *CAB*? (We know it cannot be *ACB* since *A* and *B* are 6 units apart). We can find out which of the two possibilities is correct by determining the distance between *B* and *C* in another set of two-factor testcrosses. We can predict that the gene order is *CAB* if genes *B* and *C* show 16% recombination but that the gene order would be *ABC* if we found 4% recombination between *B* and *C*, as follows:

The three sets of two-factor crosses, involving *A–B, B–C*, and *A–C*, give consistent recombination values, thereby showing that genes occur in ***fixed positions*** relative to one another, that they occur in a ***linear order*** on the chromosome, and that they are found at ***particular distances*** from one antoher. We can use these results to predict the numbers of recombinant progeny that will be found any time these three genes are involved in crosses with one another.

The chromosome map (or gene map of the chromosome) has the following features:

1. It summarizes the types of progeny obtained from particular crosses, and it summarizes all the linkage data.
2. It contains two items of information order of and distance between the genes, and it indicates consistency of recombination values in crosses.
3. It is a map of a linkage group, including all the genes found to be linked together on a single chromosome. A chromosome map can be derived for each linkage group of the genome (whole set of genes or chromosomes) in a species.
4. The same gene map will be derived no matter which alleles are carried by each parent in the crosses; the same results are obtained whether alleles are in the coupling (cis) or the repulsion (trans) phase of linkage.
5. Gene maps are reliable forecasting devices. They allow predictions about numbers and kinds of progenies in new crosses. Linkage analysis is, therefore, a powerful tool for describing the genome of a species.

By proceeding with two-factor crosses of linked genes, we can add more and more genes to the liner sequence, depending on the availability of mutant alleles for the wild-type alternatives. We cannot identify, much less map, a gene unless we know its inheritance pattern To do this we must be able to identify segregation and recombination patterns involving members of pairs of alleles. Once a number of genes have been shown to be linked–that is, all are in the same linkage group–the order of and distances between genes are found to be *consistent* in all the combinations analyzed. The genes, therefore, are indeed in a linear arrangement. As Sturtevant pointed out, such an arrangement of genes made a very strong argument in favour of their location in the chromosome, which is the only known linear component in the cell that has a hereditary function.

We can construct an actual map of the chromosome by placing the linked genes, which have been analyzed, into arbitrary locations, based on the distances between them. When a gene is discovered to have linked genes only to its right and none to its left, that gene may be positioned at locus 0.0. The genes to its right are then located in accordance with the map units of distance found from linkage analysis. In Sturtevant's study of six X-linked character in *Drosophila*, for example, the gene for body colour was put at locus 0.0 of the X chromosome and the others were placed in relation to this gene.

By putting the mutant allele on the map, we can more easily identify the gene, knowing at the same time that each of these alleles is recessive to its wild-type alternative. If a dominant mutant allele is mapped, then a capital-letter symbol is shown on the map.

Sturtevant's map differs somewhat in the specific locus designations from the standard map of the X chromosome in *Drosophila*. The standard map was constructed from numerous linkage studies; it is based on vast amounts of data and can, therefore, be more refined in its details. Note that alleles *w* (white eye colour) and w^e (eosin eye colour) are positioned at the same locus. The fact that testcrosses involving different X-linked genes and *w* or w^e consistently showed about the same percentage recombinations, was preliminary evidence that *w* and w^e were two mutant alleles of the same wild-type w^+ gene. The more critical and reliable test for multiple alleles of the same gene versus alleles of different genes verified this tentative conclusion. Since crosses between white-eyed and eosin-eyed flies always gave mutant progeny, these alleles were interpreted to be at the same locus on the chromosome. If the situation had been mutant × mutant → some wild-type segregants, then two different genes would have been implicated in the development of the same phenotypic character (eye colour, in this case), as was found for *w* and *v*.

Double Crossover Problem

The percentage of genetic recombinants is converted directly into percentage of crossover gametes or chromosomes, and this latter percentage in turn is translated into map units of distance between two linked genes. This method has a pitfall, however, because recombination and crossing over are not the same thing. Crossing over is a chromosomal process involving an exchange of homologous chromosome segments; in contrast, recombinations are genetically identified from genotypic constitutions in progeny. We *infer* that a recombination arises as the result of a crossover between two genes, but the values for genetic recombination and detectable chromosomal crossovers are not always identical and, therefore, measurement of distances between genes in map units is not always entirely accurate.

The main reason that we cannot always equate genetic recombinations and chromosomal crossovers is that crossing over is a random process that takes place along the length of the chromosome. Greater distances between pairs of genes provide *more chances* for crossing over to take place, at random, between these genes. Another way of saying this is that the probability is greater that two or more crossovers will take place between genes that are farther apart than between genes that are closer together on the chromosome.

Suppose that genes *A* and *B* are really 30 map units apart on a chromosome. What are the chances that one crossover will take place between them? The answer is 30% because 30-map unit measurement is derived from the percentage of recombinants in progenies and is translated directly into percentage of crossover gametes. What are the chances that two crossovers will take place, at random, within this distance of 30 map units, or, what is the probability of a double crossover in the region between two genes that are 30 map units apart? If the probability for one crossover is 30%, and each crossover is an independent event that takes place at random and without regard to other crossovers in the region, then the probability for two crossovers is the product of their separate probabilities. In this case, the probability for each crossover is 30% and, therefore, the probability of two crossovers between *A* and *B* is 30% × 30%, or 9% (0.3 × 0.3 = 0.09). In other words, we may expect tht out of every 30 gametes per 100 produced by crossing over between *A* and *B*, 9 of these will be double crossover. What is the consequence of double crossovers? They yield alleles in parental combinations and, therefore, lead to parental, not recombinant, phenotypes. They reduce the percentage of recombinants recovered in the progeny and, therefore, lead to an *underestimate* of crossing over. This in turn leads to an underestimate of the map units of distance between two genes.

Instead of 30% of the gametes from the heterozygous parent in a testcross having the *Ab* or *aB* recombinant genotype and producing *Ab/ab* or *aB/ab* recombinant testcross progeny, only 21% (30 – 9 = 21) of the gametes will lead to recombinants. The remaining 9% of these gametes will have double crossover chromosomes and will lead to the parental tyeps *AB/ab* or *ab/ab* in testcross progeny. On finding 21% recombinants in the two-factor testcross progeny, we would assume 21% of the gametes resulted from crossing over and put only 21 map units between genes *A* and *B* on the chromosome.

The probability is lower for two crossovers to take place in the space between two genes if the distance is smaller. If genes *C* and *D* are actually 5 map units apart, then 5% of the gametes will have crossover chromosomes. Of these, only 0.25% (0.05 × 0.05) will be double crossovers and 4.75% will be single crossovers, on the average. The discrepancy for small distances is, therefore, less significant and usually can be ignored until very fine detail and location are required for the standard map or for other purposes.

The double crossover problem may be remedied when a third gene is situated between *A* and *B*. The gene in the middle serves as a "marker" for doule crossover chromosomes. If we looks at the double crossover chromosomes, we see that the *middle pair of alleles* is reversed when compared with the parental arrangements of alleles.

For the following reasons, the *three-point testcross* is the usual method of choice for mapping linked genes on chromosomes.

1. The presence of three gene differences in testcross parents allows the identification of double crossover recombinants. This identification makes the relationship more meaningful between percentage of recombinants and percentage of gametes resulting from crossing over and provides more accurate values for determining map units of distance between linked genes.
2. A trihybrid cross yields eight phenotypic classes and each class represents a unique genotype. Eight classes is still a manageable number to analyze from reasonable numbers of progeny. We know that 2^n = the number of different phenotypic classes for alleles showing complete dominance, where n = the number of pairs of heterozygous alleles. Thus, 16 phenotypes are produced for 4 pairs of alleles (2^4), 32 different phenotypes for 5 pairs of alleles (2^5), and so forth. Four- or five-factor testcrosses would require substantially larger numbers of progeny in order to recover all the phenotypic classes in the numbers needed for reliable samplings of genetic events observed from recombinations. From a practical standpoint, three-factor testcrosses are more convenient and equally as reliable as crosses involving more factors. The situation could be different, of course, for microorganisms that can be raised in huge numbers in very brief time and in a relatively small space.
3. It is more efficient and less time-consuming to analyze three genes at one time than to analyze three genes in several two-factor testcrosses. Instead of $A \times B$, $A \times C$, and $B \times C$ to discover the map order and distances between *A*, *B*, and *C*, we need only one three-factor testcross involving *A*, *B* and *C*. This reduction of the number of crosses needing to be made is particularly important when we use organisms with a long life cycle or ones requiring a great deal of space and maintenance. Anything that saves time and money and gives the same or improved results is highly desirable.

As we analyze representative three point testcross experiments, we should try to keep these three features of experimental design and management in mind.

Mapping by the Three-Point Testcross Method

We can begin with a familiar examples to illustrate how linkage analysis is performed when the three genes are known to be in the same linkage group. Since the goal is to learn the relative order and distance between linked genes, we can start with the three X-linked genes in *Drosophila* already mentioned: body colour (gray, y^+, versus yellow, *y*), eye colour (red, w^+, versus white, *w*), and wing length (long, m^+ versus miniature, *m*).

If *ywm*/*ywm* females are crossed with $y^+ w^+ m^+$ males, which is simply abbreviated throughout the discussion to + + + for the three wild-type alleles, F_1 progeny consist of heterozygous + + +/*ymw* females and *ymw* males. We can either perform a testcross of F_1 females to triply recessive males from another stock or obtain the F_2 since these crosses would be identical. The F_2 or testcross progeny include the following:

Phenotypic class	*Maternal chromosome present*	*Number of progeny*
gray, red, long	+ + +	1087
yellow, white, miniature	*y w m*	1042
gray, white, miniature	+ *w m*	17
yellow, red, long	*y* + +	15
gray, red, miniature	+ + *m*	543
yellow, white, long	*y w* +	502
gray, white, long	+ *w* +	6
yellow, red, miniature	*y* + *m*	4
		3216

We must first determine the pair of parental phenotypic classes in this progeny. Since these genes are linked, we expect an excess of parentals, so we look for the largest two classes among the total of eight; these are + + + and *ywm*. We already know this in this example because we know the alleles that were present on each of the X chromosomes in the F_1 females, and the parental, or noncrossover, genotypes are identical with these unchanged chromosomes.

The second step is to determine the pair of classes with the double crossover chromosomes since this pair will define the middle gene and, therefore, give us the order of the three genes in the chromosome. The double crossover phenotypic classes are the least numerous of all

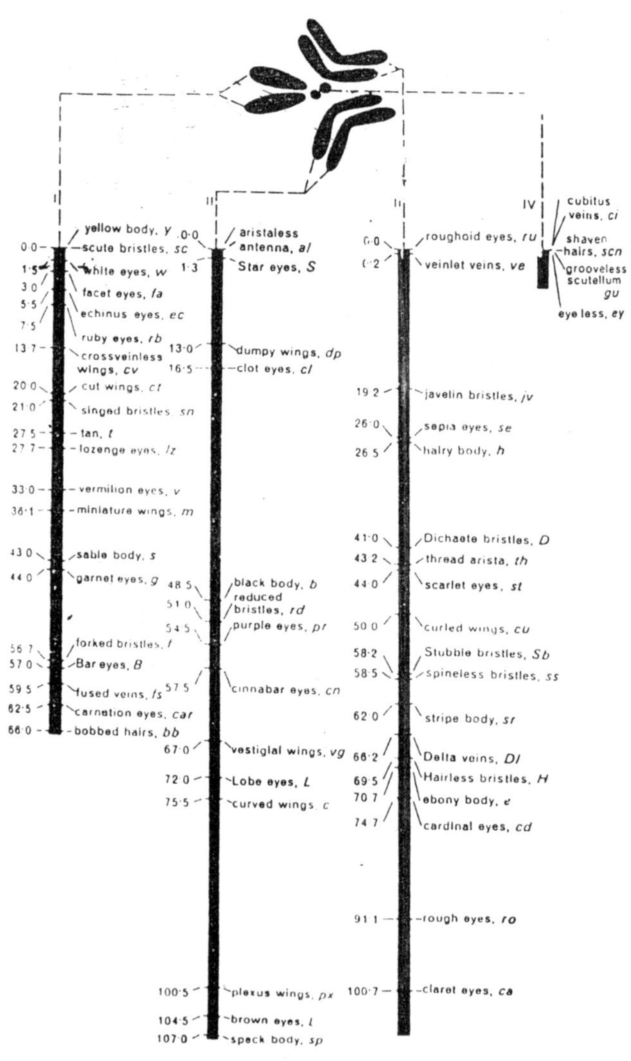

Fig. 13.12. Linkage maps of four different chromosomes of Drosophila.

because double crossovers are less frequent than single crossover events. To produce a double crossover, one crossover must occur in the space between the middle gene and the gene to its right and another crossover between the middle gene and the gene to its left. The probability of a double crossover occurring is the product of the separate probabilities for each single crossover occurring, so it is less frequent than either of the single crossover events. By comparing the parental + + + and *ywm* with + *w* + and *y* + *m* in the two phenotypic classes present in the lowest numbers, we can see that *w* must be the middle gene and that it is flanked by *y* and *m*.

It is now a simple matter to identify each of the single crossover pairs of classes. Crossing over between *y* and *w* would change the combination of y^+/y in a chromosome retaining the parental combination of *wm* or + + alleles.

The remaining single crossover classes must be those involving the space between the loci for *w* and *m*.

Notice that each of the crossover events give rise to reciprocal classes of recombinant gametes. We expect to find approximately equal numbers of each genotype arising from two reciprocal crossover chromosomes because approximately equal numbers of gametes will have each of the crossover chromosomes due to the segregation of homologous chromosomes at meiosis.

The percentages of crossovers are calculated separately for the two regions in which crossovers occurred, we usually start at the left and label the first space region I and call the second space region II. For crosses of more than three factors, additional regions are identified in sequence by roman numerals. When we calculate crossovers in region I, we must *add* the total of the double crossovers to the total of the single crossovers for this *y*–*w* region. Similarly, to obtain the total percentage of crossovers in region II, we add the double crossovers to the single crossover events between *w* and *m*.

The reason we add double crossover to each of the single crossover values is that crossing over took place in region I and region II in the double crossover classes. If we omit the double crossovers, we lose one of the advantages of the three-point testcross, namely, that we can detect region I crossovers or region II crossovers, even when a second independent crossover event has obscured the expected recombinations. When all the crossovers within a region are summed, the resulting percentage of crossovers (gametes or chromosomes) more accurately reflects the distance between two gene and translation into map units is closer to the actual distance.

The results of calculations from the three-point testcross are as follows:

Region	Gene loci	Proportion of crossovers	Percentage of crossovers	Distance in map units
I	*y–w*	$\frac{32+10}{3216}$	1.3	1.3
II	*w–m*	$\frac{1045+10}{3216}$	32.8	32.8
I + II	*y–m*˙	$\frac{32+1045+10+10}{3216}$	34.1	34.1

These three gene loci can now be mapped in their proper sequence and distances from each other on the X chromosome.

Now let us look at linkage analysis for autosomal genes in a three-points testcross. previous studies of *Drosophila* have shown that the genes for *gray* versus *black* body colour (b^+ and *b*), *red* versus *purple* eye colour (pr^+ and *pr*), and *long* versus *vestigial* wing length (vg^+ and *vg*) are located on autosomal chromosomes since they produce identical progeny in reciprocal crosses. X-linked genes show different reciprocal progeny. The gene symbols indicate that all the wild-type alleles are dominant to their mutant alternatives, as was discovered in breeding analysis. We now want to find out if these genes are in the same linkage group, each in a different chromosome, or two in one chromosome and the third in another linkage group. To do this, homozygous flies from true-breeding stocks are crossed to obtain F_1 progeny that are triply heterozygous for the three pairs of alleles. The F_1 heterozygous ♀♀ are testcrossed to triply recessive ♂♂ and the following progeny are obtained:

Phenotype	Chromosome from heterozygous parent	Number of individuals
wild type	+ + +	59
black, purple, vestigial	*b pr vg*	51
vestigial	+ + *vg*	416
black, purple	*b pr* +	402
purble	+ *pr* +	23
black, vestigial	*b* + *vg*	21
		972

The first question is: Are these genes linked? The F_2 consists of only six phenotypes instead of eight, with only two of the phenotypes present in very large numbers. These phenotypes are reciprocal (what is recessive in one phenotype is dominant in the other for each pair of alleles: + + *vg* and *b pr* +), and there is no hint of a 1:1:1:1 ratio, which means that no two genes are assorting independently. We can, therefore, assume that all three genes are linked. There is no evidence for independent assortment. The noncrossover, or parental, chromosomes from the heterozygous parent are therefore, + + *vg* and *b pr* + since they are the most numerous classes. The missing phenotypes must be the relatively more rare double crossovers, none of which was produced, for some unknown reason. (Perhaps a crossover in region I or in region II reduces the probability for antother crossover in the other region). In any event, it is clear that the missing two genotypes are + *pr vg* and *b* + + since all the other possible combinations are present.

If we assume that + *pr vg* and *b* + + are the missing double crossover phenotype classes, we can find the middle gene by comparing these classes with the parental types. The transposed pair of alleles, relative to the arrangement in the noncrossover chromosomes, is + *pr*, which is the gene in the middle [(+ + *vg*/ *b pr* +) versus (+ *pr vg*/ *b* + +)]. The sequence is, therefore, *b pr vg*, and we can now find the single crossovers in region I, between *b* and *pr*. We then calcualte the percentage recombinants as $^{44}/_{972}$ = 4.5 map units in region I. The single crossovers in region II, between *pr* and *vg*, include 110 recombinants out of a total progeny of 972, which is 11.3% recombination and thus the distance between *pr* and *vg* is 11.3 map units. The map is therefore:

We can verify the gene order by calculating the percentage recombination between *b* and *vg*. If these are the outside members of the chromosome segment and if *pr* is in the middle, then the percentage of crossover between *b* and *vg* should approximate 4.5 + 11.3. A crossover between *b* and *vg* would produce the following:

The recombinant phenotype classes showing *b* and *vg*, those showing b^+ and vg^+ in the same chromosome are + + + and + *pr* +. When these are totaled (51 + 21 + 59 + 23 = 154), we can calculate $^{154}/_{972}$ = 15.8% recombinants involving *b* and *vg*. This is exactly what is expected for the two genes that are farthest apart and at each end of the chromosome segment carrying these three linked genes.

We also could have taken genes in combinations of two to determine the percentages of recombinants in each two-factor combination. By finding 4.5% for *b*–*pr*, 11.3% for *pr*–*vg*, and 15.8% for *b*–*vg* crossovers, we would have put these genes together on the map in the same logical order. It would be very similar to the example in the discussion of mapping by two-point testcrosses for genes *A, B,* and *C.*

As a final example of the value of the three-point testcross, it is useful to examine the progeny of a testcross with F_1 ♀♀ heterozygous for ebony (*e*) body colour, rough (*ro*) eyes, and vestigial (*vg*) wings, as follows:

e rovg	210
+ + +	202
e ro +	198
+ + *vg*	206
e + *vg*	47
+ *ro* +	49
e + +	48
+ *rovg*	50
total progeny	1010

We first notice that some linkage is involved, because independent assortment of all three genes would have produced a 1:1:1:1:1:1:1:1 ratio. Since crossing over is infrequent, the four largest classes must be noncrossover genotypes. We can find out which two genes are linked by determining which two pairs of alleles do not assort in the noncrossover classes. For the genes *e* and *ro*:

e ro	210	
+ +	202	
e ro	198	408 *e ro*:408 + + =
+ +	206	1:1

Since there are only two classes, *e ro* and + +, these two pairs of alleles have not assorted to give a 1:1:1:1 ratio and must, therefore, be linked. These large classes are the noncrossover, parental classes.

We can now find out whether *vg* is linked to *e* and *ro* by checking the parental classes again. If *vg* is linked to *e* (or *ro*), there will be only two parental types, just as before.

e vg	210
+ +	202
e +	198
+ *vg*	206

Since the ratio is 1:1:1:1 for the two pairs of alleles, they must have assorted independently and are, therefore, not linked. (And *vg* is not linked to *ro* either because *ro* is linked to *e*).

We can now calculate linkage between *e* and *ro* by looking at the smaller-sized recombinant classes.

e +	47
+ *ro*	49
e +	48
+ *ro*	50

194 recombinants/1010 total progeny = 0.192

0.192 × 100 = 19.2% recombinants = 19.2 map units.

14

SEX DETERMINATION

Sex confers the great advantage of making it possible for sexually reproducing eukaryotes to reshuffle their DNA during meiosis by crossing over and independent assortment of chromosomes. This is a prime source for the tremendous variety we see in the living world.

Factors controlling male and female gamete formation have been found to be complex, involving genetic as well as other internal and external factors. Our objective in this section is to explore the process of sex determination in several plant and animal species in order to show the genetic complexity and species uniqueness involved in this aspect of differentiation and development.

Two questions with regard to sex, however, have not yet been broached: (a) What determines the sex of the individual? (b) How do the secondary sexual characters of individual develop? These two questions will be answered in the present chapter.

The question as to what determines whether an animal shall be a male or female is a very ancient one, and it is only during the present century that we have solved the puzzle.

A great many theories of sex determination have been proposed, some of which are as follows:

(a) Hippocrates and some subsequent theorists believed that the sex of the offspring depended on the relative vigor of the parents, the more vigorous parent giving his or her sex to the offspring.
(b) Thury thought that the sex of the offspring depended on the degree of ripeness of the ovum at the time of fertilization.
(c) Various writers claim that statistics show that germ cells from the right ovary produce males and those from the left ovary females.

(d) *The nutrition theory*. The egg is a much more highly nourished cell than the spermatozoan, and the idea seems natural that high degrees of nourishment of the mother produce female offspring and lower degrees of nourishment male offspring. Professor Schenk of Vienna gained a huge reputation by controlling the diet of certain royal prospective mothers and predicting the sex of the offspring accordingly. He was correct in his predictions several times, but his success was short-lived. His early predictions were merely lucky, just as one might be who could guess heads or tails correctly several times in succession.

Sex Chromosomal Mechanism of Sex Determination

In dioecious diploidic organisms following two systems of sex chromosomal determination of sex have been recognized:

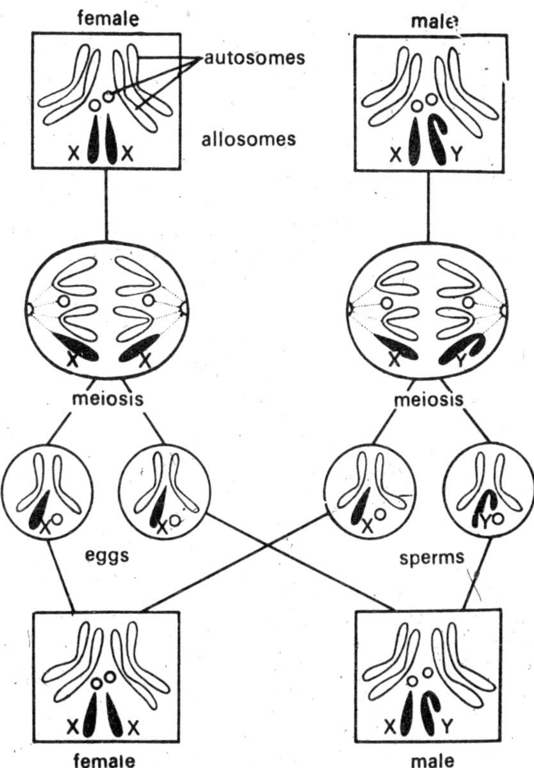

Fig. 14.1. Distribution of sex chromosomes and sex determination in Drosophila.

(a) Heterogametic males.
(b) Heterogametic females.

Heterogametic Males

In this type of sex chromosomal determination of sex, the female sex has two X chromosomes, while the male sex has only one X chromosome. Because, male lacks a X chromosome, therefore, during gametogenesis produces two types of gametes, 50% gametes carry the X chromosomes, while the rest 50% gametes lack in X chromosome. Such a sex which produces two different types of gametes in terms of sex chromosomes, called heterogametic sex. The female sex, because, produces similar types of gametes, is called, homogametic sex. The hetrogametic males may be of following two types:

(i) XX-XO system

The somatic cells of a female grasshopper contain 24 chromosomes, whereas those of the male contain only 23 chromosomes. Thus, in grasshopper (and in many other insects), there is a chromosomal difference between the sexes; females referred to as XX (having two X chromosomes) and males as XO ("X–oh", having only one X chromosome). The X chromosome is called a *sex chromosome*, the remaining chromosomes are called *autosomes.* Thus in XX-XO system, all the eggs have one X chromosome whereas the sperms are of two types, X and O, that is half the sperms have one X chromosome and the other half have none.

(ii) XX-XY system

In mammals, *Drosophila* and some plants (e.g. the angiosperm genus *Lychnis*) etc., the females are generally referred to as XX (having two X chromosomes) and males as XY (having only one X chromosome and another one called Y chromosome). In *Drosophila* there are four pairs of chromosomes; three pairs of these chromosomes (that is, two pairs of V-shaped chromosomes and a pair of small dot-like chromosomes) in both sexes are called *autosomes.* The fourth pair of chromosomes is different in the two sexes; these are *sex chromosomes.* In female *Drosophila* both the sex chromosomes are identical and each is called X chromosome, in male one of the sex chromosomes is straight (X chromosome) but the other is bent having two unequal arms (Y chromosome). In man, the females have 44 autosomes and two X chromosome (44+X.X) whereas the males have 44 autosomes and one X chromosome and a Y chromosomes (44+X+Y). For further details

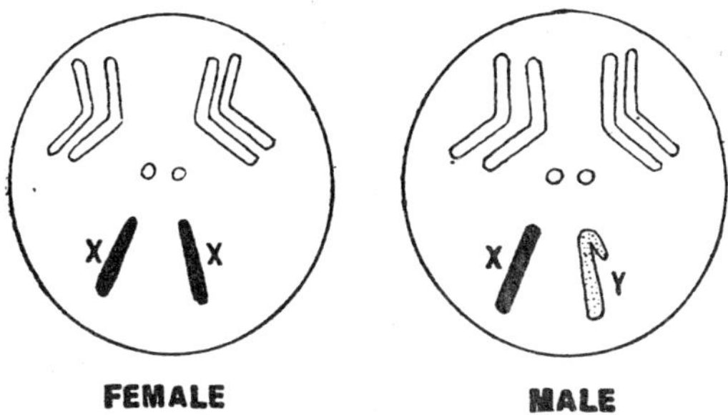

Fig. 14.2. Diagram showing the chromosomes of Drosophila melanogaster.

of the human chromosomes. In the XX-XY system, all the ova have one X chromosomes whereas the sperms are of two kinds, X and Y. In both the XX-XO (*vide supra*) and XX-XY types, the male is the *heterogametic sex* (producing two types of sperms whereas the female is the *homogametic sex* (producing only one type of ovum).

Heterogametic Females

In this type of sex chromosomal determination of sex, the male sex possesses two homomorphic X chromosomes, therefore, is homogametic and produces single type of gametes, each carries a single X chromosome. The female sex either consists, of single X chromosome or one X chromosome and one Y chromosome. The female sex is, thus, heterogametic and produces two types of eggs, half with a X chromosome and half without a X chromosome (with or without a Y chromosome). To avoid confusion with that of XX-XO and XX-XY type of sex determining mechanisms, instead of the X and Y alphabets, Z and W alphabets are generally used respectively. This kind of sex determination mechanism is called Abraxas mechanism of sex determination, (*Kuspira* and *Walker*, 1973).

The heterogametic females may be of following two types:

(i) ZO-ZZ system

This system of sex determination is found in certain moths, butterflies and domestic chickens. In this case, the female possesses single Z chromosome in its body cells (hence, is referred to as ZO) and is heterogametic, producing two kinds of eggs, half with a Z chromosome and half without any Z chromosome. The male possesses two Z chromosomes (hence referred to as ZZ) and is homogametic, producing single type of sperms, each of which carries a single Z

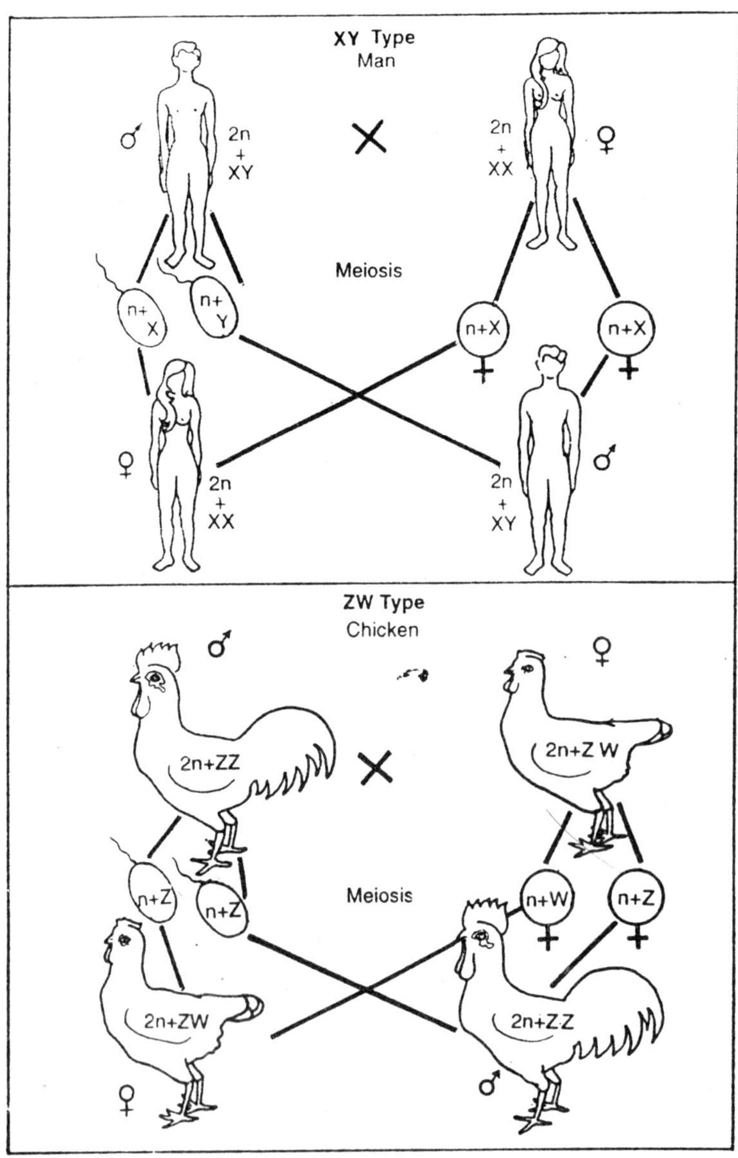

Fig. 14.3. The XX-XY and ZW-ZZ type determination of sex in man and chicken.

chromosome. The sex of the offspring depends on the kind of egg as shown below :

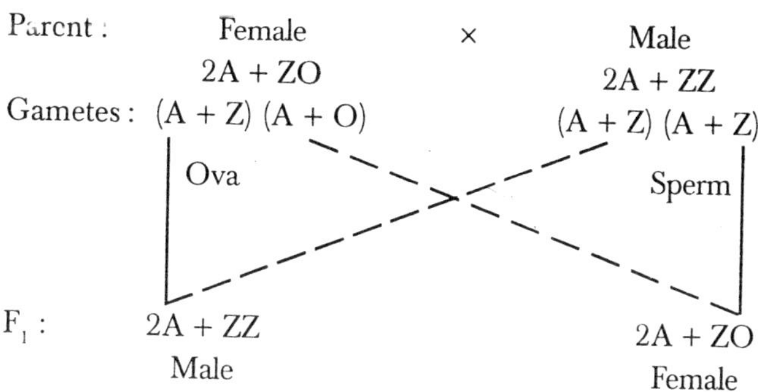

(ii) ZZ-ZW system

In birds (including the domestic fowl), butterflies, moths and some fishes, the female is heterogametic but the male is homogametic. To avoid confusion, the sex chromosomes in this case are often designated as Z (instead of X) and W (instead of Y). Thus in these cases females are ZW (or XY) and males are ZZ (or XX).

Sex chromosomes in monoploid (haploid) organisms

The sporophyte of liver-wort *Sphaerocarpos* (phylum Bryophyta) contain seven matching pairs of chromosomes plus an eighth pair in which one of the two chromosomes is much larger than the other. The larger member of this eighth pair of chromosomes is called the X chromosome, its smaller partner the Y chromosome. Meiosis terminates the diploid sporophyte generation. At meiosis, the four meiospores are produced from each meiocyte; two meiocytes receive an X chromosome and two a Y chromosome. Meiospores containing an X chromosome develop into female gametophytes whereas those with a Y chromosome develop into males. Thus in liver-worts, the females are designated as X, males as Y, and the asexual sporophytes as XY.

Table 14.1. Various types of chromosomal differences between sexes.

Female	*Male*	*Examples*
XX	XY	*Drosophila*, mammals, some dioecious angiosperms (e.g. *Lychnis*)
XX	XO	Grasshoppers; many Orthoptera and Hemiptera.
ZW	ZZ	Birds, buterflies, moths and some fishes.
X	Y	Liverworts.

Balance Concept of Sex Determination

Soon after sex chromosomes were identified it became obvious that sex determination was more complicated than preliminary observations had indicated. A more intricate mechanism than the segregation of a single pair of chromosomes was in evidence. The most fundamental contributions to the definition of this mechanism came from Bridges' investigations on Drosophila, which showed that female determiners were located in the X chromosome and male determiners were in the autosomes. More than one gene, perhaps a great many (in the X chromosome), were found to influence femaleness. Bridges also demonstrated that genes for maleness were not located in the Y chromosome of Drosophila, but were distributed widely among the autosomes. No specific loci have been identified, and the present evidence suggests that many chromosome areas are involved. Thus, it was shown that sex-determining genes are carried in certain chromosomes in Drosophila, and that all individuals carry genes for both sexes. The genic balance theory of sex determination was devised as a more detailed explanation of the mechanics of sex determination.

The XO or XY chromosome segregation was interpreted as a means of tipping the balance between maleness and femaleness, whereas more deep-seated processes were involved in the actual determination. Bridges experimentally produced various combinations of X chromosomes and autosomes in Drosophila. The first irregular chromosome arrangement resulted from nondisjunction, the failure of paired chromosomes to separate in anaphase of the reduction. X chromosomes, which ordinarily came together in pairs in meiotic phase of oogenesis and separated to the poles in anaphase of the reduction division, remained together and went to the same pole. As a result, some female gametes received two X chromosome and some received no X chromosomes.

Following fertilization by sperm from wild-type males, all zygotes had 2n autosomes. Some received two X chromosomes from the mother and a Y from the father. These XXY flies, which in appearance were normal females, were mated with wild-type (XY) males. All progeny had two sets of autosomes, whereas some received XXX, others XXX, XY, or YY sex chromosomes. XXX flies, now called meta-females, were sterile and highly inviable. The XXY combination resulted in females that were normal in appearance and reproductive function, XY equated with normal males, and YY zygotes did not survive. Experimentally produced XO males were similar in sex manifestations to XY males, but the XO males were sterile. XXY females were similar

chromosome composition	chromosome formulation	ratio of X chromosomes to autosome sets	sex
	3X/2A	1·5	metafemale
	3X/3A	1·0	female
	2X/2A	1·0	female
	2X/3A	0·67	intersex
	3X/4A	0·75	intersex
	X/2A	0·50	male
	XY/2A	0·50	male
	XY/3A	0·33	metamale

Fig. 14.4. Chromosomal composition and ratio of X chromosomes to autosomes in Drosophila.

to XX females. These results indicated that, in Drosophila, the Y chromosome is not involved in sex determination but that it does control male fertility.

Flies produced experimentally with three whole sets of chromosomes (3 genomes or 3n triploids) were then included in the studies and, later, some with four genomes (tetraploids) were added. On the basis of many experimental combinations of autosomes and X chromosomes, Bridges established a standard by which various sets of autosomes and X chromosomes were compared with reference to the relative potency of a male-and female-determining capacity. Two sets of autosomes (AA) were found to have enough male-determining potency to overbalance the female-determining capacity of one X chromosome. The chromosome combination AAX(Y) thus gave rise to a normal male. In the presence of two X chromosomes and two sets of autosomes (AAXX), a normal female was produced. Bridges showed that equal numbers of X chromosomes and sets of male-determining autosomes occurred only in females. When the number of sets of autosomes was greater, maleness was expressed. A large over-balance of X chromosomes resulted in metafemales. An overbalance of autosomes was associated with metamales. When the difference in proportion was not great, intersexes with characteristics of both males and females were produced. The various demonstrated chromosome combinations and sex expressions are summarized in Table 14.2

Table 14.2. Different combinations of X chromosomes and autosomes and corresponding sex expressions in Drosophila (after Bridges)

X Chromosomes	*Sets of Autosomes*	*Sex*
1	2	Male
2	2	Female
3	2	Metafemale
4	3	Metafemale
4	4	Tetraploid female
3	3	Triploid female
1	1	Haploid female[a]
3	4	Intersex
2	3	Intersex
2	4	Male
1	3	Metamale

[a] Determination based on patches of tissue in individual flies which show female traits.

No other animals or plants have been investigated with equal thoroughness, but indirect evidence suggests that, in many organisms, some such balance is involved. Intersexes can be produced experimentally in some animals by upsetting this balance during the developmental stages. In nature, a margin of safety exists which makes intermediates between the two sexes uncommon.

GYNANDROMORPHS

In some animals such as insects, upsets in the chromosom: behaviour result in sexual mosaics called gynandromorphs. Some parts of the animal express female characteristics while other parts express those of the male. Some gynandromorphs in Drosophila are bilateral intersexes, with male colour pattern, body shape, and sex comb on one half and female characteristics on the other half of the body. Both male and female gonads and genitalia are sometimes present.

Bilateral gynandromorphs have been explained on the basis of an irregularity in the chromosome mechanism at the first cleavage of the zygote. Infrequently, a chromosome lags in division and does not arrive at the pole in time to be included in the reconstructed nucleus of the daughter cell. If one of the X chromosomes of an XX (female) zygote should lag in the center of the spindle, one daughter cell would get only one X chromosome and the other would get XX. The baisi for a mosaic pattern would thus be established. One cell in the two-cell stage would be XX (female) and one would be XO (male). In

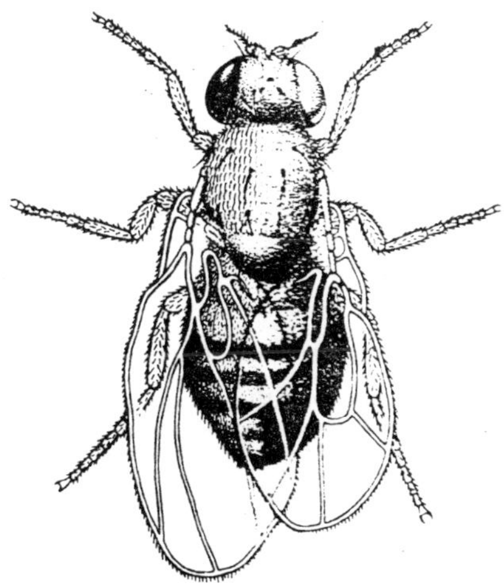

Fig. 14.5. Bilateral gynandromorph in Drosophila.

Drosophila, the right and left halves of the body are determined at the first cleavage. One cell gives rise to all the cells making up the right half of the adult body and the other give rise to the left half. If the chromosome loss should occur at a later cell division, a smaller proportion of the adult body would be included in the male segment. The position and size of the mosaic sector would be determined, therefore, by the place and time of the division abnormality.

Gynandromorphs were described in Drosophila by Sturtevant, Morgan, and Bridges beginning in 1919. Following the original descritions, a few conspicuous examples were reported in flies, but the condition was considered extremely rare. More extensive observations have since shown that a gynandromorph of some kind is produced in every 2000 to 3000 flies; many of these represent small sections of tissue involving only a few cells. Spencer Brown and Aloha Hannah-Alava devised a technique to increase experimentally the frequency of gynandromorphs by making use of a ring X chromosome first discoverd by L. V. Morgan.

1. Bilateral Gynandromorph

In a bilateral gynandromorph the body parts of one side (right or left) shows female secondary sexual characters and another side shows male secondary sexual characters. The bilateral gyandromorph have been observed in butter fly, Colias philodices and in fruit fly, Drosophila melanogaster.

2. Anterio-posterior Gynandromorph

In an anterio-posterior gynandromorph, the anterior side of animal body shows characters of one sex and posterior side of body shows the characters of opposite sex. The anterio-posterior gynandromorph has been reported in beette, *Lucanus cervus.*

3. Sex Piebalds

In sex piebalds, the body of gynandromorph consists of female tissues having irregularly distributed patches of male tissue. Such sex piebalds have been reported in *Drosophila.*

Origin of Gynandromorphs

The gynandromorphs of different organisms are originated by following methods:

1. Degenerates of X Chromosome

In insects and other invertebrates, the first mitotic division of zygote determines the future right and left halves of the embryo. A bilateral gynandromorph of Drosophila, for example, begins its

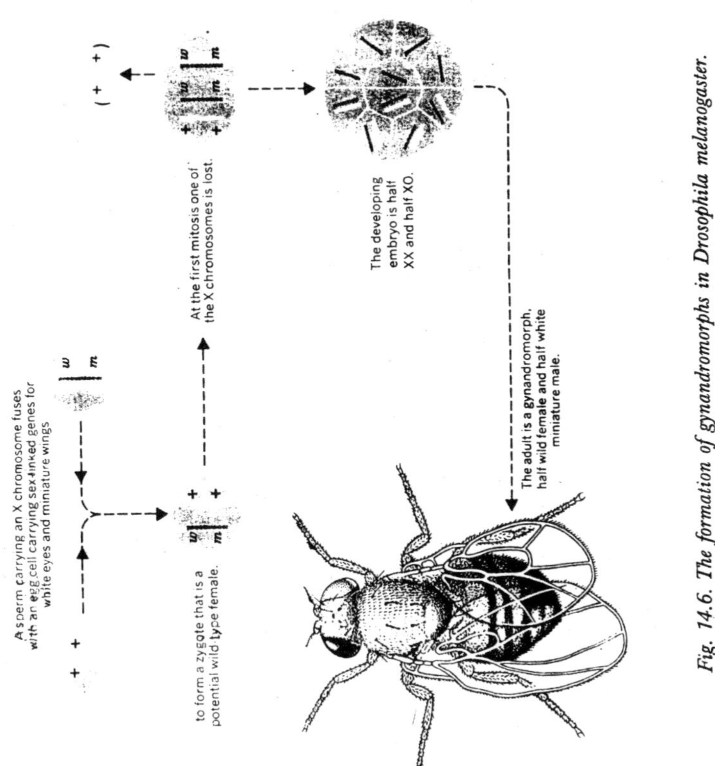

Fig. 14.6. The formation of gynandromorphs in Drosophila melanogaster.

development with a normal diploid zygote having two X chromosomes. During first cleavage due to anaphase lag two kinds of blastomeres are formed : One blastomere (XO) has single X chromosome while other blastomere (XX) has two X chromosomes. The blastomere with XO blastomere with XX chromosomal constitution develops into a female phenotype of gynandromorph.

2. Retention of Polar Nucleus

In silkworms which have a heterogametic (XY) female sex and

homogametic male sex (XX), during oogenesis the XY chromosomes normally disjoin or separate and one of them goes to the polar body while another is retained by the egg nucleus. But sometimes the polar body having one sex chromosome does not leave the egg and retained by ooplasm along with egg nucleus. The egg becomes binucleate and has one nucleus with X chromosome and other nucleus with Y chromosome. During fertilization, when X bearing sperm nuclei fuse with each of the egg nucleus to give XX and XY nuclei. The XX egg nucleus develops into male phenotype and XY egg nucleus gives origin to female phenotype.

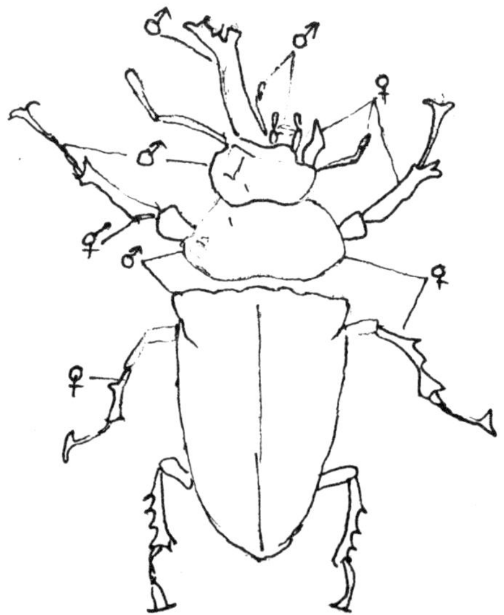

Fig. 14.7. An anterio-posterior gynandromorph of beetle Lucanus cervus.

The gynandromorphs of bees have analogously similar origin. Bees also have binucleate eggs which are originated either due to retention of polocyte nucleus by egg or pre-fertilizic (parthenogenetic) division of zygote nucleus. During fertilization only one sperm enters in the egg and its pronuclei fuses with one of the nucleus of binucleate egg. The resultant diploid nucleus develops into female tissues, while haploid (unfertilized) nucleus develops into male tissues to form a gynandromorph.

Sex Determination in Plants

In plants as also in case of human beings the sex is mainly controlled by the Y chromosome. Consequently, irrespective of the number of X-chromosomes and the number of sets of autosomes, if Y-chromosome is present, the individual would be male while if Y-chromosome is absent, it would be female. The mechanism of sex determination has been studied in a large number of plant materials.

1. Coccinia indica and Melandrium album (Lychnis)

The mechanism of sex determination in Coccinia indica, a member of family Cucurbitaceae has recently been studied in some detail by Prof. R.P. Roy and his co-workers at Patna University. They studied the sex in diploid, triploid and tetraploid plants with and without Y chromosome and observed that irrespective of the number of X-chromosomes and/or autosomes, the presence of a single Y chromosome gave a male individual.

Table 14.3. The relation between chromosome constitution and sex in Coccinia indica.

Chromosome constitution	*X/A ratio*	*Sex*
2A+XX	1.00	female
2A+XY	0.50	male
2A+XYY	0.50	male
3A+XXY	0.67	male
3A+XXX	1.00	female
4A+XXXX	1.00	female
4A+XXXY	0.75	male

A similar case earlier worked out by H.E. Warmke in 1946 refers to *Melandrium album.* In case of *Melandrium,* diploids, tiploids and tetraploids having different doses of X and Y chromosomes were studied with respect to their sex. Thus, in *Coccinia* as well as in *Melandrium* a plant is male when one or more Y chromosomes are present and is female when Y chromosome is absent. The female potentialities showed some expression only when the ratio of Y:X reached 1:4. The number of autosomes did not visibly affect the sex expression.

In *Melandrium,* the Y chromosome is longer than the X-chromosome. Plants having different individual fragments of Y chromosome were studied as a result, it was possible to divide the Y and the X-chromosomes into five different segments. These segments are known to control different stages of the development of sex organs. The X and the Y chromosomes have a common segment IV, which helps in pairing and the regular disjunction of X and Y chromosomes during meiosis. The remainder of Y chromosome has three segments, namely I, suppressing femaleness; II, initiating anther development and III, controlling the late stages of anther development. The X-chromosome also has a differential segment V, which should promote femaleness in the absence of female suppressing segment I on Y chromosome.

Table 14.4. Expression of sex in Melandrium with different numbers of X-chromosomes and autosome sets.

Chromosome constitution	*X/A ratio*	*Sex*
2A + XX	1.00	female
2A + XY	0.50	male
2A + XXY	1.00	male
3A + XXY	0.67	male (occasional ♂)
4A + XXY	0.50	male
4A + XY	0.25	male
4A + XXXY	0.75	male (occasional ♂)
4A + XXXXY	1.00	hermaphrodite (occasional ♂)

Therefore, while comparing *Drosophila* with *Coccinia* and *Melandrium*, we find that the male determining genes are present on autosomes in *Drosophila* but on Y-chromosome in plants.

2. Mistletoes (Viscum)

Several species of *Viscum* are dioecious, which are found in Africa, Madagascar, Europe, Asia and Taiwan. Sex is determined in this case through multiple translocations as determined only recently. In *Viscum fischeri*, for instance, male plants have 2n = 23, out of which, at meiosis, 14 chromosomes form seven bivalents and the remaining nine chromosomes form a ring (7" + 1.9) Seven bivalents disjoin normally but the ring of nine chromosomes disjoins in 4:5 thus giving rise to two types of gametes, one with 11 and the other with 12 chromosomes. The female plant has 2n=22 forming 11 bivalents (11") disjoining normally and giving only one type of gametes with 11 chromosomes each. Thus male is heterogametic due to translocations without any sex chromosomes involved. Two types of male gametes will give rise to two sexes. This phenomenon is also common in several other species of *Viscum*.

Table 14.5. Chromosome constitutions of two sexes in some dioecious plant species.

Mechanism		*Examples*
1. Male heterogametic (♀ XX; ♂ XY)		*Human lupulus*
		Melandrium album, *M. rubrum*, *Rumex angiocarpus*
	less well established	*Populus* spp., *Salix*, *Smilax*, and *Cannabis*
2. Male heterogametic (♀XX; ♂XO)		*Vallisneria spiralis*
		Dioscorea sinuata
3. Male heterogametic (♂ with an extra chromosome)		*Phoradendron flavescens*
		P. villosum
4. Female heterogametic (♀XY; ♂ XX)		*Fragaria elatior* and other *Fragaria species*
5. Compound chromosomes (e.g. ♀XX; ♂Y_1XY_2)		*Rumex acetosa*
		Humulus japonicus
6. No sex chromosomes (perhaps gene controlled)		*Spinacia oleracea*
		Ribes alpinum
		Vitis cinerea
		V. rupestris
		V. vinifera
		Carica papaya
		Asparagus officinalis
		Bryonia dioica
7. No sex chromosomes in male (translocation heterozygosity)		*Viscum fischeri*

Sex Determination in Humans

In early human embryonic development, the gonads are neuter, that is, neither ovary nor testes are differentiated. The neuter embryonic gonad has an outer region called the cortex and an inner region called the medulla. In an XY embryo, the cortex degenerates and the medulla forms the testes. In an XX embryo, the medulla degenerates and the cortex forms the ovaries. Thus, it is apparent that the X and Y chromosomes must contain genes that determine which of these developmental pathways will be followed.

Recently it has been determined that the human and mouse Y chromosomes have genes coding for a protein called the H–Y antigen. This protein acts upon the cells of the undifferentiated gonads and tilts them in the male direction so that they become testes.

Hormones produced by the ovary and testes also play an important role in influencing the development of sex characteristics. For example, once the testes or the ovaries have been formed,the effects of the different hormones produced by them cause degeneration of several ducts and the eventual formation of the external genital organs. Later in life, hormones play an important role in the determination and maintenance of secondary sexual characteristics, such as breast and pubic hair formation.

An unusual hormonal-cellular interaction has been found to be controlled by an X-linked recessive gene. The presence of this mutant gene in a genotypic male leads to a condition referred to as the testicular feminization syndrome. Such individuals are sterile females with male testosterone levels.

15

Sex Linked Inheritance

Sex Linkage in Man

More than fifty sex-linked traits have been reported in man and most of these are due to recessive genes. Some of the sex-linked traits in man include red-green colour blindness, hemophilia, two forms of diabetes insipidus, nonfunctional sweat glands (anhidrotic ectodermal dysplasia), certain forms of deafness, absence of central incisors, spastic paraplegia, a form of cataract nystagmus, optic atropy, night blindness, juvenile muscular dystrophy, juvenile glaucoma, etc. (Burns, 1969). Most of these traits have been clearly found to be due to recessive genes. Defective tooth enamel, which results in early wearing of the teeth down to the gums, is a sex-linked trait in man and is due to a dominant sex-linked gene.

The criteria for identifying sex-linked recessive genes from pedigree studies may be summarized as follows (1) expressions occur much more frequently in males than in female; (2) traits are transmitted from an affected man through his daughters to half of their sons; (3) an X-linked gene is never transmitted directly from father to son; and (4) because the gene is transmitted through carrier females, affected males in a kindred may be related to one another through their mothers.

If the X-linked gene should be dominant, such as the Xg gene for a rare blood type, males expressing the trait would be expected to transmit it to all their daughters but none of their sons. Heterozygous females would transmit the trait to half of their children of either sex. If a female expressing the trait should be homozygous, all of her children would be expected to inherit the trait. Sex-linked dominant inheritance cannot be distinguished from autosomal inheritance in the progeny of females expressing the trait but only in the progeny of affected males.

Incompletely Sex-linked Genes in Man

Besides the nonhomologous part of the X chromosome that carries the usual sex-linked genes, the X chromosome of man has section that is homologous with a part of the Y chromosome. The situation is similar to the case described in Drosophila for the section carrying the gene (bb) for bobbed bristles. Several genes have gene postulated for this region on the basis of pedigree studies. These include the gene for total colour blindness; that for xeroderma pigmentosum, asking disease characterized by pigment patches and cancerous growths on the body; the gene for retinitis pigmentosa, a progressive degeneration of the retina, accompanied by deposition of pigment in the eye; and that for a type of nephritis, a kidney disease. These are presumably represented in the X and Y chromosomes as allelic pairs and segregate like ordinary autosomal pairs, although they do not segregate independently of sex as do autosomal genes. Even though these genes are located on the X chromosome, the usual crisscross pattern for sex linkage is not expected because of their paired (allelic) arrangement. Questions have been raised concerning the interpretation of genetic or pedigree data for incomplete sex linkage in man but the cytological evidence is good. Chiasmata have been observed between sex chromosomes. More extensive pedigree studies will undoubtedly provide evidence for this mode of inheritance.

Y Chromosome Linkage in Man

Certain published pedigrees have indicated that the Y chromosome may have a section with genes distinctive to that chromosome. Genes located in a nonhomologous part of the Y chromosome are expected to control "holandric" inheritance because they are transmitted exclusively through the male line. The pedigree evidence for transmission from father to son is the only criterion on which they have been predicted Published pedigrees interpreted to show this pattern, for the most part, have not been substantiated, and there is reason to suspect that at least some of the most spectacular cases are not accurately reported. One example, that of "hairy pianna" of the ear reported by Dronamraju in 1960, is well substantiated. Judgment on the extend of Y-linked genes in man must be withheld until more complete evidence is available.

Sex-Linked Lethals

One of the many well-known sex-linked lethals in Drosophila produces the notched-wing effect. Appropriate test crosses and cytological observation have demonstrated that females homozygous for a gene associated with the notch phenotype die before hatching. Numerous other sex-linked lethals have been induced by irradiation in experiments designed to identify mutagenic agents and determine mutations rates under different environmental conditions. Methods of detecting sex-linked lethal mutations are described in next chapter.

Induced sex-linked lethal mutations in man have been indicated by differences in the sex ratio following irradiation of parents. W. J. Schull and J. V. Neel have analyzed the sex ratios of children born to parents who were exposed to atomic bombing in Japan during World War II. The data were grouped according to whether the father, mother, or both parents were exposed to irradiation. A trend evicting from the sex ratio in the general population was detected among the children of mothers that are exposed. Male children occurred less frequently than expected. This trend was interpreted to indicate that sex-linked lethals had been induced in the mothers and were being expressed in their sons. Data on sex-linked lethals are thus utilized for determining mutations rates that occur spontaneously and under different environmental conditions.

Sex-Influenced Traits

The end product of some gene action is influenced by hormones. For example, autosomal genes responsible for horns in some breeds of sheep behave differently in the presence of the male and female sex hormones. More than a single pair of genes is involved in the production of horns, but assuming all other genes to be homozygous, the example can be treated as if only a single pair were involved. Among Dorset sheep, both sexes are horned, and the gene for the horned condition is homozygous (h^+h^+). In Suffolk sheep, neither sex is horned and the genotype is hh. Among the F_1 progeny from crosses between these two breeds, horned males and hornless females are produced. Because both sexes are genotypically alike (h^+h) the gene must behave as a dominant in males and as a recessive in females; that is, only one gene is required for an expression in the male, but the same gene must be homozygous for expression in the female.

When F_1 hybrids are mated together, a ratio of 3 horned to 1 hornless is produced among the F_2 males, whereas a ratio of 3 hornless to 1 horned is observed among the F_2 females. Genotypes and phenotypes of the two sexes are summarized. The only departure from the usual pattern is concerned with the heterozygous (h^+h) genotype. This genotype in the male results in the horned condition, but females with the same genotype are hornless. Dominance of the gene is apparently influenced by the sex hormone.

Table 15.1. Genotypes and corresponding phenotype in male and female hybrid sheep

Genotypes	*Males*	*Females*
h^+h^+	Horned	Horned
h^+h	Horned	Hornless
hh	Hornless	Hornless

Some human traits, such as certain type of white forelock, absence of the upper lateral incisor teeth, and a particular type of enlargement of the terminal joints of the fingers, have been reported to allow the sex-influenced mode of inheritance. Other abnormalities such as harelip, cleft palate, an stuttering have hereditary bases and occur more frequently and more severely among males than females. The inheritance mechanism is complex. Environmental as well as genetic factor are involved but autosomal and not sex-linked genes have been associated with the traits. The higher incidence of affected amelus presumably indicates some sort of sex influence on gene action.

Sex-Limited Traits

Some genes can not express themselves in the presence of certain hormones and therefore are considered to be sex-limited. In most breeds of domestic poultry, plumage of the two sexes is strikingly different, but in some Sebright bantams, for example, both sexes are hen-featherd. In the Hamburgh breed, both hen-feathered and cock-feathered males may be produced, but all females are hen-feathered. Results of appropriate crosses whom that hen feathering is due to a dominant gene h^+ and cock feather in to its recessive allele h. Cock feathering, however, not only requires a particular genotype, but also is limited to the male sex. Even though females carry the proper genotype (hh) for cock feathering, they are hen-featherd. Genotypes and corresponding phenotypes that might occur in mixed breeds of chickens are summarized as shown in Table 15.2.

The Hambrugh strain, in which males are cock-feathered and females are hen-feathered, carries the homozygous genotype hh. Sebright bantams, in which both sexes are hen-feathered, carry the homozygous genotype h^+h^+. In hybrids between the two breeds, the genotype and presence or absence of the male sex hormone determine the feathering pattern. Both alleles are apparently seggregating in populations that include hen-feathered and cock-feathered males. In such flocks, all females, regardless of genotype, are hen-feathered, and males carrying the gene h^+ are also hen-feathered; only *hh* males are cock-feathered. Experimental gonadectomies have elucidated the relation between genes and hormones. Removal of the ovary in a hen-feathered female (*hh*) results in cock feathering. This indicates that the female sex hormone normally in habits the *hh* genotype from producing cock-feathered. Furthermore, in castrated males, the h^+ gene is inhibited. Thus, the hormones of both sexes limit gene action.

Table 15.2. Genotypes and corresponding phenotypes in male and female hybrid chickens

Genotypes	*Males*	*Females*
h^+h^+	Hen-feathered	Hen-feathered
h^+h	Hen-feathered	Hen-feathered
hh	Cock-feathered	Hen-feathered

Premature (pattern) baldness in man has also been explained as a sex-limited trait. Other types of baldness are associated with abnormal ities in thyroid metabolism and infectious disease. About 26 percent of the men over 30 in the United States are baldheaded. Approximately half of these became bald prematurely, in their twenties or early thirties. Baldness is known to be more common in some families than in others. Several different modes of inheritance have been associated by different investigators with this trait. An explanation for a particular type of premature baldness based on sex-limed inheritance was set forth by J. B Hamilton and supported by statistical data accumulated by H. Harris. A single dominant gene that expresses itself only in the presence of an adequate level of androgenic hormone was postulated by these investigator to account for the observed facts. The level of hormone necessary for expression of the trait is seldom if ever reached in women, but is attained in all normal men.

A pedigree illustrating a hereditary pattern of baldness in one family group is presented in Fig. Affected males, symbolized by darkened squares, became bald before they reached the age of 35. Men symbolized by light squares had thick hair, some even in old age. No women in this family group were baldheaded.

Expressions of some genes are sex-limed for more basic reasons. For example, milk production among cattle and other mammals is limited to the sex that is equipped with developed mammary glands and appropriate hormones. It is true that milk production is affected by environmental factors, but inheritance plays a part, and it is well known that milk-yielded genes are carried in the chromosomes of bulls as well as cows. Certain bulls are in great demand along dairy breeders and artificial insemination associations because their mothers and daughters have good milk-production records.

The frequency of twins and other multiple births in human families is hereditary to some extent. Mothers are immediately involved, but evidence indicates that genes from their fathers may influence the tendency toward multiple births. Genes of both parents control directly or indirectly many anatomical and physiological characteristics that express themselves only in one sex. The width of the pelvis, age of onset of menstruation, and distribution of body hair in females depend on genes common to both sexes. Genes that directly or indirectly influence the fertility of one sex or sex linked what force in *Drosophila*.

Sex Linked Inheritance in Drosophila

The first X-linked gene found in *Drosophila* was the recessive white eye mutation (*Morgan*, 1910). When a homozygous red-eyed female (dominant) is crossed with a white-eyed male (recessive), all individuals in the F_1 are red-eyed but when the cross is between a white-eyed female and red-eyed male, male offspring in the F_1 have white eyes. When heterozygous red eyed females are crossed with white-eyed males, both sexes segregate 1:1 rather than 3:1, for eye colour. These experiments demonstrate that the genes in this case are carried by the X chromosome, but not by the Y.

The inheritance of X-linked recessive gene can be understood more properly by considering the behaviour of X and Y chromosomes, such as follows:

In the cross between red-eyed female and white-eyed male *Drosophila,* the red eyed female contains the gene 'WW' for white colour of eye which remains located in X-chromosome. This one allelic condition of male is termed hemizygous in contrast to the homozygous or heterozygous possibilities in the female. The female being homogametic produces only one type of gametes or eggs each with the gene 'W' for red-coloured eyes. The male being heterogametic produces two types of gametes, 50% sperms with X chromosome

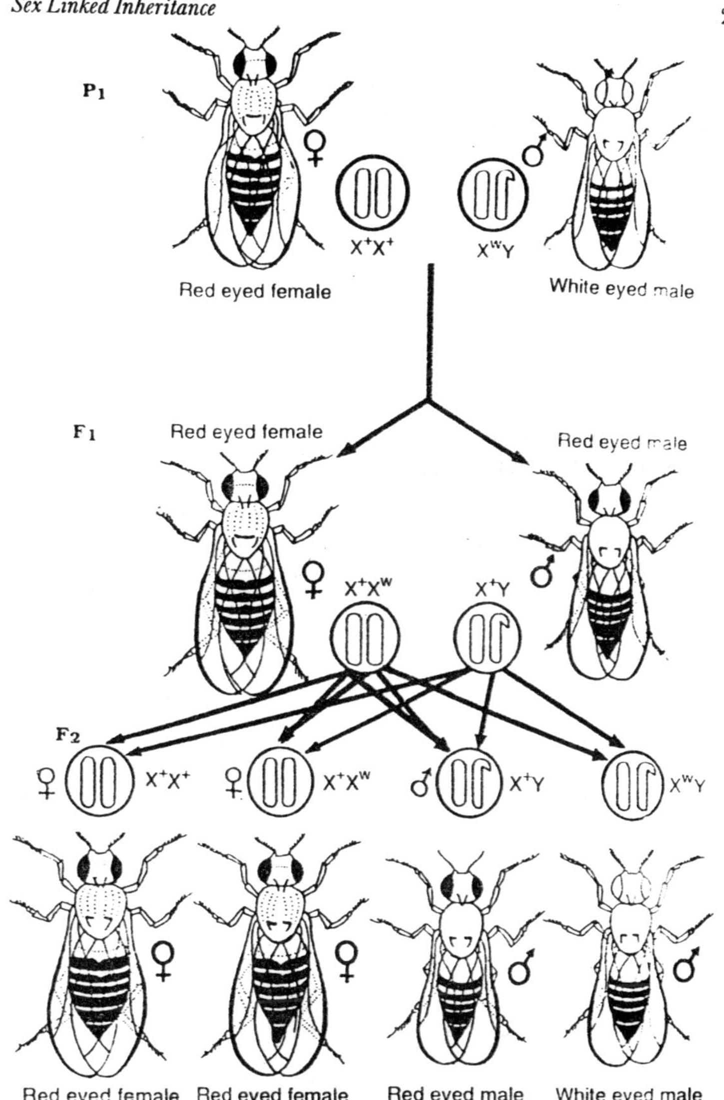

Fig. 15.1. A cross between red-eyed female and white-eyed male showing sex-linked inheritance in Drosophila.

containing 'w' gene and 50% sperms with 'Y' chromosome without any such gene. The gametes of both parents unite in fertilization to produce F_1 progeny. The F_1 hybrids which receive a X chromosome with 'W' gene from the female and a X chromosome with 'w' gene from the male becomes red-eyed female because gene 'W' is dominant over gene 'w'. Further the hybrid which happens to receive a single gene 'W' from mother and Y chromosome from father produce red-eyed males.

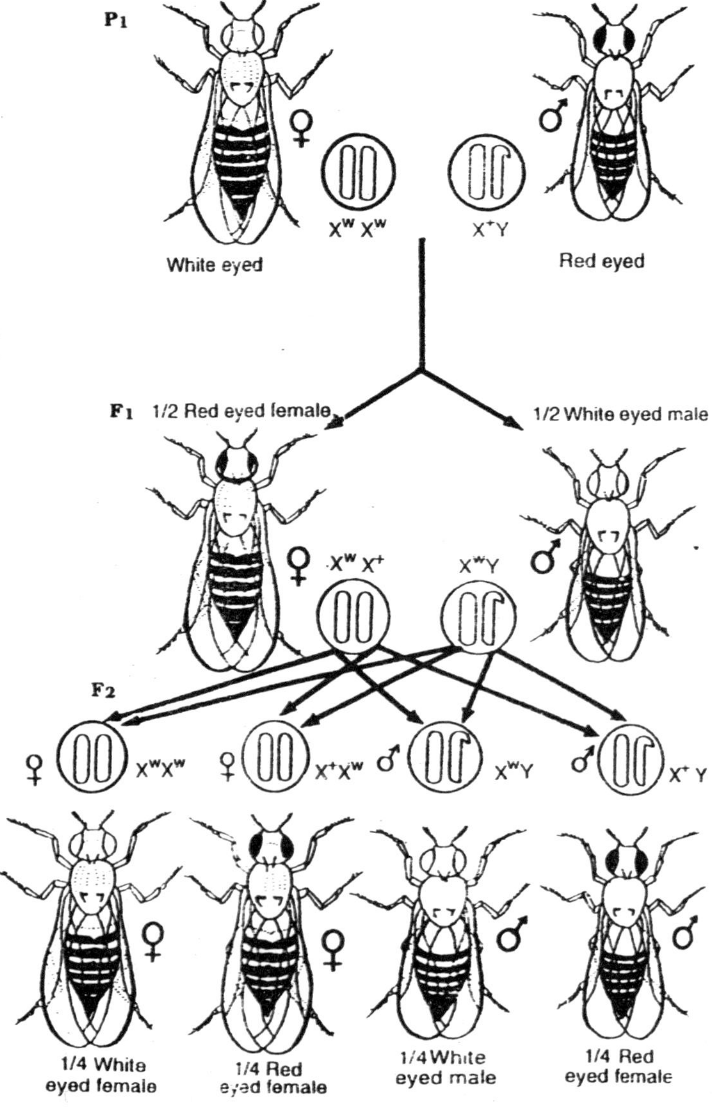

Fig. 15.2. A cross between white-eyed female and red-eyed male showing X-linked inheritance in Drosophila.

The F_1 red-eyed female with the gene 'Ww' when crossed with F_1 red-eyed male having the gene 'W'. The female hybrids produces two types of eggs, 50% eggs carry the gene 'W and the remaining 50% carry the gene 'w'. The males produce two types of sperms, half carry the gene 'W' and half carry no such gene on Y chromosomes. The union of sperm sand ova of F_1 offsprings may produce four possible types of F_2 individuals.

1. The eggs with 'W' genes if fertilized by sperms with 'W' genes produce homogametic red-eyed females.
2. The eggs with 'W' gene if fertilized by the sperms with 'Y' chromosome produce the red-eyed males.
3. The eggs with the gene 'w' when fertilized by the sperms having the gene 'W' produce heterogametic red-eyed females.
4. The eggs with the gene 'w' when fertilized by the sperms having the Y chromosome white-eyed males are produced.

Likewise in other experiment in which a white-eyed female is crossed with red eyed male. Similar X-linked inheritance of recessive gene for white eye colour is revealed. The white-eyed female contains the gene 'ww' located on both X chromosomes. The red-eyed male contains the gene 'W' located on single X-chromosome. The female being homogametic produces single type of eggs with single 'w' gene for whiteness, while the male beings heterogametic produces two types of sperms, 50% sperms carry the gene 'W' and remaining 50% sperms carry no such genes on the Y chromosomes. In the mating of both parents by the union of these eggs and sperms, two kinds of F_1 hybrids are produced. The eggs with 'w' genes by union with the sperms having 'W' genes produce heterogametic red-eyed female and the eggs with 'w' gene by union with the sperms having Y chromosome produce white-eyed male. When the F_1 brothers and sisters with the gene 'w' and 'Ww' bred together they produce four types of individuals in F_2 such as white-eyed female (ww), red-eyed female (Ww), white-eyed male (w) and red-eyed male (W).

The results of these experiments clearly indicate towards linkage of genes for red-eyes and genes for white-eyes on the X chromosomes. Further, from these experiments characteristic criss-cross pattern of inheritance of recessive X-linked from P_1 father to F_1 daughter, in which it remain phenotypically unexpressed (and such F_1 female is called carrier) and from F_1 daughter this trait is transmitted to F_2 son, in which it is expressed phenotypically.

Inheritance for Z-Linked Gene for Coloured (Black or Red) Feathers in Chicken

In case of birds, moths, butterflies, etc,, the females are heterogametic and males are homogametic quite unlike that of

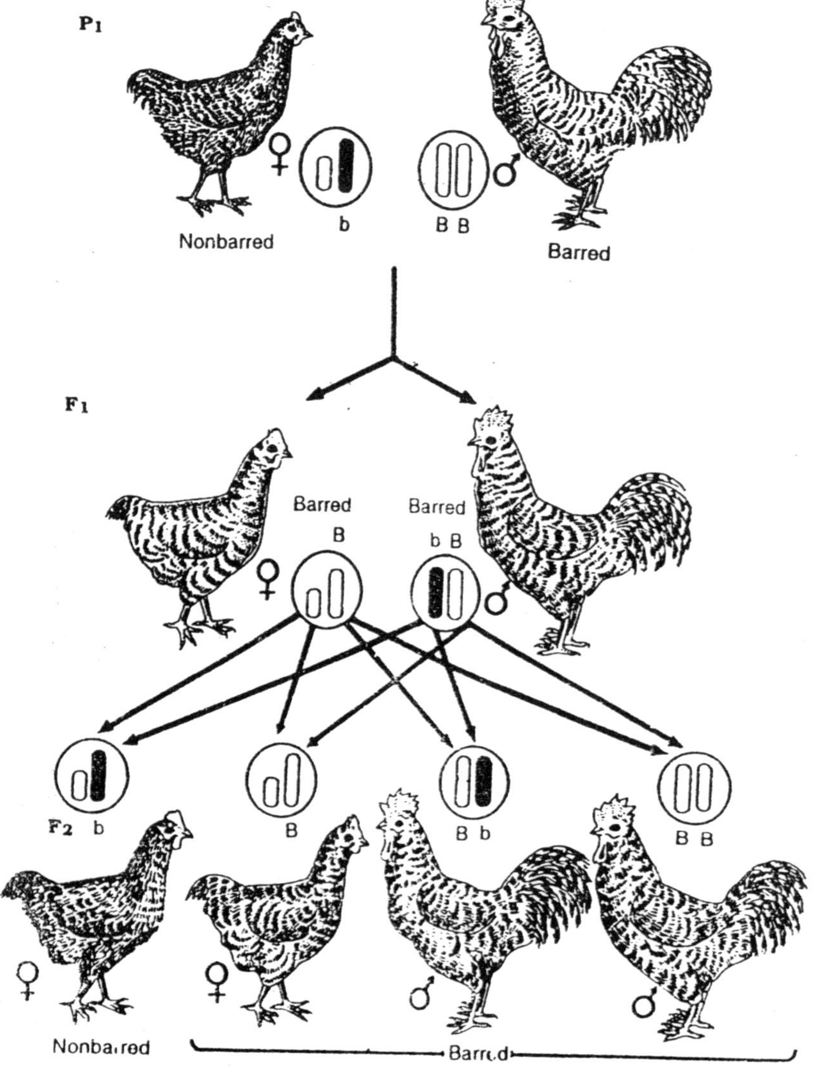

Fig. 15.3. A cross between a non-barred hen and a barred cock showing sex-linked (Z-linked) inheritance.

Drosophila and man. Here also sex-linked genes follows the "crisscross" pattern but from mother through heterozygous F_1 sons to grand-daughters of F_2. The common example of ZW-ZZ sex-linkage is Plymouth rock chicken.

In Plymouth rock chicken the gene fro barred feathers is dominant and the gene for black or red unbarred feathers is recessive. Both the genes are Z-linked. A barred male chicken contains two genes for barring because it has two sex chromosomes (ZZ). When this barred male with the gene BB is crossed with unbarred female containing single recessive gene 'b' in its Z chromosome (W chromosome contains no genes), produce in F_1 only barred males and females. These F_1 barred males and females when inbred produce in F_2, a hemizygous non-barred female, a hemizygous barred female, a heterozygous barred male and a homozygous barred male.

In another cross, when barred hens and non-barred cocks are crossed, the F_1 has half barred male and half non-barred female. These by inbreeding produce in F_2 half barred males and female and half non-barred males and females.

Besides fowl, many birds, e.g., pigeon, duck, canary, moths, butterflies, all reptiles, some fishes and amphibians which have XX male and XY or XO female type of sexes showing sex-linked inheritance of (ZW-ZZ) type.

Males Express Sex-Linked Recessive Genes

Congenital hyperuricemia (Lesch-Nyhan syndrome), characterized by excess production of uric acid, is inherited through a *sex-linked recessive gene.* This means that the mother contributes the X chromosme with the defective gene to a male zy-gote. Half of the male children of carrier mothers may be expected to inherit the disease. These are deficient for the enzyme hypoxanthine-guanine phosphoribosyl-transferase (*HGPRT*). This enzyme is involved in nucleotide synthesis. Infants who receive the gene appear normal at birth and for several months, but may show symptoms of excessive uric acid in the urine, such as orange "sand" (uric acid crystals). By about 10 months of age, they may become abnormally irritable and lose motor control. Weak and flabby muscles prevent the child from sittings walking, or speaking normally. By the second year of life, the nervous condition has progressed to a degree that self-mutations occurs, manifested by lip-bitting, finger-chewing, teeth-grinding, and marked swinging of the arms. Death, which is usually secondary to severe renal and neurological damage, usually occurs within a few years, but some victims live into there 20s. Affected features as well as sex can be detected by amniocentesis.

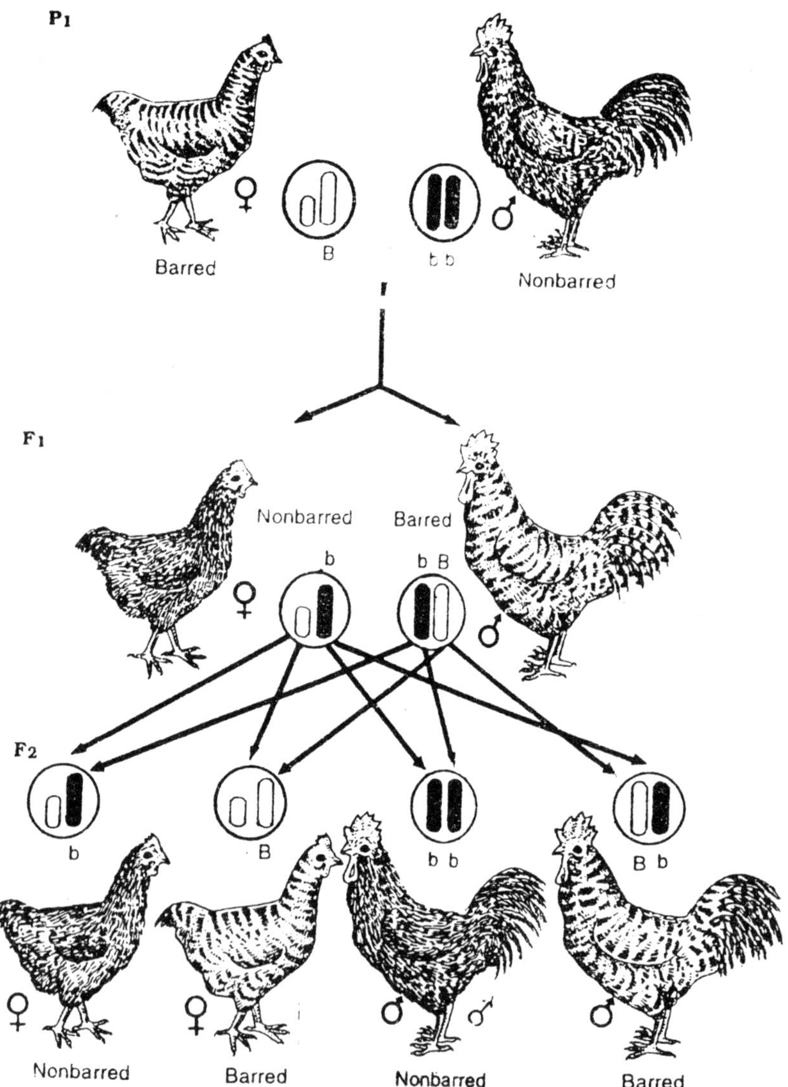

Fig. 15.4. A cross between a barred hen and non-barred cock showing sex linked (Z-linked) inheritance.

Juvenile muscular dystrophy also depend son a *sex-linked recessive gene.* If the mother is known to be a carrier for this gene, either from her pedigree or through tests that are available, about half of her male children are expected to be affected. Male fetuses can be identified by a chromosme study. Juvenile muscular systrophy affects males, usually before they reach teen age, with muscular deterioraction that progresses rapidly during the early teen years. Muscles of the legs and shoulders become stiff, and the children usually become paralyzed and crippled during their middle or late teens. Virtually all die before age 21. All female children born to a carrier mother are expected to be normal, since the possibility for their being homozygous for a sex-linked recessive gene is virtually nonexistent.

Another severe disease following the pattern of *X-linked recessive* inheritance is the *Hunter syndrome*. It is characterized by mental retardation, coarse features, hirsutism (abnormal hairness), and a characteristic facial appearance that includes a broad bridge of the nose and a large protruding tongue. Symptoms appear in early childhood. A chemical means of diagnosing this condition is being developed. Certain constituents in the amniotic fluid indicate the presence of this disease, which is associated with an abnormal processing of mucopolysaccharides synthesized in early pregnancy. Mucopolysachharides also accumulate in skin cells of persons who are monozygous for the gene for Hunter syndrome. When amniotic or skin cells are grown in culture and stained with o-toludine blue, any mucopolysachharide cell inclusions will be stained pink. It is thus possible to identify a heterozygous carried of the gene as well as an affected fetus.

The pattern of mucopolysaccharide (containing an amino sugar as well as uronic acid units) metabolism by Hunter cells is so strikingly different from the normal that it can be used along with chromosome analysis for sex determination in prenatal diagnosis, a situation in which clinical observation is obviously impossible. Of the many cell types originally present in amniotic fluid, fetal fibroblasts are the only ones to multiply in culture. Like fibroblasts from skin biopsies, they show an excessive accumulation of mucopolysaccharide or stainable cell inclusions if the fetus is affected with the Hunter syndrome.

Glucose 6-Phosphate Dehydrogenase (G6PD) Deficiency

One well-known gene on the X- chromosomes controls the production of an enzyme, *glucose 6-phosphate dehydrogenase* (G6PD), which is involved in carbohydrate metabolism and is important in maintaining the stability of red blood cells. Individuals who have abnormally low amounts of this enzymatic activity are prone to a severe anemia that occurs when many for their red blood cells cannot function normally and therefore break down and are destroyed. The anemia can be provoked by a number of environmental triggers such as inhalting pollen of the broad bean Vicia faba or eating the bean raw, in which the illness is known as *favism.* Such individuals are also sensitive to certain drugs such as naphthalene (used in mothballs), certain sulfa antibiotics (such as sulfanilamide), or the antimalarial drug primaquine. In the absence of the offending substances these individual are completely normal, and they recover from the anemia when the agents are eliminated. G6PD deficiency is found in high frequency in people of Mediterranean extractin (10 to 20 percent or more of males are affected) and among Asians (About 5 per cent of Chinese males are affected), and it occurs in about 10 percent of black American males. As noted, the locus of G6PD is X linked, and well over 50 alleles coding variant forms of the enzyme are known. However, only a few of these alleles lead to sufficiently defective form of G6PD to cause the drug sensitivity and anemia associated with G6PD deficiency.

Colour Blindness

Almost everyone is familiar with the common form of colour blindness, called *red-green colour blindness,* which is inherited as an X-

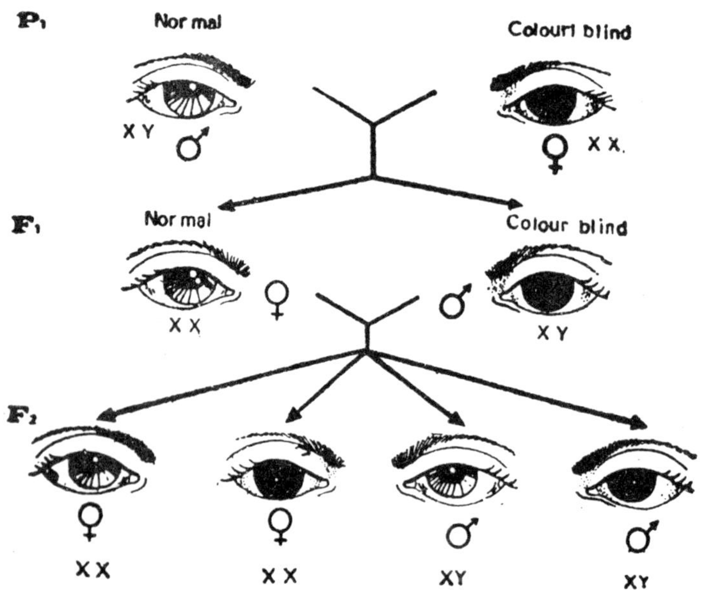

Fig. 15.5. Cross between a red-green colour blind female (XX) and a normal male (XY) showing X-linked inheritance in man.

linked recessive condition. Several other kinds of defects in colour vision are known; a they differ according to which of the three pigments in the retina of the eye–red, green, or blue–is defective or present in an abnormally low amount. The most common types of colour blindness involve the red or green pigments; these are collectively known as red-green colour blindness, but they are not a single entity. Both conditions are X linked, however, and in different Caucasian populations the frequency of red-green colour-blind males is between 5 and 9 percent. Colour blindness in females is much rare, of course, but it does occur.

Two loci on the long arm of the X chromosoe seem to be involve in red-green colour blindness. Defects in green vision are due to mutations at one of the loci; defects in red vision are due to mutations at the other locus. Among Western European males, about 5 percent have defects in green perception and another 1 percent have defects in red perception. Two loci are known to be involved in red-green colour perception because the stations that cause the red defects and the green defects exhibit *complementation.* That is to say, women who carry a red-defect allele on one X chromosome and a green-defect allele on the

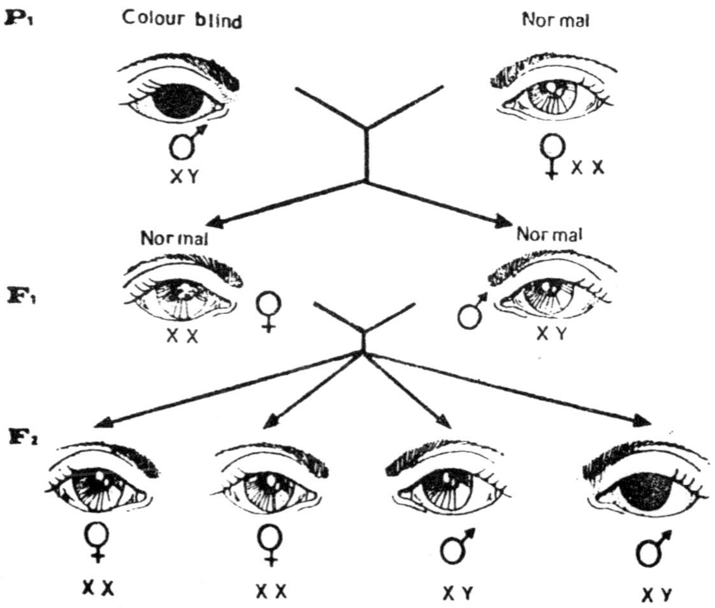

Fig. 15.6. A cross between a red-green colour blind male and normal female showing X-linked inheritance in man.

other have normal colour vision. Such complementation is expected when two loci are involved because these women are actually heterozygous for a normal allele and a recessive mutation at each of two loci. Because of segregation, nearly half the sons of these women will have the green defect and the other half will have the red defect. Because of recombination, however, a proportion of their sons will have normal colour vision and an equal proportion will have defects in both their red and green colour vision. The two loci are very close to each other on the X chromosome, so the actual proportion of sons who carry recombinant X chromosomes is very small.

HEMOPHILIA

Recall that hemophilia is a bleeding disorder due to an X-linked recessive that results from the excessively long time required for the blood to clot following an injury. Affecting about 1 in 7000 males, it is much rare than red-green colour blindness, yet is probably as well known. This is partly because it affects the blood; and even though we now know that blood has no magical or hereditary properties, we still carry a vestige of the old beliefs in such expressions as "cure blood" and "blood lines" and "blood relation." A second reason hemophilia is so well known is that it occurred in many members of the European royalty who descended from Queen Victoria of England (1819–1901).She was a carrier of the gene, and by the marriages of her carrier granddaughters, the gene was introduced into the royal houses of Russia and Spain. Ironically, the present royal family of Great Britain is free of the gene because it descends form King Edward VII, one of Victoria's four sons, who was not himself affected.

Actually, there are several forms of hemophilia, and the incidence of 1 per 7000 in males applies only to the special type exemplified in some of Queen Victoria's descendents. Blood clotting is a complex process involving more than a dozen ingredients known as *cloting factors*, and different genes are responsible for producing these various clotting factors. An abnormal form of any one of these genes can lead to an inherited form of excessive bleeding. Most hereditary hemophilias are caused by genes on the autosomes, but two forms are inherited as X-linked recessives. One of these X-linked forms, called ***hemophilia B*** or ***Christmas disease***, involves clotting factor IX, and it is extremely rare. The other X-linked form, called ***hemophilia A*** or ***Royal hemophilia***, involves clotting factor VIII. Hemophilia A is the more common X-linked form, and it is the one present tin some of the European royal families.

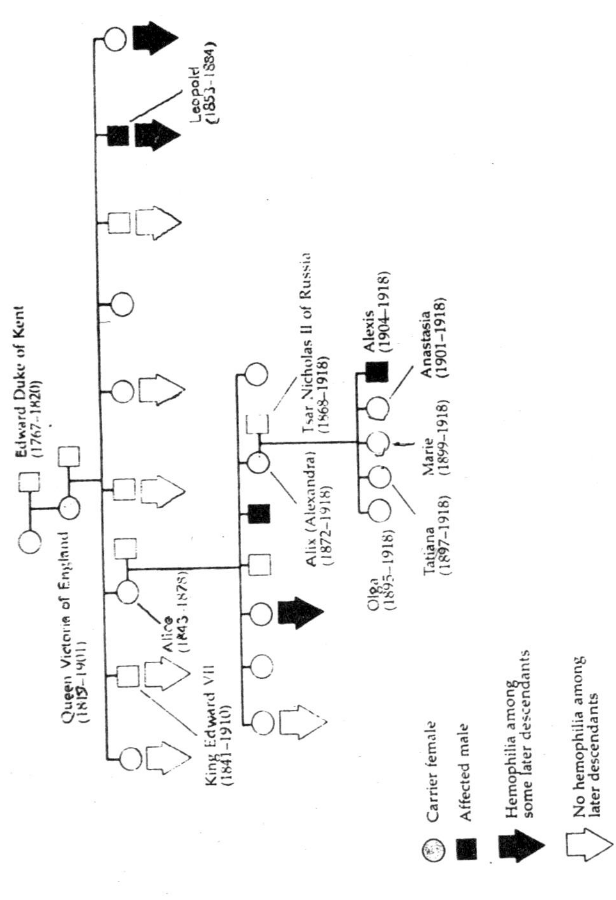

Fig. 15.7. Partial pedigree of hemophilia A among the descendants of Queen Victoria, including Alexandra, Empress of Russia, and her five children.

Where Queen Victoria's hemophilia-A mutation came from is not known. She herself was a carrier, but her father was completely normal and nothing in her mother's family suggests that the gene was present. The best guess is that either the egg or the sperm that gave rise to Queen Victoria carried a new mutation. Perhaps it was in the sperm of her father, Edward Duke of Kent, because it is thought that mutations are somewhat more likely to occur in the sperm of older men, and he was 52 when Victoria was born. However it happened,

Victoria was a carrier. She had nine children five daughters (two were certainly carriers, two were almost certainly not, and the remaining daughter may or may not have been as she left no children) and four sons (one a hemophiliac–Leopold Duke of Albany–and three normal). In the five generations since Queen Victoria, 10 of her male descendants have had the disease. Virtually all died very young. Those who survived childhood often died in their 20s or early 30s, usually from excessive bleeding following injuries.

The most famous of Victoria's affected male descendants is undoubtedly her great-grandson Alexis, the son of her granddaughter Alix (Empress Alexandra of Russia) and Tsar Nicholas II. Alexis was Alexandra's first born son and heir to the Russian throne. Unbeknown to Alexandra, she was a carrier of the hemophilia-A allele, and her son had inherited the disease. Anna Viroubova, one of Alexandra's favorite ladies-in-wating, recounts:

The hair was born amid the wildest rejoicings all over the Empire. After many prayers, there was an heir to the throne of the Romanoffs. The Emperor was quite mad with joy. His happiness and the mother's, however, was of short duration, for almost at once they learned that the child was affected with a dread disease. The whole short life of the Tsarevich, the loveliest and most amiable child imaginable, was a succession of agonizing illnesses due to this congenital affection. The sufferings of the child were more than equaled by those of his parents, especially of his mother.

The parents became preoccupied with the boy's health. At one point the Tsar observed in his diary how difficult it was to live through the worry resulting from Alexis's excessive and worry resulting from Alexis's excessive and life-threatening bleeding from minor injuries and bruises that all children unavoidably experience. The boy survived his childhood, but some historians have argued that the Tsar's preoccupation with Alexis's health contributed to the neglect of the empire that ultimately bought on the Bolshevik Revolution in 1917. During the revolution, the Tsar, Alexandra, Alexis, and his four sisters disappeared. The fate of the family is unknown, but they are thought to have been machine-gunned to death in Ekaterinburg fortress on July 17, 1918, just 13 days before what would have been Alexis's seventeenth birthday.

Lesch-Nyhan Syndrome (HGPRT Deficiency)

As a final example of X-linked recessive inheritance we consider the rare condition known as *Lesch-Nyhan syndrome*, which is due to a deficiency of the enzyme HGPRT. HGPRT refers to (*hypoxanthine guanine phosphoribosyl transferase*). Persons with this conditions, virtually all male, have two groups of symptoms. One results from excessive accumulation of *uric acid* in the blood. Recall from Chapter 4 that one normal function of HGPRT is to route certain cellular metabolities into DNA via the minor pathway of DNA synthesis. (A *metabolite* is any compound produced during the course of chemical processes in the cell; the totality of all these processes is called *metabolism*). In the absence HGPRT, these metabolites accumulate and are eventually broken down by other enzymes into uric acid. Accumulation of excess uric acid in the blood), bloody urine, crystals in the urine, urinary tract stones, severe inflammation of the joints (*arthritis*), and the extremely painful swelling and inflammation of joints in the hands and feet (particularly the big toe) known as *gout*. (Although gout is always associated with elevated levels of uric acid, the inheritance of most forms of gout is multi factorial; only aminority of patients with excessive uric acid and gout have HGPRT deficiency.)

Symptoms of excessive uric acid can be treated with appropriate drugs, but Lesch-Nyhan syndrome has another group of symptoms that are of unknown origin and are not alleviated by treatment for excessive uric acid. These symptoms involve the nervous system and are associated with jerky, unwilled, and uncoordinated muscular movement. The most bizarre symptoms of Lesch-Nyhan syndrome is extreme aggressive behaviour, most often expressed as self-mutation of the hands and arms, usually by biting. At present, such self-mutilation can be prevented only by appropriate binding of the hands and arms. Lesch-Nyhan syndrome is an extremely serious condition. Most patients die before the age of five, and almost all die before adulthood. Though rare, the Lesch-Nyhan syndrome is important because it involves HGPRT, which allows us to understand the basis of some of its symptoms. The condition also illustrates the profound influence that genes can exert on behaviour, even though the biochemical basis of this influence is not yet understood.

16

POLYGENIC INHERITANCE

POLYGENIC THEORY

Colour in Wheat Kernels

During the period between 1900 and 1910, many geneticists thought continuous variation reflected an entirely different mechanism of inheritance from discontinuous variation. However, when the idea of Mendelian inheritance had become well established, a few keen investigators began to envision a common basis for the results of Mendel and those of Kolreuter. Genes with small but cumulative effects were postulated to behave in a Mendelian fashion. An explantion for continuous variation thus emerged in the form of multiple-gene hypothesis. Experimental results and interpretations substantiating this hypothesis were obtained from the classical investigations of H. Nilsson-Ehle (1873-1949) in Sweden and E.M. East in the United States during the period 1910 to 1913.

Fig. 16.1. H. Nilsson-Ehle, Swedish plant breeder who developed the multiple gene hypothesis to explain the genetic mechanism for quantitative inheritance.

One of the these studies was based on cross between two varieties of wheat producing red and white kernels, respectively. The F_1 seed were intermediate in colour between those of the two parent. They were lighter than those of the red parent but distinctly more coloured than those of the white-parent variety. When the F_2 seeds were classified according to intensity of colour, a continuous gradation was observed from red to white, and classes were more or less arbitrary. About 1/16 of the F_2 seeds were as red as those of the red parent, and about 1/16 were white, and about 14/16 were intermediate, ranging between the colour of the red and the white original parents.

When the 14/16 of the F_2 seeds were classified further, on the basis of the colour intesity, it was shown that about 4/16 had more colour than the F_1 intermediates, about 6/16 were intermediate like the F_1's, and about 4/16 were lighter than the F_1/s. This result suggested a segregation of two gene pairs and was explained on the basis of duplicate genes acting onthe same character and producing a cumulative effect. When the results of this cross were examined criticaly,they were found to resemble those that Kolreuter had obtained many years before. A second cross between red-kernel and white-kernel varieties of wheat was carried to the F_2 and this this about 1/64 of the F_2 progeny produced white kernels and about 1/64 produced red kernels. Some 62/64 were intermediate in colour, ranging between those of the two parental varieties. Again continuous gradations were recognized between the extremes of the parents. The result of this cross resembled those of a Mendelian trihybrid cross in which three independent pairs of alleles had similar and cumulative effects. Therefore, three independent gene pairs were postulated to explain this result in contrast to the two pairs for the previous cross. Evidently, one pair, which was segregating in the second cross, was homozygous in both parents in the first cross.

The concept of multiple genes for quantitative inheritance is now one of the most important principle of genetics. It has been strengthened greatly by the use of statistical methods devised by R. A fisher in England, sewall Wright in the United States, and others. The explantion, based on the action of many genes (polygenes) usually segregatin independently but influencing the same phenotype in a cumulative fashion, has been well established, although the details of the mechanism require further investigation.

Polygenic inheritance differs from the classical Mendelian pattern in that the whole range of variation is covered in a graded series from one parental extreme to the other. Only averages of populations are considered and not values for individuals. Such factors as dominance, epistasis, cytoplasmic influences, interactions among genes and gene products, and interactions with the environment are reflected inthe averages. Polygenic inheritance is a statistical concept.

Because most characteristics of domestic plants and animals that have practical significance (including height, weight, time required to reach maturity, and qualities for human nutrition) depend on polygenic inheritance, much attention has centered around this principle. If all of the practical genetic experimental projects that are now in progress in various experiment stations throughout the world could be listed and classified, the results would probably indicate that some 80 to 90 percent of all practical studies involve quantitative inheritance. Some human characteristics of interest and significance also depend on multiple genes.

Skin Colour in Man

C. B. Davenport, one of the first to investigate the inheritance of skin colour in man, believed that two pairs of genes accounted for the difference in pigment between Negro and Caucasina people. His most significant studies were conducted among selected groups is Bermmuda and Jamaica, where intermarriages between people with different degress of pigmentation were relatively frequent, and illegitimate births, which could obscure the data, were relatively uncommon. As a part of these studies,an objective classification system was devised to detect different degrees of pigmentation. A rotating disk was prepared by which the skin colour of any individual could be matched with a standard. The validity of such a system was increased when it was discovered that the same pigment, that is, melanin, was present in varying degrees in so-called "white" people, those with mixed ancestry,and Negroes. An arbitrary scale was developed in terms of percentage, with albino people designated at 0 and the most heavily pigmented at 100 percent. Average "white" individuals were found to have about 5 percent of the menalin pigment. Objective tests showed considerable variation among whites from only white ancestry. The amount of pigment varied form practically none in the case of light blonds to about 11 percent for some brunettes. The average of individuals sampled from the American Negro population showed about 75 percent of pigment. A range from about 56 to 78 percent of pigment was observed among the American coloured people with only Negro ancestry. Other coloured people with more pigmentation than that of the average American Negroes were shown to approach the arbitrary 100 percent of pigment.

When samples were taken from the immediate families resulting from intermarriage between coloured people with only Negro ancestry and white people with only white ancestry, a range from 27 to 40 percent of pigment was observed. When individuals resulting from intermarriage in previous generations were examined, a colour range extending continuously from white to black was encounted. Different classes were arbitrarily distinguished among individuals of mixed ancestry on the basis of intensity of colour. The class with 12 to 25 percent of pigment was identified as light; 26 to 40 percent, medium (mulatto); and 41 to 55 percent dark. These three classes were intermediate between white and black.

Table 16.1. Degree of pigmentation and phenotypic classes in man

Percent Melanin Pigment	*Phenotype*
0-11	White
12-25	Light
26-40	Medium (mulatto)
41-55	Dark
56-78	Negro

In the entire population resulting from intermixture, a continuous gradation was recongized. The first generation hybrids, that is, mulattoes, were found to be intermediate between their black and white parents. When large numbers of individuals known to have come from marriages between first generation hybrids were classified, it was shown that some were as light as the white parent and some were as dark as the original black parent. Most are somewhere between the extremes of the parents. Davenprt's analysis of the data showed a proportionof about 1 black, 4 dark, 6 medium, 4 light, and 1 white. This proportion was based on a model in which two pairs of alleles were involved, each producing about the same amount of pigment. No appreciable dominance existed between the alleles, and the action of the genes was cumulative. By using gene symbols a and b, the cross based on this hypothesis may be reconstructed.

The well-established pattern of quantitative inheritance based on multiple genes is amply demonstrated in the production of gradations of pigmentation. The common belief that dominance favours the pigmented condition over the white is left without a foundation, and the frequently repeated statement that white parents can have black children is unsubstantiated by scientific fact. If one parent is white and genetically homozygous (aabb), the children would not be genetically darker than the other parent. If, however, each parent carries one or two of the genes for pigmentation, combinations producing more intense pigment than shown by either parent might occur. One serious difficulty in such phenotypic analyses comes form the influence on pigmentaton of environmental factors, both those from natural causes such as the sun and wind and those supplied at the cosmetic counter. It is sometimes difficult, indeed, to distinguish on the basis of colour alone between an individual genetically white (aabb) but representing a brunette type, and an individual with one gene for pigment (Aabb or aaBb) but representing the lower range of expression for that genotype.

Comparisons become even more difficult because the amount of pigment increases from birth to maturity and decreases from maturity to old age. Undoubtedly, modifiers of the basic pigment genes are involved. Thus, minor genetic variations may account, part at least, for the small differences among individuals with small differences among individuals with similar ancestry and presumably comparable genotypes. Different racial groups with different intensities of pigmentation may be explained on the basis of past mutations that influenced the degree of pigmentation.

Although quantitative inheritance of skin colour is well established, questions have arises concerning the number of gene pairs involved. Davenport's model based on two pairs is too simple to acount for the data now available. A more recent model by R. R. Gates is based on three pairs of genes. Curt Stenr's mathematical analysis shows that 4, 5 and 6 gene pairs agree better with the observed data than a higher or lower number, but the actual number of genes involved is not known. Various investigators have estimated numbers ranging from 2 to 20 pairs. Variation in the degree of pigmentation exists among coloured individuals from only coloured ancestry. Likewise, different Caucasian peoples show individual variation. The relatively slight differences that exist among white persons from only white ancestry and coloured persons from only coloured ancesry may be attributed to minor genetic variation and environmental influence.

Ear Length in Maize

One of the classical studies on quantitative inheritance, which did much to establish the multiple gene hypothesis, was made by R. A. Emerson (1873–1947) and E. M. East on the inheritance of ear length in maize. A cross was made betwen varieties with samll ears and those with large ears, and the F_2 results were critically analyzed. The small-eared parent was variety of popcorn called the Tom Thum, with ears averaging 6.6 cm in length and ranging from 5 to 8 cm. The other parent was Black Mexican, a sweet corn with ears averaging 16.8 cm and ranging from 13 to 21 cm. The F_1 progeny were intermediate in ear length, averaged 12.1 cm, and ranged form 9 to 15 cm.

Fig. 16.2. R.A. Emerson, pioneer American plant breeder who did basic work on maize genetics.

The F_2 represented a widere range of variation than the F_1, with some of the ears extending into the range of both parents. This is characteristic pattern for the results of crosses involving traits dependently on only a few genes. When many genes are involved, the extremes represented by the parent occur infrequently in the F_2. The histograms represent the maize data and illustrate graphically the characteristic pattern of multiple genes inheritance. Since the parental lines were inbred and relatively homozygous, the variation within these lines and that shown for the F_1 were mostly environmental, or a result of sequential development in time. The F_1 represented the fully heterozygous condition between the genotypes of the two original parents. The results of genetic segregation were illustrated by the wider variation in the F_1.

As in other examples involving quantitative inheritance, the environmental influence is the most confusing factor in analysis, and it must be considered and is adequately controlled as possible. The environment can produce results similar to those of the genes with respect to size differences between large and small varieties. Plants raised in unfavourable environments–that is, without sufficient water, sunlight, or soil nutrients–will be smaller than others of the same genotype that enjoy a more satisfactory environement. On the other hand, under ideal environmental conditions, organisms with inferior genotypes may develop phenotypes equivalent or superior to those with better genotypes but inferior environments. Even minor environmental variations affect the expression of quantitative traits. The environmental influence must be controlled as completely as possible with experimental design.

If, in the maize example, two gene pairs are active, the variation is all hereditary, and each gene produces an equal effect on size of the ear above the size dependent on the residual genotype (6.6 cm), the individual contributions would be

$$\frac{16.8 - 6.6}{4} = 2.55 \text{ cm per gene}$$

Each active allele thus would produce 2.55 cm in addition to the 6.6 made possible by the residual genotype for the small variety. When F_2 plants were classified according to phenotype a, ratio of 1:4:6:4:1 was obtained. This is a modification of the 1:2:1:2:4:2:1:2:1 ratio, which may be changed to 9:3:3:1 by dominance.

Size in Rabbits and Corolla Length in Tobacco

Another example of multiple gene or polygenic inheritance may be taken from the work of W.E. Castele on the inheritance of weight of rabbits. Rabbits from the large Flemish Giant breed, which, when fully mature, averaged about 13 pounds, were crossed with those from a small Polish variety averaging about 3 pounds. The F_1 hybrids were intermediate between the two parents averaging 7 to 8 pounds, and the F_2 progeny ranged from a size nearly as small as the Plish breed to a size approaching that of a Flemish Giant. Several litters of the F_2 progeny were obtained, but no rabbits reached the extremes of the parental strains, presumably because a large number of gene pairs were acting in the cross. If enough F_2 individuals were raised, the parental extremes would be expected to occur eventually. On the other hand, the extremes of the parents might never be obtained, even with large numbers of hyrids, because further work of Castle showed that the effects of the genes were not proportional. Modifiers and maternal size effects were involved along with the effects of polygenes.

In examples that involved several thousand F_2 progeny from plant crosses, none even approached the parental phenotypes and many gene pairs were presumed to be involved. As many as 200 pairs of alleles were postulated to be active in some traits that have been reported. E.M. East studied the coraolla length in a cross between two varieties of tobacco, Nicotiana longiflora, and found none of the several hundred

Individual contribution of genes:

(A or B) 2.55cm

size produced by residual genotype : 6.6 cm Genotypes of parent:

Black Mexican		Tom Thumb
AABB	×	aabb

F_1 × Aabb (12.1 cm)

F_2 Geno type	Frequency	Phenotype (cm)	Phenotypic Ratio
AABB	1	16.8	1
AaBB	2	14.2	4
AABb	2	14.2	
AaBb	4	11.7	6
aaBB	1	11.7	
AAbb	1	11.7	
aaBb	2	9.1	4
Aabb	2	9.1	
aabb	1	6.6	1

Fig. 16.3. Diagrammatic analysis of Emerson's maize data based on the multiple gene hypothesis with two pairs of active alleles, each contributing equally to the phenotype.

F_2 plants to resemble the parents (P). Many gene pairs were therefore postulated.

Estimating the New Number of Gene Differences

The contributions of individual genes to a quantitative character can be evaluated roughly from the results of some of the foregoing detailed crosses. In the maize data, for example, each extreme of the parents occurred in the proportion of about 1 in 16 in the F_2. Two pairs of alleles, then, might be assumed to be operating. Determining the number of genes involved in a given cross is usually more difficult, however, mainly because environmental as well as genetic variations are represented by the same measurements. In nature, the genes may not all influence the phenotype in the same way or to the same extent. Models devised to estimate the number of genes are oversimplified if based on the assumption of equal effects.

One of the problems confronting the geneticist when he begins to study a trait known or suspected to depend on quantitative inheritance is the estimation of the number of genes involved. In some organisms, a rough estimate has been made by determining the frequency of occurrence in the F_2 population of the extremes representing the parental phenotypes. If the extremes occur with a frequency of 1 in 4, one pair of genes may be assumed to be operating. If the phenotype of each pairs are involved; in 64, three pairs; 1 in 256, four pairs; and 1

Table 16.2. Probability of occurrence of F_2 individuals as extreme as either parent

Pairs of Segregating Alleles	*Fraction of F_2 as Extreme as Either Parent*
1	$^1/_4$
2	$^1/_{16}$
3	$^1/_{64}$
4	$^1/_{256}$
5	$^1/_{1024}$

in 1024, five pairs. Obviously, this is an inadequate approximation of the number of active genes, but the method is useful for preliminary genetic analysis. More complicated mathematical procedures are employed when sufficient F_1 and F_2 data are available. This method is based on the assumption that all genes produce comparable effect on the phenotype and that random assortment occurs among gene pairs. Gene combinations and interactions may become complex, and simple conclusions are not always possible.

Complementary Genes

"The complementary genes are two independent pairs of genes which interact to produce a trait in such a way that neither dominant can produce its effect unless the others is present too." Presence of at least one dominant gene from each pair of complementary genes produces one trait whereas the alternative character results from the absence of either dominant gene or of both dominant genes.

Example. Several white-flowered varieties of sweet pea, *Lathyrus odoratus* are known. Mating of most white-flowered plants produce only white-flowerd offspring but it was found that when plants from two particular white-flowered varieties were crossed, all the F_1 plants had purple flowers. When two of these purple flowered F_1 plants were crossed, or when they were self-fertilized, plants of F_2 generation were produced inthe ratio of 9-purple flowerd plants : 7 white-flowered plants. It has been found that two pairs of genes located on different chromosomes are involved; one gene (C) controls some essential step in the production of a raw material, and the other gene (E) controls the formation of an enzyme which converts the raw material into pigment of a purple colour. Plants having homozygous recessive cc are unable to synthesize the raw material, whereas those having homozygous recessive *ee* lack the enzymes to convert the raw material into purple pigment.

	White flowers	X	White flower
P :	CC ee		cc EE
	↓	↓	↓
P gametes :	(Ce)		(cE)
F_1 :	Purple flower	X	Purple flower
	Cc Ee		CcEe

F_1 Male gametes →
F_1 Female gametes ↓

F_2 :

	CE	Ce	cE	ce
CE	CC EE Purple	CC Ee Purple	Cc EE Purple	Cc Ee Purple
Ce	CC Ee Purple	CCee White	Cc Ee Purple	Cc ee White
cE	Cc EE Purple	Cc Ee Purple	cc EE White	cc Ee White
ce	Cc Ee Purple	Cc ee White	cc Ee White	cc ee White

F_2 phenotypic ratio : 9/16 purple : 7/16 white or 9 : 7.

Fig. 16.4. Complementary genes : production of 9 : 7 phenotypic ratio in sweet peas due to complementary genes.

When CCee plants were mated with ccEE, the F_1 plants all were CcEe and had purple flowers because they had both raw material and enzyme for the synthesis of the purple pigment. In F_2 generation, seven plants with genotypes CCee, Ccee, ccEE, ccEe, Ccee. CcEe and ccee were all white.

Epistasis

When one gene interferes with or masks the phenotypic expression of another, the interaction is called epistasis. The complementary genes, cn and u, described for *D. melanogaster* are an example of epistasis, since if either is homozygous mutant the effect of the other is masked. The term epistasis is used in a more general way by some geneticists to describe any kind of gene interaction particularly when dealing with quantitative characters.

Epistasis can give rise to a variety of different kinds of ratios depending on the relationship of the genes involved. That the sex-linked white gene, w^+, is involve in the production of both the brown and red pigments. When its mutant allele, w, is present homozygous, none of the genes involved in the synthesis of these pigments will be expressed. Consequently, no matter how many of the other genes involved in pigment synthesis are segregating, the ratio obtained from a cross ww × wy will produce only white-eyed offspring. The white allele is thus epistatic to the genes involved in red and brown pigment production.

Note that the gene bw^+ (brown) is involved in the synthesis of the red pigments. When the mutant allele, bw, is homozygous and all other eye colour genes are represented by their + alleles, the fly's eyes will be brown in colour, since only the brown pigment will be formed. If a fly carries both u and bw, its eyes will be white and phenotypically indistinguishable from ww flies. The same result will occur with bw bw flies that are also homozygous for any one of the mutant alleles cn, cd, or st.

Numerous examples of these kinds of phenomena are also found in plants. In the snapdragon, *Antirrhinum majus*, the genes N and P interact in the formation of the pigments that give colour to flowers. They each have recessive alleles, n and p. The homozygous NN PP plant has the same colour flowers as the heterozygote Nn Pp, which in our example here is yellow-orange. The results of a dihybrid cross indicate the genotypes, and the phenotypes expected. The phenotypic ratio is 9 yellow-orange : 4 pure white : 3 ivory. This is because when n is homozygous no pigment had ever is formed, and when p is homozygous with N present, a different pigment colour, ivory, is produced.

If it is assumed that these two genes directly control different reactions in the synthesis of the flower pigments, and that their recessive alleles are inactive, then the results can be easily explained. As shown in the figure, N controls the formation of a precursor to all the pigments. When its recessive allele is homozygous, however, precursor 1 is not formed and hence the flower have no pigment and are white. When both N and P are present, all the pigments are present, and this mixture gives the yellow-orange colour. P controls the conversion of precursor 5 to the pigments kaempferol and pelargonidin, and when its allele, p is homozygous, these are not formed. Only apeginin and aureusidin

P :	White Leghorn	X	White Plymouth Rock	
	CC II		cc ii	
	↓	↓	↓	
P gametes :	(CI)		(c i)	
F_1 :	White	X	White	
	Cc Ii		CcEe	

F_1 male gametes → / **F_1 female gametes ↓**	CI	Ci	cI	ci
CI	CC II White	CCii White	Cc II White	Cc Ii White
C i	CC Ii White	CCii Coloured	Cc Ii White	Cc ii Coloured
cI	Cc II White	Cc Ii White	cc II White	cc Ii White
ci	Cc Ii White	Cc ii Coloured	cc Ii White	cc ii White

F_2 : (the checkerboard above)

F_2 phenotypic ratio : 13/16 white : 3/16 coloured breeds or 13 : 3.

Fig. 16.5. A cross between two white coloured breeds of fowls to get 13 : 3 F_2 dihybrid ratio.

are formed, and the flowers are ivory coloured. There are other genes involved, of course, but we have considered only two here so as not to make matters too complicated. N and P interact in the sense that both must be present to form all these pigments. But if the plant is nn, it makes no difference what the allele at the P locus is.

Dominant Epistasis in Dogs

Among dogs, the colours of cats depend upon the action of two genes. One gene locus has dominant epistatic inhibitor allele(I) of coat colour pigment. The allele I prevents the expression of colour allele at another independently assorting, hypostatic gene locus (B or b) and produces white coat colour. The alleles of hypostatic gene locus (BB, Bb, or bb) express only when two recessive alleles (ii) occur on the epistatic locus, i.e, ii BB or ii Bb produces black and ii bb produces brown. When, two such white coat colours appear in 12 : 3 : 1 ratio as shown in the following figure :

		White (Male) Ii Bb	X		White Female Ii Bb
P :					
P Male gametes →		IB	Ib	iB	ib
P Female gametes ↓	IB	II BB White	II Bb White	Ii BB White	Ii Bb White
	Ib	II Bb White	II bb White	Ii Bb White	Ii bb White
F_1 :	iB	Ii BB White	Ii Bb White	ii BB Black	ii Bb Black
	ib	Ii Bb White	Ii bb White	ii Bb Black	ii bb Brown

F_1 phenotypic ratio : 12/16 White : 3/16 Black : 1/16 Brown or 12 : 3 :1.

Fig. 16.6. Checkboard derived from a cross between two heterozygous white coated dogs showing 12 : 3 : 1 ratio due to dominant epistatic inhibitor genes.

Recessive Epistasis in Mice

In mice various types of epistatic genetic interactions have been reported. The most interesting case is of recessive epistasis in coat colours. The common house mouse occurs in a number of coat colours, i.e., agouti, black and albino. The agouti colour pattern is commonly occurring one (wild type) and is characterized by colour banded hairs in which the part nearest the skin is grey, then a yellow band and finally the distal part is either black or brown. The albino mouse lacks totally in pigments and has white hairs and pink eyes.

The agouti coat colour is controlled by a gene A which is hypostatic to recessive allele c. Thus, cc AA, cc Aaa and cc aa genotypes produce albino phenotypes. Further, dominant allele C in the absence of A gives coloured (black mice, i.e., CC aa and Cc aa genotypes give black coat colours. Moreover, in the presence of dominant epistatic allele C, the hypostatic dominant allele A gives rise to agouti coat. Thus CCAA, CcAA, CcAa and CC Aa genotypes produce agouti coat. When black mice (CCaa) are crossed with albino (ccAA), agouti mice (CcAa) appear in F_1. In F_2, 9 agouti, 3 black, and 4 albino individuals are obtained due to recessive epistasis, as shown in the following figure:

P_1 :	Black CCaa	X	Albino ccAA	
P_1 gametes :	(Ca)	↓	(cA)	
F_1 :		Agouti CcAa		

F_1 male gametes → / F_1 female gametes ↓	CA	Ca	cA	ca
CA	CC AA Agouti	CC Aa Agouti	Cc AA Agouti	Cc Aa Agouti
Ca	CC Aa Agouti	CC aa Black	Cc Aa Agouti	Cc aa Black
F_2 : cA	Cc AA Agouti	Cc Aa Agouti	cc AA Albino	cc Aa Albino
ca	Cc Aa Agouti	Cc aa Black	cc Aa Albino	cc aa Albino

F_2 phenotypic ratio : 9/16 Agouti : 3/16 Black : 4/16 Albino.

Fig. 16.7. Recessive epistasis : a cross between black (CC aa) and albino (cc AA) mice, in which dihybrid ratio of 9 : 3 : 3 : 1 becomes modified into 9 : 3 : 4 due to recessive epistatic genes.

Modifying Factors

In cattle both solid colour and spotting are found. The solid colour of cattle is due to a dominant genes S and the spotting due to its recessive alleles. However, the size and distribution of spots is ss individuals depends on a large series of multiple genes. The multiple genes (polygenes) which affect the degree of expression of another gene are called *modifying factors.*

17

MULTIPLE ALLELES

Although much has been learned about the chemistry of genetic material and the mechanism of genetic material and the mechanism of gene action in microorganisms and in higher forms, the basic principle of allelism discovered by Mendel but explained and namely by Bateson remains substantial and valid. For this discussion, alleles will be considered in the traditional sense as alternative genes at the same locus in a chromosome. Examples will be cited from several animals and plants. Human blood-group and serum-protein genes that illustrate various combinations of alleles and nonalleles will then described.

A few definitions will first be reviewed in their historical setting. The word "Gene as used by Johannsen in 1903 was a convenient invention to cover such terms s *unit factor, genetic element,* or *allele* as represented in the gamete. Johannsen also coined the words "genotype" to identify the genes of a particular individual and "phenotype" to denote the trait or expression of the genes. Mendel had suggested a physical element and Boveri and Sutton has observed a parallel between the behaviour of such units and that of chromosomes, but the physical nature of genes was not yet appreciated. Proof of their physical existence was established when Morgan, Sturtevant, Bridges, and others demonstrated that genes were spatial entities arranged in linear order in chromosomes. The term "gene" was then extended to mean a locus, that is, a discrete unit or location in a chromosome; and alleles, which has previously been considered by Bateson only as hypothetical partners involved in Mendelian segregation, were defined as alternative genes at the same locus. The ultimate source of new alleles was recognized as gene change or mutation in an individual organism or somewhere in the ancestry of that organism.

Since alleles are located in corresponding parts of homologous chromosomes, only one member of a pair can be present in a give chromosome and only two are ordinarily present in a cell of a diploid organism. The traditional criteria for identifying mutant genes as members of allelic pairs are summarized as follows: (1) alleles must segregate into separate gametes, and (2) the heterozygote usually expresses the phenotype of the parent or an intermediate between those of the two parents. The first criterion is indicated by the simple monohybrid ratios (1:2:1 or 3:1) that can be expected from crosses between diploid individuals heterozygous for the alternative genes. In haploid organisms such as Neurospora, where each strain carries a single allele, the expected ratio from a cross between strains with different allels is 1:1. Examples given thus far in this book have involved only single pairs of alleles, usually with one member wild-type and the other a mutant type.

Multiple Alleles

Many examples have been found which more than two alternative alleles or multiple alleles are present in a population. In such cases, two or more different mutations must have ocurred at the same locus but in different individuals or at different times. Multiple allels are thus alternative states at the same locus. The different members of a series are conventionally represented by the same basic symbol. Superscripts and subscripts are used to identify different members of a series of alleles.

Most alleles produce variations or gradations of the same character, but some produce very different phenotypes. In Drosophila melanogaster, for example, the gene ss makes the bristles small in contrast to the effect of the wild-type gene ss^{+}. No effect has been observed on the legs or antennae. Another allele in the series, ss^{a} (aristapedia), reduces the bristles

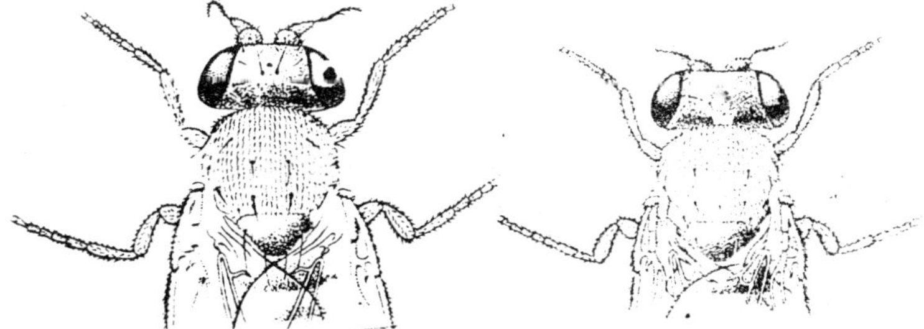

Fig. 17.1. Different phenotypes of Drosophila antennae produced by different alleles.

slightly, but the flies also undergo a more conspicuous phenotypic alteration: legs develop on the head in place of antennae.

Coat Colour in Rabbit

A classical example of multiple alleles was discovered many years ago in rabbits Albino (white) rabbits were known to occur occasionally in wild (bariously coloured) populations. It was shown by the monohybrid results (for example, 3:1 ratio in F_2) from crosses between coloured and white rabbits that members of a pair of alternative gene, c^+ or c, were responsible for the production to either coloured or albino rabbits, respectively. When crosses were made between homozygous coloured (c+c) and albino (cc) rabbits, all the F_1 progeny were coloured and the F_2 were about 3 coloured to 1 albino. This ratio indicated that only one single pair of alleles was involved (one wild type, c+ and one mutant allele, (c) and c^+ was dominant over c.

Other rabbits, called chinchilla had a gray appearance because of mixed black and white pattern. When, crosses were made between fully coloured and chinchilla rabbits, all of the F_1 were about 3 coloured, whereas the F_2 were about 3 coloured to 1 chinchilla. This ratio indicated that these genes were also alleles that c^+ was dominant over the chinchilla gene cch.

Table 17.1. Phenotypes and corresponding genotypes for alleles of c locus in rabbits

Phenotypes	*Genotypes*
Full colour (agouti)	c^+c^+, c^+c^{ch}, c^+c^h, c^+c
Chinchilla	$c^{ch}c^{ch}$, $c^{ch}c^h$, $c^{ch}c$
Himalayan	c^hc^h, c^hc
Albino	cc

Another fur pattern was characterised by a white coat and black tips on the ears, nose, and feet. When crosses were made between himalayan and fullly coloured rabbits, all the F_1 progeny were coloured. In the F_2 the proportion of about 3 coloured to 1 himalayan was recognized. Crosses between chinchilla and himalayan resulted in all chinchilla in the F_1 and about 3 chinchilla to 1 himalayan in the F_2. This result indicated that cch and ch were also alleles and that cch was dominant over ch. Finally, crosses between himalayan and albino produced only himalayan in the F_2 and about 3 himalayan to 1 albino in the F_2. The consistent monohybrid ratios indicated that all for genes were members of the same series of alleles. Gradation in dominance was recognized in the following order : C^+, cch, ch and c. Presumably,

Fig. 17.2. Coat colours in rabbits dependent on members of a series of multiple alleles. A–full colour, B–chinchilla, C–himalayan, and D–albino.

cch, ch, and c originated somewhat in the ancestry as mutations from the wild-type gene (c^+).

Coat Colour in Mice

In another example from rodents, colour pattern as well as viability could be shown to depend on different members of a series of multiple

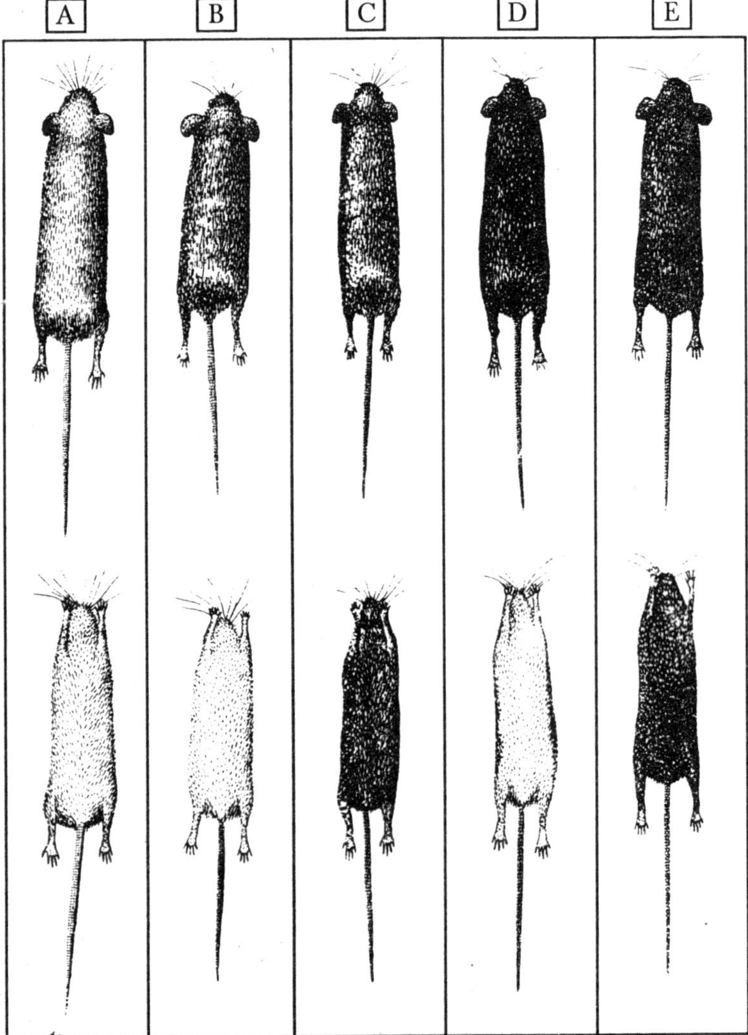

Fig. 17.3. Mouse skins illustrating the expressions of a series of multiple alleles at the A locus. (Upper) Dorsal surface : (lower) ventral surface of the same specimens. A, yellow, B, agouti light belly, C, agouti, D, black and tan, E, black.

alleles. The gene A controlling the agouti pattern in mice (and other mammals) has several alleles. Among the members of the series a gradation exists in dominance, similar to that of the c series in rabbits, but in this example two members, A^y and A^L, are dominant over the wild-type gene A^+. Studies on yellow mice carried out in the early 1900's showed that it was impossible to obtain a pure yellow strain. Yellow mice mated with non-yellows produced half yellow and half nonyellow, whereas yellows mated to yellows produced 2 yellows to 1 nonyellow. The gene A^y for yellow mice thus behaved like the gene C for creeper chickens as a homozygous lethal. In heterozygous condition, the gene A^y removes nearly all the black pigment from the fur but does not affect the colour of the eyes.

Table 17.2. Agouti (A) series of alleles in mice, listed in order of dominance

Phenotype	*Allele*
Yellow	A^y
Agouti light belly	A^L
Agouti	A+
Black and tan	a^t
Nonagouti (black)	a

Wing Size in Drosophila

A series of multiple alleles was discovered to account for an interesting group of wing characteristics of *D. melanogaster.* In this example, a progressive series of wing abnormalities was recognized, ranging in size from no wing at all to a normal wing. Graduations in the amount of wing present were associated with certain gene combinations. Normal wings are dependent on the vg^+ allele. The extreme expression with no wings is the result: of one allele vg^{nw} in homozygous condition. A small stump wing is associated with the allele vg^{st}, notched with vg^{no}, and nicked with vg^{ni}. Each of these phenotypes results from the homozygous arrangement of a particular allele. Heterozygous combinations usually produce intermediate phenotypes, but the wild-type allele (vg^+) is dominant over the other members of the series, except in the presence of certain modifiers.

Self Sterility in Nicothana

Multiple alleles have been associated with self-incompatibility in several groups of plants. As early as 1764 Kolreuter described self-sterility in tobacco, Nicotiana. The reason for the incompatibility was

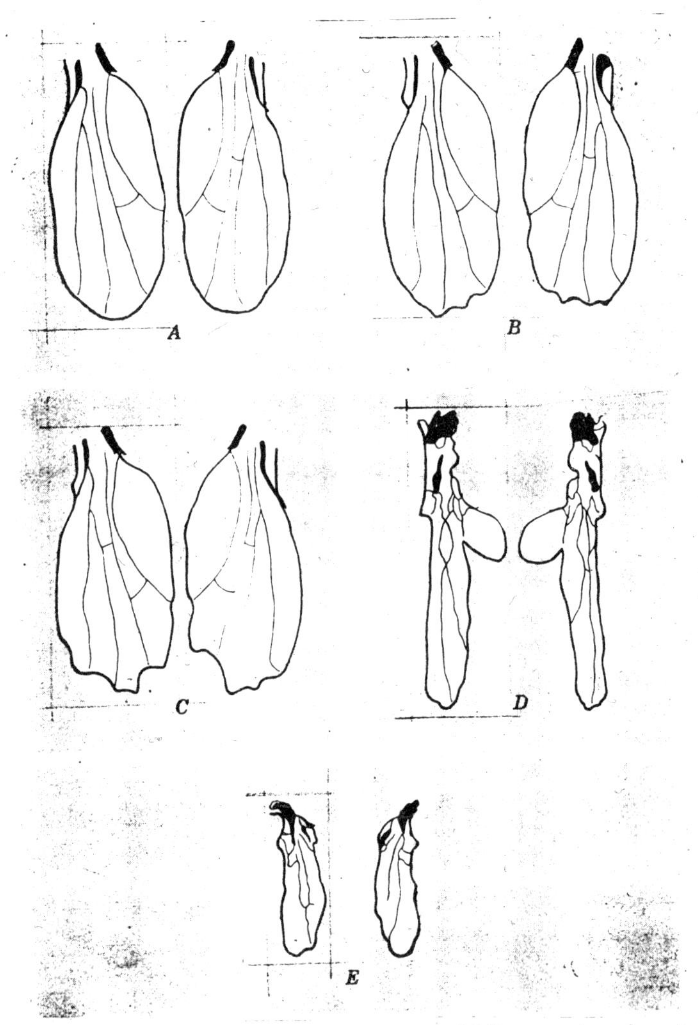

Fig. 17.4. Wings of the fruit fly, illustrating expressions of different members of a series of multiple alleles at the vg locus. A–normal wings (vg^+); B–nicked (vg^{ni}); C–notched (vg^{no}); D–strap (vg^{st}); E–vestigial (vg).

discovered in 1925, when it was shown that pollen grains produced by certain plants failed to germinate, or that the pollen tubes grew too slowly to be effective in fertilization when placed on the stigmas of the same plant or those of certain others plants. Results of appropriate crosses showed that the genes associated with the pollen tube inefficicncy were members of a series of multiple alleles, and the letter S was used as the base symbol for their identificaion. Subscripts were used to identify the different alleles. Thus, one plant was found to have S_1S_2 and another S_3S_4. None of the cross-fertilization tobacco plants was homozygous, for example S_1S_1 or S_2S_2. When crosses were attempted between different S_1S_2 plants, it was observed that the pollen tubes did not develop normally, but pollen from S_1S_2 plants was effective on stigmas of plants with other alleles, for example, S_3S_4.

When crosses were made between seed parents with S_1S_2 and pollen parents with S_2S_4, two kinds of pollen tubes were distinguished. Pollen grains carrying S_2 were not effective, but the pollen grains carrying S_3 were capable of fertilization. Thus, from the cross $S_1S_2 \times S_2S_3$, two kinds of progeny, S_1S_3, were produced. From a cross S_1S_2 and S_2S_3, were produced. From a cross $S_1S_2 \times S_3S_4$, all the pollen was effective and four kinds of progeny resulted: $S_1S_3 \times S_3S_4$, all the pollen was effective and four kinds of progeny resulted: S_1S_3, S_1S_4, S_2S_3, and S_2S_4. Several other S alleles were found, and many combinations occured in different populations of tobacco plants.

In a rare species of the eventing primrose, Oenothera, 37 members of an allelic series of incompatibility genes were detected among 500 plants collected at random. A similar series of incompatibility alleles was found in red clover. In one particular study, 41 were distinguished and symbolized S_1 to S_4. Exhaustive investigations would undoubtedly uncover even larger series of multiple alleles affecting the fertility of plants.

The ABO Blood Groups

The most widespread kind of tissue transplant is an ordinary blood transfusion. Transfusions are not usually thought of as transplants, perhaps partly because they are so common place. But in addition, the immune barrier to transfusions does not involve the major histocompatibiity locus (HLA), since histocompatibility antigens are not found on red blood cells. Other antigens predominate on these cells, and these antigens dcfine the familiar ABO and Rh blood groups as well as dozens of other blood groups that are less widely known. The ABO and Rh blood groups were discussed briefly in next chapter as examples of simple Mendelian inheritance in humans. Here a somewhat more detailed discussion is in order.

The *ABO blood groups* are the most important ones in transfusions. The antigens are determined by one locus on chromosome 9 at which there are three alleles: I^A, I^B, and I^O. (Actually there are several kinds of I^A alleles, but this is a detail we need not cover here.) Individuals who are genetically I^AI^A have on their red blood cells a particular antigen called the A antigen. People who are genetically I^B/I^B have on their red blood cells an antigen called B. Heterozygotes of genotypes I^A/I^B have both the A and B antigens on their red blood cells. The I^O alleles does not produce a specific antigen, and the red blood cells of I^O/I^O individuals carry neither the A nor the B antigen. A person's ABO blood type is determined by which antigens are present on the red blood cell. I^A/I^A and I^A/I^O individuals have *blood type A* because only A antigens are present on their red blood cells: I^B/I^B and I^B/I^O individuals have *blood type B*; I^A/I^B individuals have *blood type AB*; I^O/I^O individuals are said to have *blood type O*, which means the absence of both A and B antigens. the frequencies of the ABO blood types vary from population to population. Among Caucasians (Londoners), the frequencies are O (48%), A (42%), B(8%), and AB (1%). Among Chinese, the frequencies are O (34%), A (31%), B (28%), and AB (7%). As with HLA, certain disease associations involve the ABO blood groups. Examples are stomach cancer (20 percent greater risk in type A that in type O), doudenal ulcers (30 percent greater risk in type O than in other blood groups), and obstructive blood clots (thrombo-embolic disease–60 percent greater risk in types A, B and AB than in type O).

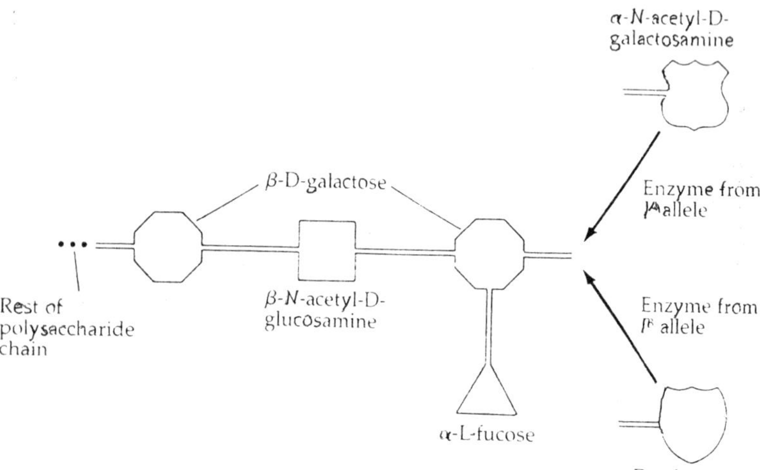

Fig. 17.5. Function of I^A and I^B alleles in producing ABO blood group antigens.

Tabl3 17.3 Important properties of ABO, Rh, and MN blood groups

Blood group and genotype	*Antigens on red blood cells*	*Antibodies present in*	*Antibodies present in saliva and other body fluids?*
ABO blood group:			
I^A/I^A or I^AI^O	A	Anti-B	Yes, if secretor (Se/Se or Se/se)
I^B/I^B or I^B/I^O	B	Anti-A	Yes, if secretor (Se/Se or Se/se)
I^A/I^B	A and B	None	NO
I^O/I^O	None anti-B	Anti-A and	Yes, if secretor (Se/Se or Se/se)
Rh blood group:			
DD	Rh–	None	
Dd	Rh–	None	
dd	None	Anti-Rh+ if exposed to Rh+ antigen	
MN blood group:			
MM	M	None	
NN	M and N	None	
NN	N	None	

From a chemical point of view, the A and B antigens correspond to allele-specific modifications of a complex carbohydrate found on the surface of all red blood cells. The I^A and I^B alleles code for alternative forms of an enzyme that attaches one or another sugar to the end of this molecule; the I^A form of the enzyme attaches one sugar, and the I^B form of the enzyme attaches the alternative sugar. The I^O form of the enzyme causes no extra sugar to be attached to the basic carbohydrate, and the unmodified form is often called the *H substance.* Under certain genetic conditions, the A and B antigens not only are present on red blood cells but are also secreted into the saliva, sweat, semen, and other body fluids. The secretion of these antigens is controlled by another locus–the secretor locus. The secretor locus has two alleles, Se and se, and the allele leading the secretion, Se, is dominant. Se/Se and Se/se genotypes are therefore secretors, whereas se/se genotypes are not. Thus, in secretors, the ABO phenotype of a person can be determined from the body fluids alone. Among

Caucasians, about 77 percent of all individuals are secretors (i.e., Se/Se or Se/se).

Another locus involved in the ABO blood groups should be mentioned. Certian rare individuals are homozygous for a recessive mutation associated with the absence of α-L- fucose from the carbohydrate precursor of the ABO substances. With such a precursor, the I^A and I^B encoded enzymes cannot modify the precursor, so such individuals cannot express the ABO antigens on their red blood cells, whatever their ABO genotype may be; they are said to have the *Bombay phenotype.*

The A and B antigens are highly antigenic; the antibodies against them are very potent. Unlike most antigens, an individual need not have been exposed to A or B antigens on red blood cells to produce antibodies. That is, a person not carrying the A antigen (and therefore of blood type B or O) will have so-called naturally occurring antibodies against A even though he or she was never exposed to red blood cells carrying the A antigen. How this happens is not clear, but it may result from the "A-like" or "B-like" antigenicity exhibited by several classes of microorganisms. Therefore, although a person may never have been exposed to type A blood, repeated exposure to substances that have antigens similar to A will induce the production of anti-A antibodies The same goes for the B antigen. Remember, however, that individuals do not make antibodies against their own antigens because of immunological tolerance. The result of all this is that a person will have circulating in the blood antibodies against the A or B antigens not present on his or her own red blood cells. Therefore individuals with type A red blood cells with possess anti-B antibody; people with type B antbodies; and those with blood type AB will have neither anti-A nor anti-B antibodies. (These naturally occurring antibodies are of the IgM type).

If a blood transfusion is performed using red blood cells against which the recipient has antibodies, the antibodies in the recipient will immediately attack the introduced cells and cause them to clump together (*agglutinate*) in various parts of the body. These clumps block the tiny capillary vessels, and parts of the brain, heart, and other vital organs are deprived of their oxygen supply. The patient goes into shock and may die, so blood to be transfused must not carry any major antigens that the recipient does not also possess. Conversely antibodies in the donor blood may clump the cells of the recipient. Therefore best medical practice requires a two-way crossmatch of donor and recipient. However, one can transfuse limited amounts of blood that carries antibodies against the recipient's antigens; for example, people with type AB blood have A and B antigens but they may receive blood from type O people even though it contains antibodies against both A and B antigens. The reason is that the small volume of transfused antibody is diluted so rapidly in the large volume of the recipient's blood that it does no harm. The main transfusion rule that must be observed, then, is that transfused cells can have no antigen unless the recipient has the same antigen. Therefore, individuals of blood type AB can receive blood of type A, B, AB, or O; type A people can receive A or O blood; type B people can receive B or O blood; and type O individuals can receive only type O blood. (Type AB people were formerly called *universal recipients* because they can receive blood of any type; type O individuals were known as *universal donors* because they can donate blood to anyone.)

The Rh Blood Groups

The one other blood group system that is of significant medical importance is the *Rh system.* The genetic basis of the Rh system is rather complex, and it is not clear whether it involves a single locus on chromosome 1 with many alleles or a tightly linked cluster of loci. For our purpose, we may treat it as a single locus with two alleles, usually called D and d, although the actual situation regarding the number of loci and the number of alleles is still unknown. The D allele produces on the surface of red blood cells an antigen, chemically unrelated to the A and B antigens, called the Rh$^-$ antigen; the d allele does not produce a distinct antigen. Therefore, the red blood cells of both D/D and D/d genotypes carry the Rh$^-$ antigen; they are said to be Rh$^+$ or *Rh positive.* The red blood cells of d/d genotypes do not carry the Rh$^+$ antigen; they are said to be Rh$^-$ or *Rh negative.* Individuals who are Rh$^-$ are capable of producing antibodies against Rh$^-$ individuals will not produce anti-Rh$^+$ antibodies unless they have previously been exposed to Rh$^+$ cells.

The medical significance of Rh system arises in those cases in which an Rh$^-$ mother who is producing anti-Rh$^+$ antibody due to previous exposure to Rh$^+$ antigen becomes pregnant with an Rh$^+$ fetus. (This can occur when the mother is d/d and the father is D/d or D/d.) In such a case, some of the anti-Rh$^+$ antibodies from the mother can traverse the placenta, since these antibodies are of the IgG class. The antibodies that invade the fetus's blood begin to attack the fetus's red blood cells, leading to a serious blood condition called *erythroblastosis fetalis*, sometimes called *hemolytic disease of the newborn*. In the most severe cases, the fetus cannot replace its red cells quickly enough and dies. Less severely affected infants may be born in apparent good health but soon become jaundiced due to toxic by-products from damaged red cells. If the infant lives, brain damage and mental retardation may result. Emergency treatment of the child may require an exchange transfusion– removing virtually all the child's blood and replacing it with Rh$^-$ blood. Gradually the Rh$^-$ red blood cells break down and are replaced with new Rh$^+$ cells made by the child, but the child does not produce anti-Rh$^+$ antibodies and therefore no subsequent problems arise. The time factor in these transfusions is critical because the untransfused child suffers from a shortage of oxygen. In the absence of treatment, three-quarters of the affected infants would die.

Actually, an Rh$^-$ woman will not produce anit-Rh$^+$ antibodies unless she has been exposed to Rh$^+$ antigens. This exposure usually requires one or more pregnancies during which she carries Rh$^+$ embryos. With each pregnancy, some of the Rh$^+$ cells from the fetus seep across the placenta into the mother's blood-steam, especially at the time of birth. The mother produces antibodies against these cells. Because of the need for prior exposure, the first Rh$^+$ child is often not at risk, but it serves to immunize the mother. Often the mother does not produce sufficient antibody in response to the first exposure to cause problems with even the second Rh$^-$ child. But third and subsequent Rh$^+$ children incur a substantial risk of erythroblastosis fetalis. (Only 5 percent of infants with Rh hemolytic disease are first-born, roughly 45 percent are second-born, and about half are the third or later births. At one time the frequency of the condition in the United States was 1 in 150 births, but today's smaller families have reduced the incidence.) In matings of d/d mothers with D/d fathers, half the children will be d/d; these d/d children do not have the Rh$^+$ antigen and they are never at risk, nor do they sensitize the mother.

The incidence of erythroblastosis fetalis in a population depends on the frequency of marriages of Rh$^+$ women with Rh$^+$ men. Among U.S. whites, the frequency of Rh blood is about 15 percent and that of Rh+ blood is 85 percent. Thus the type of mating in question occurs in about 13 percent of white marriages (i.e. $0.15 \times 0.85 = 0.13$). Rh hemolytic disease is virtually nonexistent in Japan, where more than 99 percent of the people are Rh$^+$. Among the Basques, a genetically semi-isolated group that lives in the Pyrenees Mountains between France and Spain, the incidence of Rh$^+$ is exceptionally high–about 43 percent.

(A similar situation exists among Basque descendants who live in norther Nevada.) When Rh^+ women from these semi-isolates marry outside their own cultural group, in the majority of cases they marry Rh+ men and therefore they run the risk of erythroblastosis fetalis among their children.

For Rh^- women who are presently coming into their reproductive years and have never been transfused with Rh^+ blood, the risk of Rh hemolytic disease has all but been eliminated by a preventive treatment. The treament involves injecting anti-Rh^+ antibodies into the Rh^- mother soon after the birth of her first Rh^- child. Any Rh^+ cells from the child that make it into the mother's blood are destroyed by these antibodies before they can immunize the mother. By the time the next pregnancy occurs, the injected antibodies have disappeared. (All proteins, including antibodies, break down or are destroyed in time.) Neither can a second Rh^+ child immunize the mother if, as before, she is injected with anti-Rh^+ antibody. This treatment is very effective provided that an Rh^- woman is not already producing anti-Rh^+ antibody. The injected anti-Rh^+, by destroying any Rh^+ cells that might be present, prevents the mother's immune system from being stimulated to produce its own anti-Rh^+ antibody. Strangely enough, a similar kind of prevention sometimes occurs naturally. If red blood cells from the fetus carry an A or B antigen not present in the mother in addition to the Rh^+ antibodies will destroy the cells before they can stimulate the production of anti-Rh^+ antibody. In this way ABO incompatibility prevents Rh disease.

One may wonder why hemolytic disease of the newborn can come about from Rh incompatibility but not form ABO incompatibility. For example, why is it that hemolytic disease of the newborn does not ordinarily occur in type A or AB children whose mothers produce anti-A antibody?) One part of the answer is that a type of ABO hemolytic disease does occur, but rarely–affecting perhaps 0.1 percent of all newborns. One reason the condition is so rare is that the antibodies responsible are not the same ones involed in ABO transfusion incompatibility. The usual anti-A and anti-B antibodies important in transfusion are of the IgM class, which are unable to cross the placenta. The antibodies responsible for ABO hemolytic disease are of the IgG class, which can traverse the placental barrier. Under ordinary circumstances, this IgE class of ABO antibody is not produced. Nevertheless, there is a smaller than expected frequency of I^A/I^O children arising from matings of I^O/I^O mothers with I^A/I^O fathers.

The amount of the deficiency varies from study to study. (In one Japanese study the deficiency was more than 10 percent of the number of I^A/I^O children expected.) Thus deficiency has been attributed to very early spontaneous abortion brought about by ABO incompatibility between mother and fetus.

Other Blood Groups

Blood groups have application extending beyond genetics and medicine. Besides the ABO system and the Rh system, there are literally dozens of blood group systems. (A *blood group system* refers to the several antigens specified by the different alleles at a single locus.) Other than the ABO and Rh systems, most of these antigens are not recognized as being very antigenic by human lymphocytes; they therefore do not usually play a significant role in blood transfusions or pregnancy. They are, however, antigenic in other animals such as rabbits. Red blood cells from one person can be injected into an animal and at some later time antibodies can be isolated. These antibodies are then tested against the red blood cells from other people, and in some instances there will be an antigen-antibody reaction (the red blood cells will clump) but not in other instances. The people whose blood cells are recognized by the antibody and clump must have the same antigen as the person whose cells were originally used to immunize the animal, but those whose red cells are not recognized by the antibody must lack this antigen. Studies of this kind are then combined with family studies to determine the mode of inheritance of the antigenic difference.

One blood group that was discovered in almost the way just described is the MN systm. The MN blood group antigens are due to two alleles at a single locus on the long arm of chromosome number 2. (The true genetic situtaion is known to be slightly more complicated but is not important here). By injecting red blood cells of the appropriate kind into rabbits, anti-M antibody and anti -N antibodies can be obtained. Anti-M antibody reacts only with cells carrying the M antigen; anti-N antibody reacts only with cells carrying the N antigen. The alleles that produce the M and N antigens are also called M and N. Three genotypes at the MN locus can be found. Some people are genetically M/M (their red blood cells react only with anti-M), others are N/N (their red blood cells react only with anti-N), and still others are M/N (their red blood cells react with both anti-M and anti-N.) In the MN system, the antigens carried on the red blood cells can be used to identify the genotype of an individual at this locus. Because of this fact, the Mendelian law of genetic segregation can be observed directly. Mating of MM × MM produce only MM children; MN × MM matings produce one-fourth MM, one-half MN, and one-fourth NN offspring. One can easily work out the offspring distribution of matings involving MM × NN, MN × NN, and NN × NN. It should be realized that these numbers are averages, that they summarize what is found in a large number of matings between the given genotypes. In any one family, because the number of children is small, chance fluctuations will often occur, and these ratios will not always be obeyed exactly.

Applications of Blood Groups

The dozens of different blood group system used collectively can identify a person almost as uniquely as fingerprints. Although many people have the same ABO blood type in common, or the same Rh blood type in common, or the same MN blood type in common, when these blood groups are considered separately, many fewer people have the same ABO, Rh, and MN blood types simultaneously. This is simply to say that there are fewer peoples with the precise blood type O; Rh^-; MN than there are people with blood type O with no regard to their Rh and MN types. Now if one considers, say, 30 blood groups instead of only three, then the odds would be exceedingly high that any two people would have different genotype with respect to one or more of the 30 gropus. Every human being is genetically unique, different from all who have come before and all who will follow; the only exceptions are identical twins, triplets, and quadruplets.

The property of blood groups to characterize individuals and groups of people has been used widely by anthropologists to study the origin and migrations of various human populations. In these kinds of studies, blood groups alleles that are common in one population but less common or rare in others are very important. For example, the B type antigen is virtually absent in American Indians (with a few exceptions, such as the Blackfeet). Another example: Among the many alleles of the Rh locus (the previous discussion divided these alleles into only two classed, D and d), one allele is highly prevalent in American Negroes but rare elsewhere; a different allele is frequent in European Caucasians but rare elsewhere. Provided sufficient supporting information can be gained from, for example, similarities in language, alleles of this sort can sometimes be used as markers of ancestry. In a similar vein, comparisons of humans and other primate species have been pursued using blood group similarities to measure genetic relationship.

The many blood group systems have found legal application in such litigations as paternity suits. The principle behind these applications is that a person falsely accused of fathering a child can sometimes be exonerated on the grounds that, based on genetic evidence, he could not possibly have fathered the child. For example, a man of blood type AB cannot have a type O child, except in the extremely unlikely event of mutation. As another example, a man of blood type O;M cannot have been the father of an A;MN child whose mother is AB:M, because although the ABO system works out satisfactorily (the child could be genotypically I^A/I^O), having received the I^O gene from the father), the MN system does not work out. Such a child would have to receive the gene for the N antigen from its father because the mother does not have it. But the accused father does not have this gene either, and therefore the real father must be somebody else. The same kind of reasoning applies to matching infants to their natural mothers in the unlikely event of a mixup in the hospital nursery. An O; Rh^+; M woman could not, for instance, be the mother of an A; Rh^-; N child.

The ABO antigens are important in certain kinds of police work. These antigens are extremely stable and persistent, so minute quantities of dried blood (or body fluids, in the case of secretor) can be typed. Indeed, the ABO blood types of some Egyptian mummies can even be determined! A minute amount of an assilant's blood under a victim's fingernail can sometimes be decisive in excluding innocent suspects; knowledge that the assailant's blood type is aB will exclude 99 percent of innocent white suspects in London, for example.

It seems appropriate to end this section with an account of one of the earliest criminal cases involving the use of blood typing for evidence:

It was Japanese who took the lead. In 1928 K. Fujiwara, Director of the Institute of Forensic Medicine in Nigata, reported a crime which had been solved by means of blood group determination from seminal stains. A sixteen-year-old gird, Yoshiko Hirai, who went from village to village selling fortune-telling slips, had been found strangled. Witnesses had seen the girl at 7:30 P.M. on her way from the village to the railroad station. She must have been raped and killed shortly afterward Fujiwara found whitish spots which contained well-preserved semen on the girl's body. Her blood group was O. Fujiwara tested the semen spots by the absorption method, adding anti-A and anti-B serums, and hus determined that the semen had the group characteristic A. Meanwhile the police had arrested two suspects. One of them was a twenty-four-yearold mentally retarded beggar named Mochitsura Tagami, who straightway confessed that he had committed the murder. The second suspect, Iba Hoshi, denied all guilt. The police were on the point of accepting Tagami's confession and releasing Hoshi when Fujiwara asked for time to determine the blood groups of the prisoners. The results proved that Tagami, who had made the confession, could not possibly be the criminal; his blood group was O. Hoshi, on the other hand, had group A. Confronted with the data, Hoshi confessed; that Fujiwara knew his blood and his semen had something in common impressed (him as) witchcraft.

18

EXTRANUCLEAR INHERITANCE

Other than the occurrence of non-Mendelian patterns of inheritance, no criterion is universally applicable to distinguish extranuclear from nuclear inheritance. In higher organisms, transmission of a trait through only one parent. Uniparental inheritance, usually indicates extranuclear inheritance. Genetic transmission of cytoplasmic factors, such as mitochondria or chloroplasts, is determined by maternal and paternal contributions at the time of fertilization, by mechanisms of elimination from the zygote, and by the irregular sorting of the elements in cell division. Uniparental transmission through the mother constitutes maternal inheritance; uniparental transmission through the father is paternal inheritance.

Organelle inheritance is higher plants is often uniparental, but the particular type of uniparental inheritance depends on the organelle and the organism. For example, among angiosperms mitochondria typically show maternal inheritance, but chloroplasts may be inherited maternally (the predominant mode), paternally, or from both parents, depending on the species. In confirm, chloroplast DNA is usually inherited paternally, but the chloroplast DNA in a few offspring results from maternal transmission. The redwood Sequoia semipervirens shows paternal transmission of both mitochondrial and chloroplast DNA. Whatever the pattern of cytoplasmic transmission, most species have a few exceptional progeny, indicating that the predominant mode of transmission is usually not absolute.

In higher animals, mitochondria are typically inherited through the mother because the egg is the major contributor of cytoplasm to the zygote. One genetic consequence of maternal inheritance is that reciprocal crosses produce different phenotypes of progeny; that is, the progeny from a mutant mother and normal father are mutant, whereas progeny from a normal mother and a mutant father are normal.

A maternal pattern of inheritance usually indicates extranuclear inheritance, but not always. The difficulty is in distinguishing between maternal inheritance and maternal effects. A maternal effect occurs when the genotype of the mother influences the phenotype of the progeny, either through substances present in the egg that affect early development of through effects of nature–for example, the intrauterine environment of female mammals or their ability to produce milk. The distinction is highlighted in the following:

1. *Maternal inheritance* occurs when the hereditary determinants of a trait are extranuclear and genetic transmission is only through the maternal cytoplasm.
2. *Maternal effect* occurs when the nuclear genotype of the mother determines the phenotype of progeny. The hereditary determinants are nuclear genes transmitted by both sexes, and in suitable crosses the trait undergoes Mendelian segregation.

Extranuclear Inheritance Patterns

The nature of extranuclear inheritance can be demonstrated by the pattern of transmission from parents to progeny and by one or more additional criteria. Although the criteria vary from one system to another, the basic extra-nuclear pattern can be demonstrated. A number of useful experimental tests or observations can be applied to identify extranuclear patterns: we will briefly describe five of these.

1. Reciprocal Crosses Yield Different Progenies

When wild-type and presumptive extra-nuclear mutants are crossed reciprocally, progeny of each cross may resemble only the female parent. For example, in *Neurospora crassa*, the progeny of *poky* ♀ × wild type ♂ are *poky*, but wild-type ♀ × *poky* yields wild-type progeny. In this example of *uniparental inheritance,* only maternal zygotes are produced. In other species genes may be transmitted from both parents to produce biparental zygotes or, rarely, from only the male parent to produce paternal zygotes. Whatever the nature of the progeny, reciprocal crosses give different results. By and large these differences are seen as different proportions of maternal, biparental, and parental zygotes in different progenies due to different transmission frequencies of alleles from the two parents.

One reason for transmission mainly through the female parent is that the female gamete supplies all or most of the cytoplasm to the zygote, whereas the male often contributes only a nucleus. This situation appears to be case in Neurospora and in many plants and animals that produce a large egg and a relatively small sperm. The relative contributions by parents is probably not the explanation for uniparental inheritance in other species, such as the isogamous alga Chlamydomonas reinhardi. We have not yet identified all of the mechanisms responsible for uniparental inheritance or for variable proportions of different kinds of zygotes in a single progeny.

In the budding yeast Saccharomyces cerevisiae, extranuclear traits may be transmitted through either the *a*- or α-mating type parent or from both. Biparental zygotes are produced in relatively high

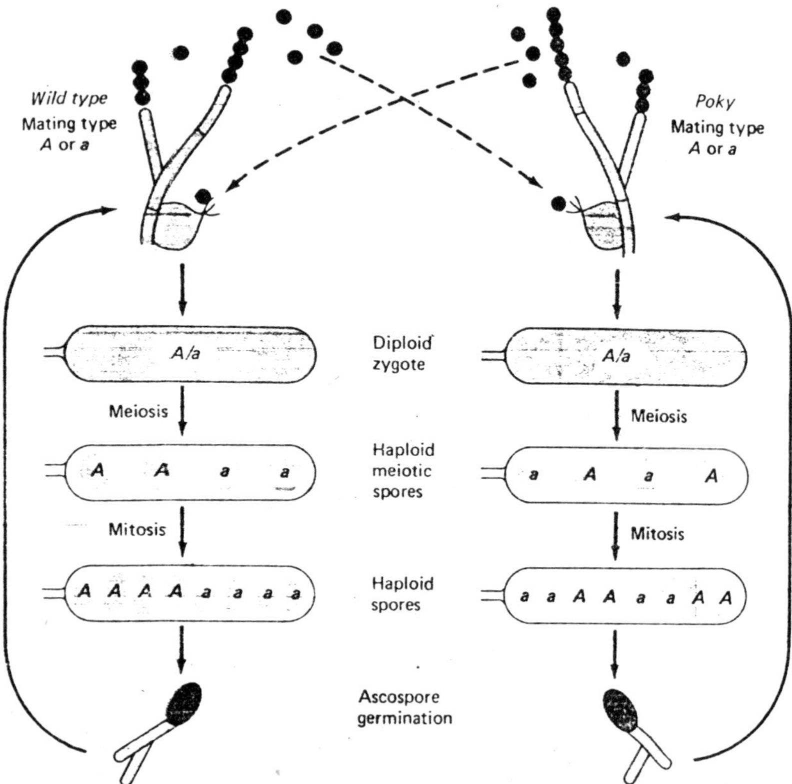

Fig. 18.1. Reciprocal crosses usually yield different progenies in relation to extranuclear traits, but nuclear-encoded traits show the expected Mendelian pattern of segregation in the same or in different crosses. The mating type alleles A and a segregate 4:4, but all the ascospores are either wild type or poky, depending on the female parent or strain of Neurospora crassa.

frequencies from parents that differ in alleles of structural genes. Uniparental zygotes are usually produced in crosses between wild-type respiration-sufficient, grande (rho-plus, or ρ^+) strains and rho-minus (ρ^-) petite strains which carry deletions that are responsible for the extranuclear trait of respiration deficiency. Some petites give no petite zygotes in crosses with grande; these are called *neutral petites.* Most petites will produce some percentage of petite zygotes in crosses with grande, and up to 99% of the progeny may be petite in cases involving *suppressive petite* strains. We will discuss these mutants more fully later in the chapter.

The pattern of transmission through only one parent is fairly typical, but it has no consistency with regard to the particular parent that transmits; transmission may be always or usually through the same mating type or sex or through either mating type or sex. Because of this variability, no rule of inheritance or universal prediction will serve for all patterns of extranuclear transmission. Each case must be analyzed by its own features.

2. Progeny show a Non-Mendelian Segregation Ratio in a Tetrad of Meiotic Products

In each of the three species we have mentioned, the zygote undergoes meiosis to produce a tetrad of spores. When the four products of a single meiotic cell are examined, all for cells often have the same phenotype, that is, the phenotypic ratio of 4 : 0 (all like one parent) or 0 : 4 (all like the other parent). Pairs of alleles for nuclear-gene markers, however, segregate 2 : 2 in these same tetrads. These results indicate at least two features of the system: (1) the nucleus behaves normally and meiosis is not aberrant since known nuclear pairs of alleles segregate in the expected 2 : 2 ratio and (2) extranuclear alternatives may not segregate at meiosis. From these results we can infer that extranuclear factors are not located on chromosomes in the nucleus; otherwise they would behave in the same way as pairs of marker alleles, which we know are located on chromosomes. We cannot, however, infer where the extranuclear factors are located until more experimental evidence is obtained.

3. Extranuclear Factors cannot be Mapped on any Chromosome in the Nuclear Genome

Conforming evidence can be obtained to verify that extranuclear factors are not situated in the nuclear genome, by tests showing that these factors are not linked to any known genes in any chromosome in the nucleus. Extranuclear factors assort independently of nuclear genes in every chromosome, and they therefore cannot be a part of any nuclear linkage group. Extra-nuclear genes will, however, show linkage to each other within an extranuclear genetic system.

4. Extranuclear Factors are Transmitted through the Cytoplasm and not through the Nucleus

The heterokaryon test can provide evidence for a cytoplasmic rather than a nuclear location of extranuclear factors in species that can maintain genetically different marker nuclei in a common cytoplasm. When haploid, multi-nucleate mycelia of *Neurospora* wild-type and *poky* strains fuse together, the heterokaryon that forms will have two genetically different kinds of haploid nuclei which remain separate in the mixed cytoplasm. When spores develop asexually from different hyphae of the mycelium, a single haploid nucleus is enclosed along with some cytoplasm. When these spores are incubated and allowed to develop, the mycelium of these sexual progeny will be either wild type or *poky*. Either one of the two kinds of haploid nucleus may be present as determined by gene markers. This shows that the source of the nucleus has no influence on poky pheno-typic development; either nucleus can be found in poky progeny. Since the nucleus does not control the poky phenotype, the cytoplasm is the only other possible source of the inheritance factor transmitted from the heterokaryon to its asexual spore progeny.

5. Extranuclear Genes in General, and Organelle Genes in Particular, usually Segregate Rapidly during Mitotic Divisions

Such *vegetative segregation* can be seen quite clearly in variegated plants, in which green, white and mixed cell lineages occur in variable amounts in different plants of a single progeny.

By taking these separate lines of evidence into consideration, we see that extranuclear factors apparently are not located on chromosomes in the nucleus. They probably are situated in the cytoplasm, which has been demonstrated to influence inheritance. The exact cytoplasmic location can be determined by other tests, which we will discuss.

Extranuclear Genes

Are extranuclear factors equivalent to genes that are located on chromosomes? If these factors are genes that happen to be located somewhere in the cytoplasm, we would predict that they would have genetic characteristics: (1) genes should have the capacity to mutate to alternative allelic forms; (2) different alleles of the same gene should segregate, and allelic types should breed true in the segregant populations; and (3) such genes should be physically identified with nucleotide sequences of DNA or RNA, the only known genetic molecules.

Although extranuclear mutants had been described according to their non-Mendelian patterns since 1909, two major problems had prevented their further analysis for more than 40 years. First, only an occasional, single kind of extranuclear mutant had been described in various species. Second, and most importantly, only one alternative was transmitted by one of the parents to the progeny in each cross; alternatives did not segregate at meiosis. There was no way to determine if these extranuclear alternatives were unit factors or not. The situation can be compared to the problem Mendel would have faced if he had crossed only tall plants and short plants and the progenies were either

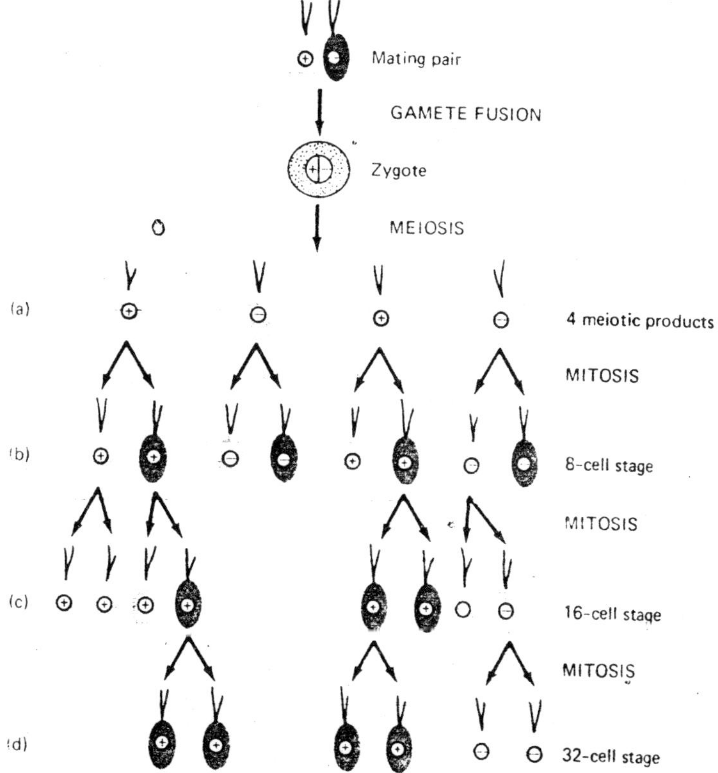

Fig. 18.2. Analysis of biparental zygotes in Chlamydomonas revealed that segregation of extranuclear factors took place in postmeiotic mitotic divisions.

all tall or all short in every case. The principles of genic inheritance probably could not have been deduced from such inheritance patterns.

Ruth Sager began the first systematic analysis of extranuclear inheritance in the early 1950s, using *Chlamydomonas reinhardi.* She discovered that streptomycin acted as a mutagen, producing extranuclear mutations that affected resistance, sensitivity, and dependence on streptomycin itself in the growth medium. She also isolated extranuclear mutations influencing cell response to other antibiotics, as well as mutations affecting photosynthesis. By fluctuation test analysis streptomycin was shown to be a mutagen in this system.

Sager was able to conduct genetic analysis of these mutants when she found that some biparental zygotes appeared among the vast majority of uniparental zygotes, that is, some of the zygotes had received extranuclear factors through both parents. When these biparental zygotes underwent meiosis, extranuclear alternatives did not segregate. But segregations did take place during mitotic divisions of the haploid meiotic products. Segregant progeny types were shown to be true breeding. These results showed that extranuclear factors existed as pairs of allelic alternatives, which *segregated postmeiotically* during mitotic divisions of the haploid progeny. Extranuclear factors, therefore, behaved like genes, and they could be studied by available genetic methods.

But where were these extranuclear genes located? Sager proposed that the extranuclear genes she studied were located in the chloroplast of *Chlamydomonas* because some of the mutations affected chloroplast characteristics. The evidence in support of this inference was not particularly convincing, and many years elapsed before certain of these extranuclear genes were generally acknowledged to be located in the chloroplast.

The development of extranuclear genetics was also aided substantially by studies of the petite extranuclear mutant in yeast, primarily by Boris Ephrussi and Piotr Slonimski in Paris. They found that the acridine dye *proflavin* acted as a specific mutagen, and up to 99% of a grande respiration-sufficient strain could mutate to become petite. The petite mutant was respiration deficient since it lacked some of the cytochrome enzymes required for aerobic respiration. This condition is not lethal in yeast because the organism can gain enough energy through glycolysis to sustain growth and reproduction. It grows more slowly, however, and produces petite (small) colonies on plates that also contain grande (large) wild-type colonies that can metabolize sugars through aerobic respiration, a far more efficient process than glycolysis.

Petites were recognizable by colony growth, enzyme analysis, respiration tests, and by electron microscopy. Mitochondria in the mutant cells were generally abnormal in appearance because the inner mitochondrial membrane was not organized into the typical invaginated cristae conformations. Although many differences could be recorded to identify petites and their inheritance was clearly extranuclear, no other mutants were available and genetic tests for recombination could not be conducted as they had been in Chlamydomonas. When it was discovered in the early 1960s that mitochondria and chloroplasts contained DNA, Slonimsky analyzed mitochondrial DNA (mtDNA) in grande and petite yeast. He showed that petite mtDNAs and grande mtDNAs were very different in base composition and therefore in sedimentation characteristics in CsCl. This study in 1966 provided the first evidence that related an alternations extranuclear DNA with mutation of a non-Mendelian inherited characteristic. The alternation in DNA paralleled the alternations in inheritance and in phenotype for petite mutants. This marked the beginning of studies of the physical basis for extranuclear inheritance.

Variegation in Leaves of Higher Plants

One of the first convincing examples of extranuclear inheritance was found in higher plants. In 1909, Carl Correns reported some surprising results from this studies on four-o'clock plants (Mirabilis jalapa). The blotchy leaves of these variegated plants showed patches of green and white tissue, but some branches carried only green leaves and others carried only white leaves.

Flowers appear on all types of branches. They may be intercrossed in a variety of different combinations by transferring pollen from one flower to another. Table shows the results of such crosses. Two features of these results are surprising. First, there is a difference between reciprocal crosses: for example, white ♀ × green ♂ gives a different result from green ♀ × white ♂. Recall that Mendel never observed differences between reciprocal crosses. In fact, in conventional genetics, differences between reciprocal crosses are normally encountered only in case of sex-linked genes. However, the results of the four-o'clock crosses cannot be explained by sex linkage.

The second surprising feature is that the phenotype of the maternal parent is solely responsible for determining the phenotype of all progeny. The phenotype of the male parent appears to be irrelevant,

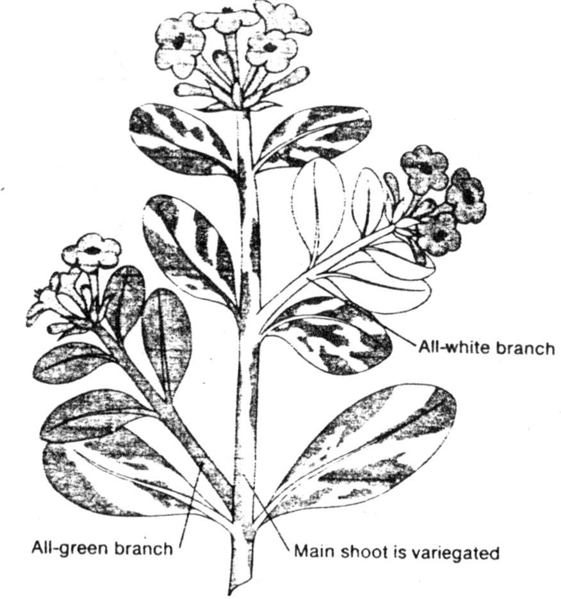

Fig. 18.3. Leaf variegation in Mirabilis jalapa, the four-o' clock plant. Flowers may form on any branch (variegated, green, or white), and these flowers may be used in crosses.

and its contribution to the progeny appears to be zero! This phenomenon is known as maternal inheritance. Although the white progeny plants do not live for long because they lack chlorophyll, the other progeny types do survive and can be used in further generations of crosses. In these subsequent generations, maternal inheritance always appears in the same patterns as those observed in the original crosses.

How can such curious results be explained? The differences in leaf colour were known to be due to the presence of either green or colourless chloroplasts. The inheritance patterns might be explained if these cytoplasmic organelles are somehow genetically autonomous and furthermore are never transmitted via the pollen parent. For an organelle to be genetically autonomous, it must have its own genetic determinants that are responsible for its phenotype. In this case, the chloroplasts would carry their own genetic determinants responsible for chloroplast colour. Thus, the suggestion arises that this organelle has its own genome. The failure to be transmitted via the pollen parent is reasonable because the bulk of the cytoplasm of the zygote is known to come from the maternal parent via the cytoplasm of the egg. Figure diagrams a model that formally accounts for all the inheritance pattern in Table 18.1.

Table 18.1. Results of crosses of variegated four-o'-clock plants

Phenotype of branch bearing egg parent(♀)	*Phenotype of branch bearing pollen parent (♂)*	*phenotype of progeny*
White	White	White
White	Green	White
White	Variegated	White
Green	White	Green
Green	Green	Green
Green	Variegated	Green
Variegated	White	Variegated, green, or white
Variegated	Green	Variegated, green, or white
Variegated	Variegated	Variegated, green or white.

Variegated branches apparently produce three kinds of eggs: some contain only white chloroplasts, some contain only green chloroplasts, and some contain both kinds of chloroplasts. The egg type containing both green and white chloroplasts produces a zygote that also contains both kinds of chloroplasts. In the subsequent mitotic divisions, some

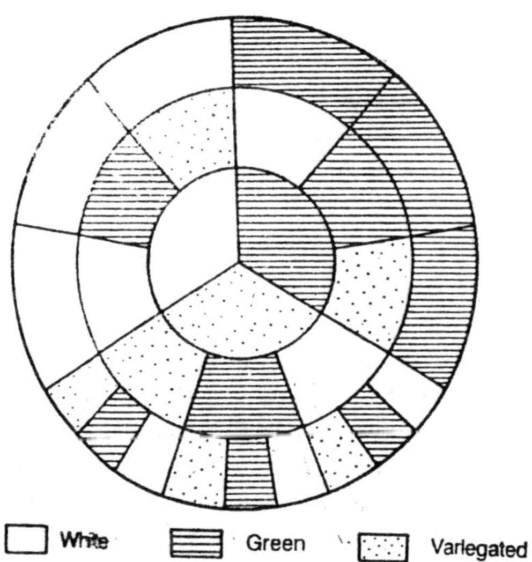

Fig. 18.4. Plastid inheritance of Mirabilis jalapa. The central circle represents the type of branch that produces flowers which are pollinated. Intermediate circle represents branch from which pollen is used and outer circle shows the progeny.

form of cytoplasmic segregation occurs that segregates the chloroplast types into pure cell lines, thus producing the variegated phenotype in that progeny individual.

This process of sorting might be described as "mitotic segregation." However, this is an *extranuclear* phenomenon, not to be confused with mitotic segregation of nuclear genes, so a new term has been invented. This term–representing a hypothetical "black box" in which the segregation and (as we shall see) recombination of organelle genotypes occur–is cytoplasmic segregation and recombination, referred to in this book by the convenient acronym CSAR. Throughout this chapter, CSAR crops up quite often; it appears to be a common behaviour of extranuclear genomes. In one sense. CSAR is the cytoplasmic equivalent of meiosis, because meiosis is the process whereby segregation and recombination of nuclear genes regularly occur. However, CSAR is purely hypothetical–no physical counterpart is known for this process.

We might suspect that random chloroplast assortment could explain the green/white segregation in variegated plants, but this hypothesis has not been proved. Furthermore, the larger numbers of chloroplasts in cells make this an unlikely possibility. In a cell with, let's say, 40 chloroplasts, consisting of 20 green and 20 colourless chloroplasts, the production by random assortment of one daughter cell containing only green and one daughter cell containing only colourless chloroplasts would be at best a very rare event. To be on the safe side, we shall treat the CSAR process as hypothetical.

Iojap in Maize

A phenotype similar to *Mirabilis* but with a different pattern of inheritance has been analyzed in maize by Marcus M. Rhoades. In this case, green, colourless, or green-and-colourless striped leaves are under the influence not only of the cytoplasm but, in addition, of a nuclear gene located on the seventh linkage group. This locus is called iojap, after the recessive mutation ij which is located there. The wild-type allele is designated Ij. Plants homozygous for the mutation ij/ij may have green-and-white striped leaves. However, when reciprocal crosses are made between plants with striped leaves and green leaves, the results are seen to vary, depending upon which parent is mutant. If the female is striped (ij/ij) and the male is green (Ij/Ij), plants with colourless, striped, and green leaves are observed as progeny. If the male parent is striped and the female green, only green plants are produced! In

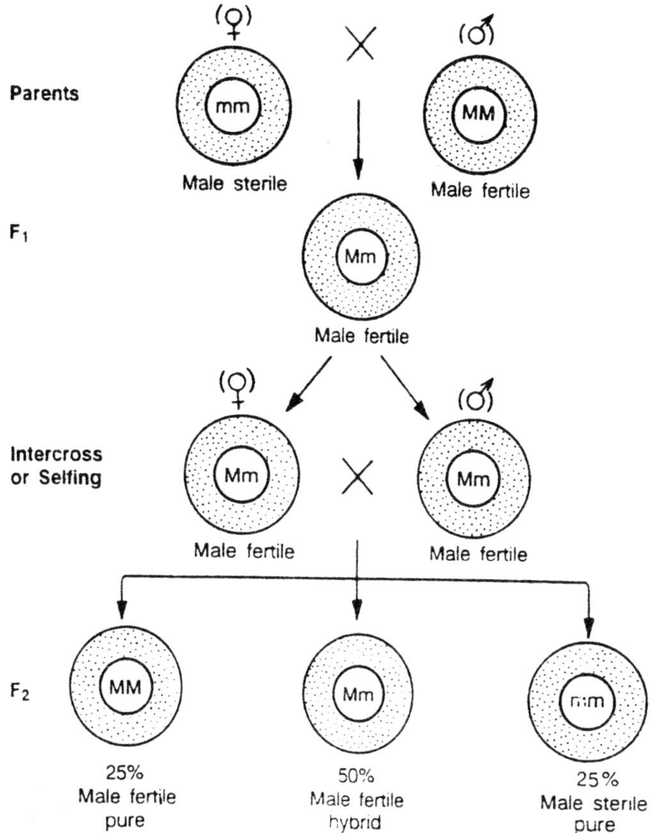

Fig. 18.5. Inheritance pattern of genetic male sterility.

both types of cross, all offspring have identical genotypes (Ij/ij). We may conclude that although a nuclear gene is somehow involved, the inheritance pattern is influenced maternally.

We can understand this pattern better by examining the offspring resulting from self-fertilization of these heterozygous plants. The striped plant gives rise to progeny with colourless, striped, and green leaves, regardless of their genotype. The green plants give rise to only green offspring. The results of these self-fertilizations substantiate that the mutant phenotype is controlled solely through the female cytoplasm, regardless of the plant's genotype.

The role of the nucleus in the origin of the mutant phenotype is unclear. However, the colourless areas of the leaf are due to the lack of the green chlorophyll pigment in chloroplast. Once acquired, colourless chloroplasts are transmitted through the egg cytoplasm, establishing the phenotypes of leaves of progeny plants.

Nonchromosomal Genes in Chlamydomonas

Very good evidence for the storage of genetic information in the chloroplast has been provided by studies of Sager with the unicellular green alga Chlamydomonas. The life cycle of this species is very simple and is comparable to that of yeast. Again two mating types, + and –, are recognized. Mating between cells of opposite mating type produces a zygote. When the zygote matures, meiosis occurs, and four zoospores (the motile products of the meiotic division) arise. Half of these are mating type + and half are type –. The zoospores may undergo mitotic division to form clones of cells. Again, the mating type is seen to depend on a pair of nuclear alleles inherited in typical Mendelian fashion.

Many characteristics in Chlamydomonas are known to depend on nuclear genes: spore colour and requirements for certain metabolities,

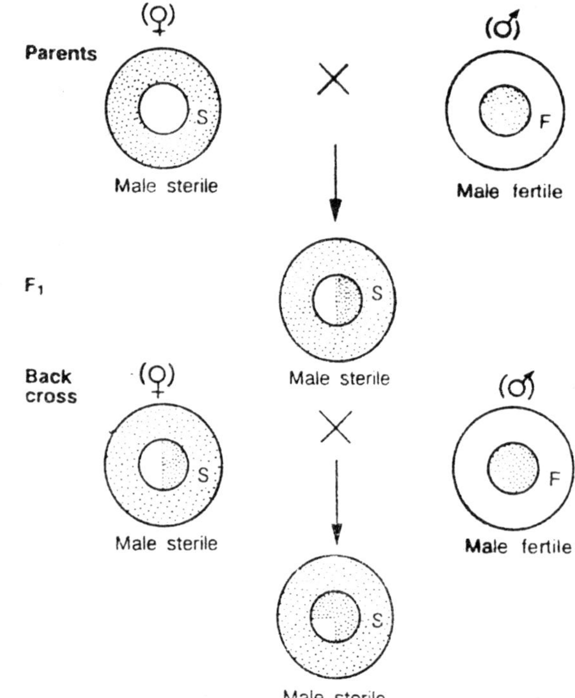

Fig. 18.6. Maternal inheritance of cytoplasmic male sterility.

for example both linkage and independent assortment have been demonstrated for these genes associated with Mendelian traits. There are, however, some genetic determinants in Chlamydomonas that are definitely nonchromosomal. Sager has worked extensively with certain streptomycin-resistant strains that exhibit non-Mendelian inheritance. As Figure shows, the results of a cross between streptomycin-resistant and streptomycin-sensitive strains depends on the phenotype of the mating type + parent.

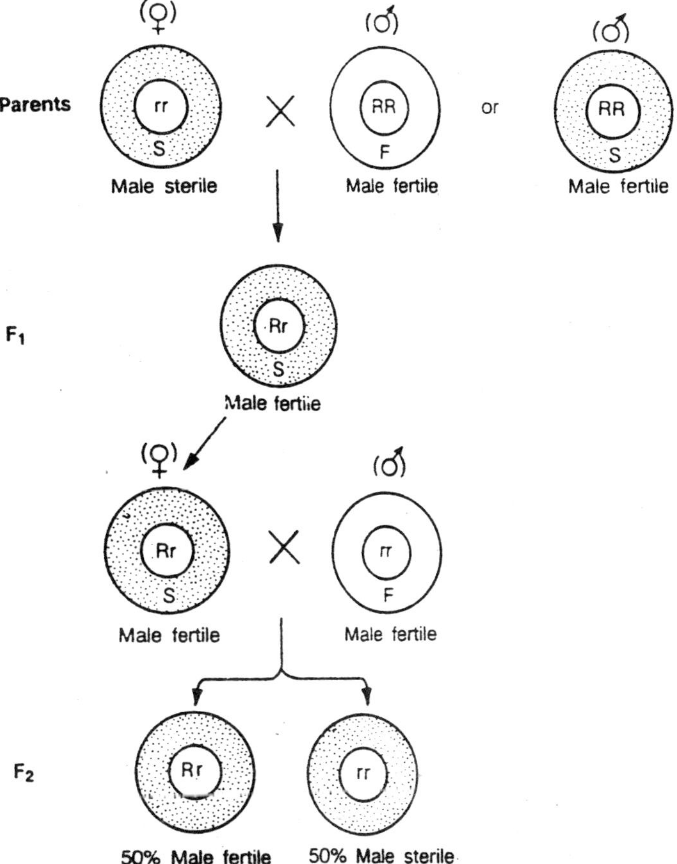

Fig. 18.7. Inheritance of cytoplasmic male sterility and effect of restorer gene in making male steriles into male fertiles.

Although mating type itself segregates among the zoospores in the Mendelian 2:2 fashion, the zoospores always have the streptomycin trait shown by the mating type + parent. This suggests maternal inheritance, but here, the + and the – cells are identical. Neither class of gamete contributes more cytoplasm and hence more cytoplasmic genes than the other. Something is operating to prevent the transmission of the nonchromosomal genes of the mating type – parent from the zygote to the meiotic products. Since the discovery of the non-chromosomal streptomycin determinants, many others have been found. Some of these were induced by mutagens, particularly by streptomycin itself, which in Chalmydomonas is able to induce mutations in nonchromosomal genes but not in those of the nucleus.

It was later found that rare exceptions occur to this mating type + pattern of inheritance. In fewer than 1% of the zygotes, exceptions arise in which both sets of nonchromosomal genes, the set from the mating type – as well as the one from the mating type + parent, are transmitted. Sager also found that if the mating type + parent is treated with a dose of ultra-violet light before mating, 50% of the zygotes exhibit hiparental inheritance. All of the offspring arising from such a zygote contain a complete set of non-chromosomal genes from each parent. Such cells are called cytotech, indicating that they are heterozygous for cytoplasmic genetic determinants.

It is possible to follow more than one pair of cytoplasmic markers in biparental inheritance. The cytoplasmic markers do not segregate when meiosis takes place. The haploid zoospores are therefore heterozygotes for the cytoplasmic genes. When a heterozygous zoospore divides mitotically and forms a clone, the alternative forms of the cytoplasmic genes segregate and undergo recombination. This segregation and recombination continues with each additional mitotic division of the zoospores until there are no more heterozygous markers to detect. The recombination frequency of these non-Mendelian genes has been measured. This has permitted the construction of a map and the assignment of several nonchromosomal genes to definite positions. It appears that the cytoplasmic genes are all part of a single circular linkage group.

Sager has presented several lines of evidence that support the hypothesis that the physical basis of the nonchromosomal genes (and hence the linkage group to which they belong) is the DNA of the chloroplast. For example, most of the cytoplasmic determinants that have been mapped were caused to mutate by streptomycin a drug known to exert its influence almost exclusively on the development of the chloroplast, with no direct effect on other parts of the cell. Other work with Chlamydomonas continues to provide greater insight into the overall significance of nonchromosomal genes, as well as into the mechanism of uniparental inheritance.

Poky Neurosopra

In 1952, Mary Mitchel isolated a mutant strain of Neurospora that she called poky. This mutant differs from the wild type in a number of ways: it is slow growing, it shows maternal inheritance, and it has abnormal amounts to cytochromes. Cytochromes are mitochondrial electron-transport proteins necessary for the proper oxidation of foodstuffs to generate ATP energy. There are three main types: cytochromes *a*, *b*, and *c*. In poky, there is no cytochrome *a* or *b*, and there is an excess of cytochrome *c*.

How can maternal inheritance be expressed in a haploid organism? It is possible to cross some fungi in such a way that one parent contributes the bulk of cytoplasm to the progeny, and this cytoplasm-contributing parent is called the female parent, even though no true sex is involved. Maternal inheritance for the poky phenotype was demonstrated in the following crosses:

poky ♀ × wild-type ♂ → all poky
wild-type ♀ × poky ♂ → all wild-type

However, in such crosses, any nuclear genes that differ between the parental strains are observed to segregate in the normal Mendelian manner and to produce 1 : 1 ratios in the progeny. All poky progeny behave like the original poky strain, transmitting the poky phenotype down through many generations when crossed as females. Because poky does not behave like a nuclear mutation, it was termed an extranuclear mutation. To hammer the point home, it also has been called a cytoplasmic mutation because it seems to be carried in the cytoplasm of the female parent.

But where in the cytoplasms the mutation carried? The important clue is that several aspects of the mutant phenotype seem to involve mitochondria. For instance, the cells' slow growth suggests lack of ATP energy, which is normally produced by mitochondria. There are abnormal amounts of cytochromes in the mutants, and cytochromes are known to be located in the mitochondrial membranes. These clues led to the conclusion that mitochondria are involved and inspired several interesting experiments designed to investigate mitochondrial autonomy.

David Luck labeled mitochondria with radioactive choline, a membrane component, and he then followed their division

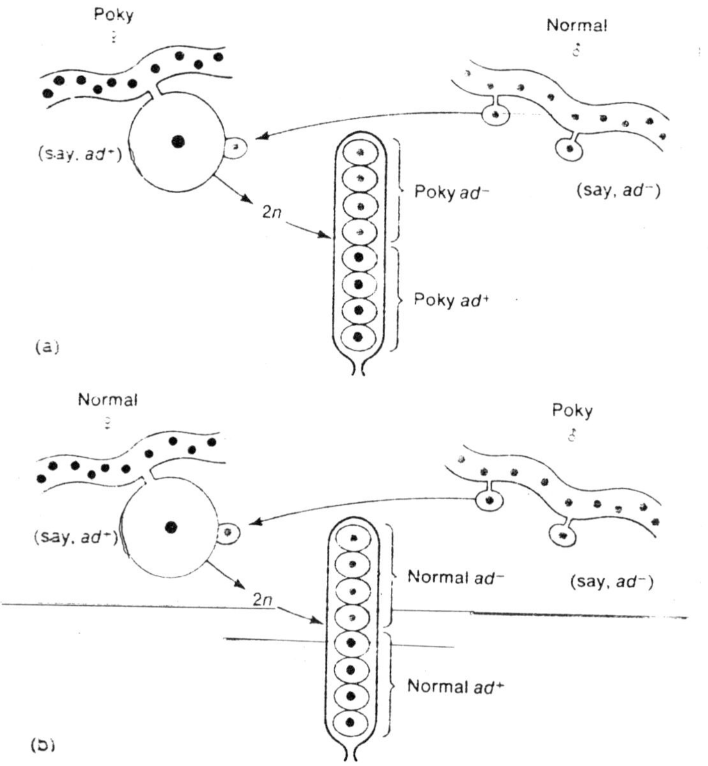

Fig. 18.8. Explanation of the different results from reciprocal crosses of poky and normal Neurospora. Most of the cytoplasm of the progeny cells comes from the parent called female.

autoradiographically in unlabeled medium. He found that even after several doublings of mass, the mitochondria were still evenly labeled. Luck concluded that mitochondria can grow and divide in a fashion that seems to be autonomous rather than nucleus direct. If mitochondria were synthesized anew, or de novo, some of the resulting population of mitochondria should be unlabeled, whereas the original ones would remain heavily labeled.

In 1965, a research group led by Edward Tatum extracted purified mitochondria from a mutant similar to poky, a mutant called abnormal. With an ultrafine needle and syringe, the experimenters injected the mitochondria into wild-type recipient cells, using appropriate controls. The recipient cells were cultured. After several of these subcultures, the abnormal phenotype appeared! In transferring the mitochondria, the experimenters had transferred the hereditary determinants of the abnormal phenotype. Presumably, then, the extranuclear genes involved in this phenotype are located in the mitochondria. These inferences gained credibility with the discovery of DNA in the mitochondria of Neurospora and other species.

The Heterokaryon Test

Maternal inheritance is one criterion for recognizing organelle-based inheritance. Another test has been applied in filamentous fungi such as Neurospora and Aspergillus. In principle, this test could be applied to other systems involving heterokaryons. A heterokaryon is made between the prospective extranuclear mutant (say, a slow-growth mutant) and a strain carrying a known nuclear mutation. If the phenotype of the nuclear mutation can be recovered from the heterokaryon in combination with the phenotype of the slow-growth mutant being tested, then the growth mutant is a good candidate for an organelle-based mutation. This is because no diploidy occurs in a heterokaryon, so that there is no genetic exchange between nuclei. The slow-growth phenotype must have been transferred solely by cytoplasmic contact. The segregation of the pure extranuclear type from the mixed cytoplasm of the heterokaryon presumably involves the CSAR process.

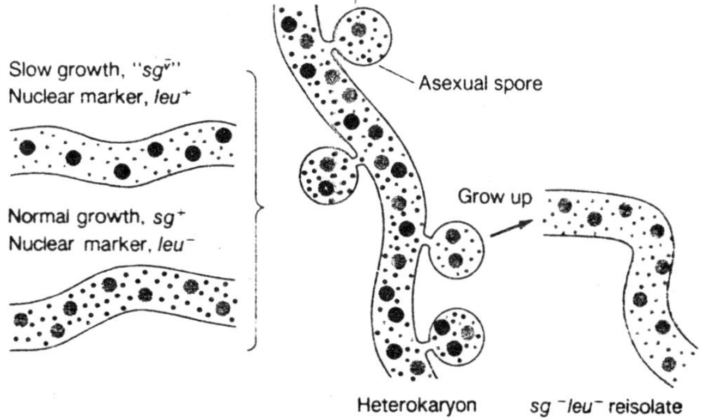

Fig. 18.9. The heterokaryon test is used to detect extranuclear inheritance in filamentous (threadlike) fungi.

Shell Coiling in Limnaea

In the gastropods of phylum Mollusca, the shell is spirally coiled. In majority of cases the direction of coiling of the shell is clockwise, if viewed from apex of the shell. This type of coiling is called *dextral* or right coiling. However, in some snails the coiling of shell may be counter clockwise or *sinistral* or left coiling. Both types of coilings are produced by two different types of genetically controlled cleavages, one being dextral cleavage, another being sinistral cleavage.

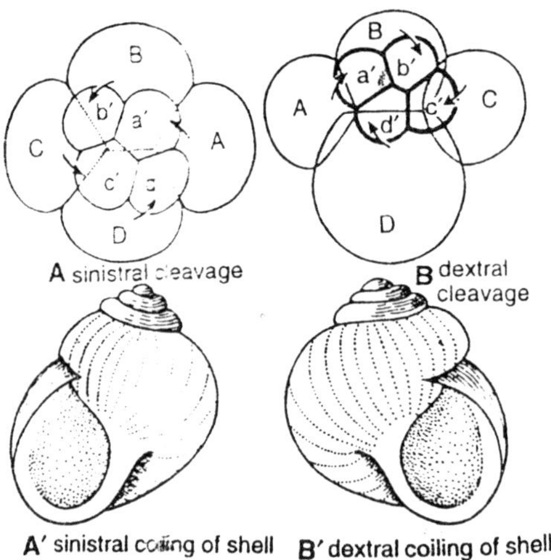

Fig. 18.10. Correlation of sinistral (A) and dextral (B) cleavage with sinistral (A') and dextral (B') coiling of the shell in gastropods.

There are some species of gastropods in which all the individuals are *sinistral* but the main interest attaches to a species in which sinistral individuals occur as mutants among a population of normal *dextral* animals. Boycott (1930) discovered such a mutant in the fresh-water snail--*Limnaea peregra.* Breeding and cross breeding of dextral and sinistral snails showed that the difference between the two forms is dependent on a pair of allelmorphic genes, the gene for sinistraility being recessive (S), and the gene for the normal dextral coiling being dominant (S^+). The two genes are inherited according to Mendelian laws, but the action of any genetic combination is manifested only in the next generation after the one in which a given genotype is found.

When a cross between a homozygous sinistral individual (SS) and a dextral individual (S^+S^+) is made, the eggs cleave sinistrally and all the snails of this F_1 generation show a sinistral coiling of the shell. Thus, the genes of sperm do not manifest themselves, although the genotype of the F_1 generation is S^+S. If a second generation (F_2) is bred from such F_1 sinistral individuals, it is all dextral, instead of showing

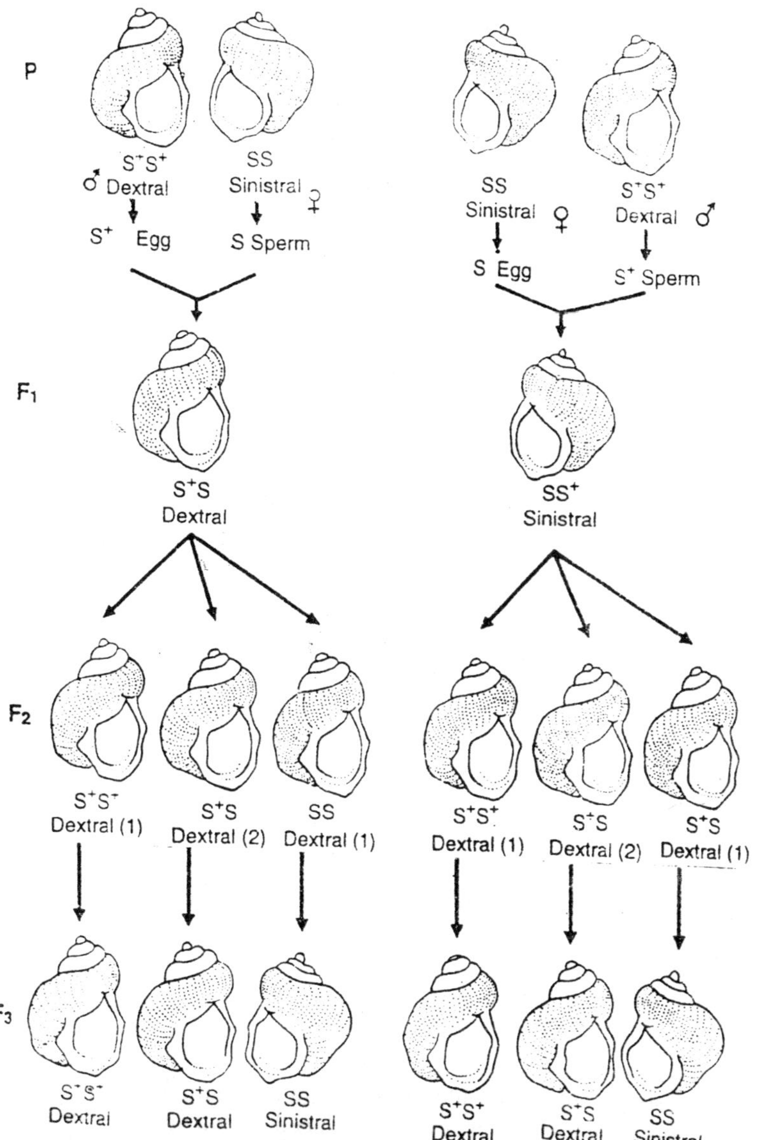

Fig. 18.11. Maternal effect in direction of coiling of the shell in Limnaea.

segregation as would be expected in normal Mendelian inheritance. In fact, segregation does take place in the F_2 generation so far as the genes are concerned, but the new genic combinations fail to manifest themselves, since the coiling is determined by the genotype of the mother. The genotype of F_1 mother being S^+S, the gene for dextrality dominates and is responsible for the exclusively dextral coiling of the second generation. Only in the F_3 generation does segregating in the ratio of 3 : 1 become apparent. Since the individuals of the F_2 generation had the genotypes –1S^+S^+, 2S^+S, 1SS, three quarters of them, on the average, produce eggs developing into sinistral individuals. Thus the results of a reciprocal cross- that is, of the fertilization of the eggs of a homozygous dextral individual (S^+S^+) by the sperm of a sinistral individual (S^+S)–will lead to a somewhat different type of pedigree: the F_1 generation will be dextral (with genotype S^+S) and the F_2 generation again all dextral (with genotypic ratio of 1S^+S^+ : 2S^+S : 1SS). The F_3 generation will show segregation among broods, just as in the cross examined first.

Now it becomes clear if it is realized that the type of cleavage (sinistral or dextral) depends on the organization of the egg which is established before the maturation divisions of the oocyte nucleus. The type of cleavage is, therefore, under the influence of the genotype of the material parent. The sperm enters the egg after this organization is already established. Lastly, the direction of coiling of shell depends upon the orientation of the mitotic spindle of first cleavage of the zygote. If the spindle is *tipped toward the left* of the median line of the egg cell, the sinistral pattern will develop, conversely if the mitotic spindle is *tipped toward the right* of the median line of the cell, the dextral pattern will develop. The spindle orientation is thus controlled by the organization of ooplasm which becomes established during oogenesis.

Maternal effect in Ambystoma

In the axolotl larva of *Ambystoma mexicanum* a lethal gene "O" has been discovered which has a maternal effect. The gene 'O' is recessive, so that heterozygote individuals (with genotype+O) are completely normal. Homozygote animals (with genotype OO) produced by a cross between two heterozygote parents (+O) develop normally in the early stages but in later life show a slight retardation of growth. Their regeneration capacity is severely reduced so that amputated legs are regenerated as would occur in normal axolotls. The homozygote (OO) males are sterile–their testes are poorly developed and spermatogenesis does not go beyond the spermatogonial stages. The homozygote (OO) females on the other hand, produce eggs capable of fertilization and normal cleavage, but at the onset of gastrulation the development is retarded and embryos normally die. This abnormal course of development is in no way affected by the genotype of the spermatozoan: the spermatozoan may carry the normal + gene or a mutant ethal O gene (if the male is heterozygote +O). The arrest in development is in both cases exactly the same : this shows that the particular type of abnormal early development does not depend on the gene present in the cells of the developing embryo (+O), but exclusively on the OO genes of the mother (maternal effect). Briggs and Justus (1967) found that eggs of a OO female lack a protein-like substance the "corrective factor" which is necessary for normal development and is synthesized by normal genes ++ of homozygous or heterozygous normal females (++ or +O) during oogenesis.

Male Sterility in Plants

Another example of cytoplasmic inheritance is associated with pollen failure. This occurs in many flowering plants and results in male sterility. In maize, wheat, sugar beets, onions, and some other crop plants, fertility is controlled, at least in part by cytoplasmic factors. Male sterility is also controlled in some plants by nuclear genes. Critical observations and tests must be made to determine the mechanism of inheritance in individual cases.

A classical example of maternal inheritance of male sterility was discovered by M. M. Rhoades in maize. A particular male sterile variety produced only male sterile progeny when fertilized with pollen from normal maize plants. The male sterile seed parent plants were then backcrossed repeatedly with pollen fertile lines until all chromosomes from the male sterile line had been substituted for those of the male fertile line. Male sterility persisted, indicating that inheritance was maternal and was not controlled by chromosomal genes. As the investigation progressed, a small amount of pollen was obtained from the male sterile line making reciprocal crosses possible. Inheritance was maternal regardless of the direction in which the cross was made. Male sterility, in this case, was attributed to cytoplasmic plasmagenes transmitted by female gametes.

The cytoplasmic effect is not entirely independent, however, because specific nuclear genes are now known to suppress maternally inherited male sterility in maize. A single dominant chromosomal gene will restore pollen fertility in the presence of cytoplasm which ordinarily would result in sterility. In one experiment, pollen abortion occurred only when a specific kind of cytoplasm was present along with a dominant gene for male sterility and the homozygous recessive allele was present at a suppressor locus.

Male sterility is of practical importance in making crosses on a large scale. Hybrid plants have certain advantages over inbred plants. They are produced commercially in maize, cucumbers, onions, wheat, and other plants for the purpose of obtaining hybrid vigour. In plant breeding programs involving plants capable of self-fertilization, the tedious process of removal of pollen (emasculation) can be avoided if it is possible to obtain a male sterile line for use in crosses.

Streptomycin Resistance in Chlamydomonas

In 1954, when Ruth Sager placed green algae (Chlamydomonas) cells on a culture medium containing the antibiotic streptomycin, most of the cells were killed but about one per million survived and multiplied, each to form a streptomycin-resistant colony. In other words, a mutant form of this alga with resistance to streptomycin was being selected from the predominantly streptomycin-suspectable cells. Subsequently, many mutations were identified in Chlamydomonas, but most of these were shown by appropriate test crosses to be changes in nuclear genes. Such mutations were merely being demonstrated by associations with the antibiotic. The approximately 10 percent that were maternally inherited and nonchromosomal, however, were found to be induced by sublethal concentrations of the drug.

Eventually, nonchromosomal mutants could be recovered from almost every colony. Streptomycin was not specific for a particular change but induced several kinds of alternations in the treated colonies. Nonchromosomal changes gave the same phenotypes as chromosomal changes, but their frequency of occurrence was much greater than the frequency of chromosomal gene changes.

Reciprocal crosses demonstrated that antibiotic resistance, when controlled by nonchromosomal factors, was maternally inherited. Mating types in this sexual unicellular algae are controlled by chromosomal genes, which were identified by the investigators as mt^{+} and mt^{-} or simply + and – (instead of female and male). All progeny from each mating were like the + mating type with respect to relative streptomycin resistance. When the + mating type was resistant, all progeny were resistant and when the + mating type was nonresistant, all progeny were nonresistant. These results of reciprocal crosses demonstrated maternal inheritance and a single contrasting pair of traits. Other nonchromosomal genes, *sr* for streptomycin resistance and *sd* for streptomycin resistant and *sd* for streptomycin dependence, were postulated to control these two alternative characteristics. The *sr* gene behaves as a dominant. Cells containing both *sr and sd* genes could survive either with or without streptomycin. Occasionally, the Chlamydomonas non-chromosomal genes do not show strict maternal inheritance but are transmitted from both parents to the progeny. When this occurs, subsequent segregation permits the investigator to follow the pattern of inheritance. Under these conditions, the genes *sr* and *sr* seemed to segregate as would members of a pair of chromosomal alleles. The two genes present in the same parent cell segregated to different daughter cells.

Another pair of nonchromosomal genes, ac_1 and ac_2, was induced in the same strain of Chlamydomonas. These mutations blocked photosynthetic activity. The mutants required acetate in the medium for growth. These genes were also found to behave as a pair of alleles. Segregation began in the first few cell divisions after meiosis, indicating that the number of copies of each gene is very low, perhaps only one.

With two pairs of nonchromosomal genes available in the same system, it was possible to check for evidence of linkage as constrasted with that for independent combinations. Cross of the dihybrid type $ac_1sd \times ac_2sr$ were allowed to grow for a few vegetative multiplications. Each cell was then classified for its segregating markers, both nonchromosomal and chromosomal (that is, mating type and others known to be chromosomal). both the ac_1/ac_2 and the *sr/sd* pairs of genes were observed to segregate early but not always in the same division. After four or five mitotic doublings, equal numbers of parental (ac_1sd and ac_2sr) combinations had been obtained. The results indicated that the two pairs of nonchromosomal genes were carried on different particles. This was a demonstration of recombination of non-chromosomal genes.

Extranuclear DNA has been detected in cellular organelles such as chloroplasts and mitochondria of Chlamydomonas and other organisms. These organelles thus could carry primary genetic information. But even so, the presence of DNA in bodies identified with genetic functions does not preclude the possibility that cells may also contain additional genetic systems in the cytoplasm which may or may not depend on DNA.

Petite in Saccharomyces

Another extensive study of mitochondrial mutations has been performed with the yeast Saccharomyces cerevisiae. The first such mutation, petite, was described by Boris Ephrussi and his coworkers in 1956. The mutant was so named because of the small size of the yeast colonies. Many independent petite mutations have since been discovered and studied. They all have a common characteristic: deficiency in cellular respiration involving abnormal electron transport. Fortunately, this organism is a facultative anaerobe and can grow by fermenting glucose through glycolysis. Thus, although colonies are small, the organism may survive the loss of mitochondrial function by generating energy anaerobically.

The complex genetics of *petite* mutations is diagrammed in Figure. A small proportion of these mutants exhibit Mendelian inheritance and are called *segregational petites*, indicating that they are the result of nuclear mutations. The remainder demonstrate cytoplasmic transmission, producing one of two effects in matings. The *neutral petites*, when crossed to wild type, produce ascospores which give rise only to wild-type or normal colonies. The same pattern continues if progeny of this cross are back-crossed to neutral petites. These results are due to the fact that the majority of neutrals lack mtDNA completely or have lost a substantial portion of it. Thus, the wild-type gamete is the effective source of normal mitochondria which are capable of reproduction. Neutral petites have now been shown to lack most, if not all, mitochondrial DNA (mtDNA).

A third type, the *suppressive petites*, behaves similarly to *poky* in *Neurospora*. Crosses between mutant and wild type give rise to mutant diploid zygotes which, upon undergoing meiosis, immediately yield all mutant cells. Under these conditions, the *petite* mutation behaves "dominantly" and seems to *suppressive petites* also represent deletions of mtDNA, but not nearly to the extent of the *neutral petites*. Buoyant density studies show a lower G-C content in suppressive mtDNA compared with normal mtDNA.

Suppressiveness remains unexplained. Two major hypothesis have been advanced. One explanation suggests that the mutant (or deleted) mtDNA replicates more rapidly, and thus mutant mitochondria "take over" or dominate the phenotype by numbers alone. The second explanation suggests that recombination occurs between the mutant and wild-type mtDNA, introducing errors into or disrupting the normal mtDNA. It is not yet clear which, if either, of these explanations is correct.

Molecular Genetics of Mitochondria

Extensive information is now available about the molecular aspects of mitochondrial gene function. This information generally parallels what is known about chloroplasts, including the similarities to prokaryotes.

It is known that mtDNA is a naked duplex, circular in most all organisms studied, and somewhat smaller than chloroplast DNA. Molecular weights vary from 10×10^6 in animals to 70×10^6 in higher plants. There appear to be no gene repetitions, and replication is dependent upon nuclearly derived enzymes. Genes have been identified which code for the ribosomal RNAs, over 20 tRNAs, ribosomal proteins, and numerous products essential to the cellular respiratory functions of the organelle.

As with chloroplast, the protein-synthesizing apparatus as well as the molecular components for cellular respiration are jointly derived from nuclear and mitochondrial DNA. Ribosomes found in the organelle are different from those present in the cytoplasm. In contrast to chloroplasts, the sedimentation coefficient of these particles is highly variable in different organisms. The data in Table show that mitochondrial ribosomes can be either lower in their coefficients than in prokaryotes (55S to 60S in vertebrates), the same general coefficient (about 70S in some algae and fungi), or as high as eukaryotic cytoplasmic ribosomes (80S in certain protozoans and fungi).

Table 18.2. Sedimentation coefficients of mitochondrial ribosomes.

Kingdom	*Example*	*Sedimentation Coefficient (S)*
Animals	Vertebrates	55–60
	Insects	60–71
Protists	*Euglena*	71
	Tetrahymena	80
Fungi	*Neurospora*	73–80
	Saccharomyces	72–80
Plants	Maize	77

Many nuclear-coded gene products are essential to biological activity within mitochondria: DNA and RNA polymerases, initiation and elongation factors essential for translation, ribosomal proteins, aminoacyl tRNA synthetases, and some tRNA species. As in chloroplasts, these imported components are generally regarded as distinct from their cytoplasmic counterparts, even though both sets are coded by nuclear genes. For example, the synthetases essential to charging tRNA molecules (a process essential to translation) show a distinct affinity for the mitochondrial tRNA species as compared with the cytoplasmic tRNAs. Similar affinity has been shown for the initiation and elongation factors. Furthermore, while bacterial and nuclear RNA polymerases are known to be composed of numerous subunits, the mitochondrial variety consists of only one polypeptide chain. This polymerase is generally susceptible to bacterial antibiotics specific to RNA synthesis but not to eukaryotic inhibitors.

There is a surprising diversity between mtDNA derived from mammalian (including human) and yeast mitochondria. The mammalian DNA is compact, is repeated in a simple orderly process, and is free of intergene (noncoding) regions. Yeast mtDNA, on the other band, demonstrates multiple origins of replication and extensive intergenic DNA sequences. Human mtDNA is free of introns, while yeast mtDNA contains such intragenic sequences. Introns are nucleotide sequences within genes that are not reflected in the protein following translation.

That genes have been exchanged between mitochondrial and nuclear DNA during evolution is established by the case of ATPase subunit 9 gene. In yeast this gene is found in the nucleus, and in Neurospora it is present in the mitochondrion. It has also been observed that the genetic code used in mitochondria contains several minor variations compared with that used in the cytoplasmic translation of nuclearly derived mRNA.

Because of the distinctive genetic systems of mitochondria and chloroplasts, there has been much speculation on the evolutionary origin of these organelles. Many of the organellar molecular components are similar to those found in prokaryotes, so it has been proposed that both organelles have been derived from free-living cells. According to the endosymbiont theory, chloroplasts have descended from free-living blue-green algae, and mitochondria were once aerobic bacteria. This hypothesis suggests that these prokaryotes independently invaded primitive eukaryotic cell long ago in evoltuionary history. Symbiotic existences were established which subsequently led to permanent associations. Through evolution, eukaryotic cells became more specialized and eventually dependent on the photosynthetic and respiratory functions. A great deal of evidence supports this exciting prospect.

Infectious Heredity

There are numerous examples of cytoplasmically transmitted phenotypes in eukaryotes that are due to an invading microorganism or particle. The foreign invader coexists in a symbiotic relationship, is usually passed through the maternal ooplasm to progeny cells or organisms, and confers a specific phenotype which may be studied. We shall consider several examples which illustrate this phenomenon.

Kappa in *Paramecium*

First described by Tracy Sonneborn, certain strains of *Paramecium aurelia* are called Killers because they release a cytoplasmic substance called paramecin which is toxic and sometimes lethal to sensitive strains. This substance is produced by particles called kappa which replicate in the Killer cytoplasm, contain DNA and protein, and depend for their maintenance on a dominant nu-multiplication, they produce the toxic products that are released and kill sensitive strains.

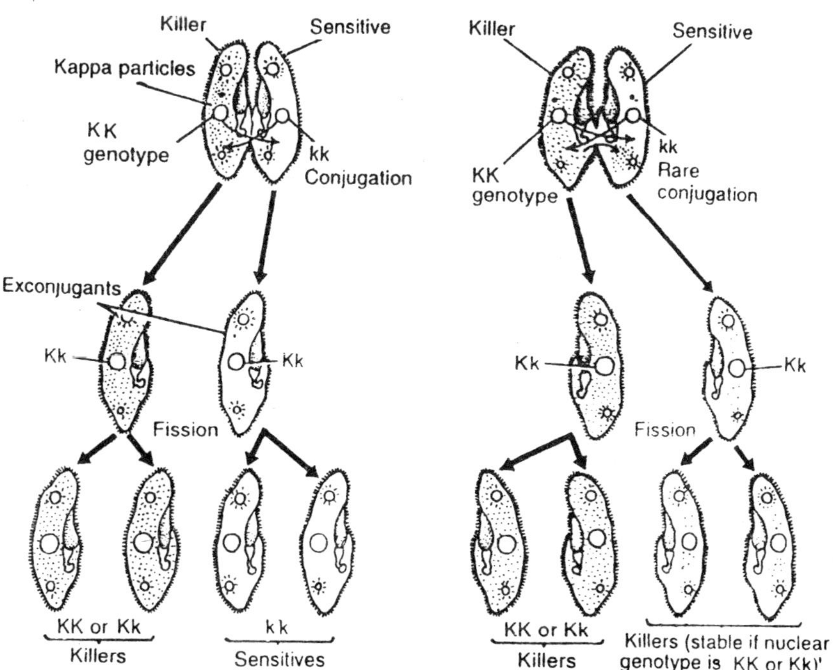

Fig. 18.12. Conjugation in Paramecium and the killer trait.

Infective particles in *Drosophila*

Two examples of similar phenomena are known in *Drosophila:* CO_2 sensitivity and sex-ratio. In the former, flies that would normally recover from carbon dioxide anesthetization instead become permanently paralyzed and are killed by CO_2. Sensitive mothers pass this trait to all offspring. Furthermore, extracts of sensitive flies induce the trait when injected into resistant flies. Phillip L'Heritier has postulated that sensitivity is due to the presence of a virus, sigma. The particle has been visualized and is smaller than kappa. Attempts to transfer the virus to other insects have been unsuccessful, demonstrating that specific nuclear genes support the presence of sigma in *Drosophila.*

A second example of infective particles comes from the study of *Drosophila bifaciata.* A small number of these flies were found to produce predominantly female offspring if reared at 21°C or lower. This condition, designated sex-ratio, was shown to be transmitted to daughters but not to the low percentage of males produced. This phenomenon was subsequently investigated in *Drosophila willistoni.* In these flies, the injection of ooplasm from sex-ratio females into normal females induced the condition. This observation suggested that an extrachromosomal element is responsible for the sex-ratio phenotype. The agent has now been isolated and shown to be a protozoan. While the protozoan has been found in both males and females, it is lethal primarily to developing male larvae. There is now some evidence that a virus harbored by the protozoan may be responsible for producing a male-lethal toxin.

Plasmids, Insertion Sequences, and Transposons

It has already been described that an extrachromosomal hereditary unit in bacteria is–the F factor. When existing gautonomously in the cytoplasm, the F factor takes the form of a double-stranded closed circle of DNA and is self-replicating. As such, it meets the criteria which define the more general category of molecules called plasmids. Plasmids, such as the F factor, which can also integrate into the bacterial chromosome are called episomes. Other plasmids maintain an independent existence in the bacterial cytoplasm, and multiple copies may be present in a single cell. Each may contain as few as three or four genes or as many as several hundred genes.

Plasmids are generally classified according to the genetic information specified by their DNA. The F factor, as you will recall, is a fertility factor and contains genes essential for sex pilus formation as well as for DNA replication. Other plasmids contain information conferring resistance to antibiotics (the R plasmids), heavy metals, and ultraviolet radiation, and leading to the production of toxins (the Col plasmids).

While plasmids as extrachromosomal units are valid topics for genetic study, two recent discoveries have dramatically increased the interest in their investigaion. The first involves segments of DNA called insertion sequences (IS). They consist of specific nucleotide sequences and are capable of moving in and out of DNA molecules. Insertion sequences can combine with other genes to form a vehicle, the transposon (Tn), which allows for transfer of genetic information from one DNA source to another. For example, transposons can move between plasmids and bacterial and viral chromosomes. As a result, they have been sometimes called "jumping genes."

Interest in plasmids has also been enhanced by the phenomenon of recombinant DNA–one aspect of the more general topic of genetic engineering. In genetic engineering, specific genes from a variety of organisms may be inserted into plasmids by experimental manipulations. As such, these genes may be cloned as the plasmid replicates and copies of it are distributed to progeny cells.

Insertion Sequences and Transposons in Bacteria

In the early 1970s, a number of independent workers, including Peter Starlinger and James Shapiro, discovered a unique class of mutations in E. coli. These mutations affected several genes in different strains of bacteria. For example, transcription of a cluster of genes (or operon) controlling galactose metabolism in E. coli was repressed or shut off as a result of one of these mutations. This phenotypic effect was heritable, but was found not be caused by a nucleotide change characteristic of conventional mutations. Instead, it was shown that a short, specific DNA segment had been inserted into the bacterial chromosome at the beginning of the galactose gene cluster. When this segment was spontaneously excised from the bacterial chromosome, wild-type function was restored.

It was subsequently revealed that several other distinct DNA segements could behave in a similar fashion, inserting into the chromosome and affecting gene function. These DNA segments are relatively short, not exceeding 2000 base pairs (2 kb, or 2 kilobase), and are now called insertion sequences (IS). The genetic information contained within an IS unit appears to be involved only in the insertion process itself.

R plasmids consist of two components: the RTF (resistance transfer factor) and one or more r-determinants. The RTF contains information essential to transfer of the plasmid between bacteria, and r-determinants are genes conferring antibiotic resistance. While the DNA of RTFs is quite similar in a variety of plasmids from different bacteria species, there is a wide variation in r-determinants. Each is specific for resistance to one antibiotic. If a bacterial cell contains r-determinant plasmids but no RTF is present, the cell is resistant but not infectious; that is, resistance cannot be transferred. However, the most commonly studied plasmids consist of the RTF as well as one or more r-determinants. Resistances to tetracycline, streptomycin, ampicillin, sulfonamide, kanamycin, and chloramphenicol are most frequently encountered.

Another unique observation is that there seems to be a modular construction of complex R plasmids which depends upon IS unit integration and excisions. As diagrammed in Figure, r-determinant genes are part of transposons, which contain inverted repeat termini of the IS unit. As a result, re-determinants move readily between plasmids as well as integrate into bacterial or viral chromosomes. Figures shows one such complex R plasmid thought to arise in this way.

The implications of these findings are great. Not only has a new and unique recombinational mechanism been revealed, but the way in which prokaryotic genomes have evolved may possibly have been discovered. Furthermore, it appears that similar events may occur eukaryotic systems.

19

Mutation

Mendel's work in 1900. In 1901 he published his accumulated data in a book entitle. The Mutation Theory. De Varies was careful to distinguish between hereditary and environmental variations, but his mutations are now known to have included several kinds of changes in the hereditary material. It took nearly half a century to clarify the situation in Oenothera and discover the causes of the variations that he observed. Chromosome changes, both structural and numerical, were eventually found to cause the phenotypic variation that de Vries had grouped under the heading of mutations. One complex alteration is described later on in this chapter in connection with the discussion of permanent heterozygotes. It was de Vries' concept of discontinuous variation rather than his precise observations and example that became significant. None of his examples would now be classified as gene mutations.

Mutations and Phenotypic Change

Because genes are chemical entities that cannot be observed and compared directly, some phenotypic alteration must be associated with the gene change (mutation) or it will go unrecognized. Minor changes presumably occur in genes as a matter of course without producing any phenotypic alterations. Some gene changes are known to be associated with only slight effects on the fertility or viability of the organism. These alterations ordinarily go undetected unless critical comparisons are made. They may, however, influence natural selection and thus represent a factor in evolution.

Mutations also represent in evolution, by which the presence of particular wild-type genes can be postulated. Normal development of any organism is influenced by numerous genes. An observable characteristic of the adult may be altered by an individual gene substitution, but because the entire organism develops as a unit, the action of individual wild-type genes is not always apparent. The existence of mutant and wild-type genes is substantiated when a gene is changed so that its influence on developmental reactions causes a visible difference from that exerted by the unmutated gene. Mutations thus provide the basis for postulating wild-type genes.

Mutated as well unmutated genes tend to give rise to other genes exactly like themselves. Once established, a mutated gene is as stable as the original gene from which the mutation occurred. This process of duplication is associated with mitotic cell division and occurs repeatedly as the number of cells increases. Duplication among genes has a crude parallel in the procedure of printing. After type is set and placed on the printing press; similar copies can be made repeatedly. If the type is changed slightly with an altered template replacing the original, the new pattern reproduces itself faithfully. Instead of a physical change, such as the substitution of one piece of metal for another, the gene in the process of mutation undergoes a chemical change. A part of the gene unit is altered by the chemical modification in such a way that it henceforth behaves differently from its unmutated ancestral gene.

Not all mutations are immediately detectable because many, perhaps the great majority, are recessive and must become homozygous before they can express themselves. Because the phenotypic effect is the only readily observable evidence of mutations, however, a sudden phenotypic change that subsequently proves to be heritable is an accepted indication that mutation has occurred in an organism. In everyday language, the phenotypic change is synomous with the gene change or actual mutation. This is quite natural because the term"mutation" was coined and used before the present gene concept was established. Historically, therefore, the word has been used to describe the perceptible change alteration. It is now known that most mutations that are not detectable by visible phenotypic change influence the viability of the organism in some indirect way.

Classification for Mutations

Dominance and recessiveness provided convenient, although perhaps superficial, criteria for early classifications of mutations. Mutations also may occur in either the autosomes or the X chromosomes, and thus may be classified as autosomal or sex-linked. Viability mutations have also been classified on the basis of their effect on the organism. Various mutations of this nature in Drosophila have been expressed in the egg, larva, pupa, or adult. When disadvantageous, the mutations are called deleterious; when disastrous to the individual, they are called lethals. Relatively few mutations occur as dominants and even fewer are advantageous in the environments where they occur. The great majority of changes are recessive and detrimental under usual environmental conditions. In a changing environment, a mutation may happen to coincide with new environmental situations and be favourable.

Pleiotropy

A given mutation may alter the organism severely or have such slight effect that it can be detected only when associated with other genes, through cumulative action. The name and gene symbol associated with a mutation are usually taken from the most conspicuous phenotypic alteration that the mutation produces. Thus, white eye (w) is used in contrast to the wild type, red; and ebony (e) body designates the deviation of dark from wild type, gray body colour. This method of designation perpetuates an overly simplified concept of the effects of mutations. Because they exert their influence on basic chemical reactions in the developmental period, single mutant genes affect more than a single trait and may modify in some way every trait of the organism.

The term *pleiotropy* refers to the situation in which a gene is known to influence more than a single trait. For example, in the presence of the white-eye mutation in *Drosphila melanogaster*, the ocelli, malpighian tubules, and testicular envelopes are colourless, whereas these structures in wild flies are coloured. The balancers and wing muscles are altered and the general viability and fertility of the flies are changed. Sheldon Reed has demonstrated a discrimination against white-eyed flies in mating. Such discrimination indicates that the gene causes some effect on the cuticle that influences the mating stimulus. White eye is thus only one several effects of the single mutant gene (w). Many examples of genes with more than single effect have been discovered. All genes (mutants and nonmutants) may be pleiotropic, even though their various effects are not recognized at present. Even though a gene may have many end effects, probably it influences only one primary function in the chemistry of the developing individual.

Congenital hydrocephalus in the mouse, cited by Gruneberg, illustrates the manifold effects of a single gene, ch, on the developing mouse. The gene occurred as a spontaneous mutation and was found by preliminary studies to influence cartilage formation in early development of the mouse. Mice that carried the gene in homozygous condition were born alive but died immediately after birth because the lungs did not inflate properly. Other conspicuous abnormalities also occurred: the skull, forehead, and face were out of proportion; large protuberances filled with fluid extended out from the cerebral hemispheres because the skull bones were abnormal and only skin covered the forehead; eyelids were always open; sensory hairs of the face were abnormal; and the sternum was abnormal with little or no bone formation. Developmental studies showed that abnormalities occurred as early as the thirteenth day in the embryo. The manifold effects were traced back to a single cause, the abnormal cartilage formation. Various skeletal abnormalities were involved directly; physiological disturbances occurred secondarily. A single primary reaction concerned with cartilage formation was controlled by the mutant gene.

Somatic and Germinal Mutation

Mutations may occur in any cell and at any stage in the cell cycle. The immediate effect of the mutation and its ability to produce phenotypic change is determined by its dominance, the type of cell in which it occurs, and the stage in the cycle of the cell. If the mutation occurs in a somatic cell that can reproduce other cells like itself but not the whole organism, the mutant change would be perpetuated only in somatic cells that descended from the original cell in which the mutation occurred.

The "Delicious" apple and the naval orange have resulted from mutations that occurred in somatic tissues. Changes that gave these two fruits their desirable qualities apparently followed spontaneous mutation in single cells, which constituted only a very small part of the body of the individual apple and orange trees that were involved. In each case, the cell carrying the mutant gene gave rise to more of its own kind, eventually producing an entire branch on its respective tree which had the charactristics of the mutant type. Fortunately, it is possible to propagate desirable types of many plants by vegetative methods such as budding and grafting. Vegetative propagation was feasible for both the delicious apple and the navel orange, and today numerous progeny from grafts and buds have perpetuated the original mutation. Descendants of the mutant types are now widespread in apple orchards and orange groves.

Results of somatic mutations have been detected in animals as well as in plants. In Drosophila, white sectors are infrequently observed in an otherwise red eye. Sectors in some male flies have been demonstrated to be composed of descendants from a single cell in which a mutation had occurred (at the w locus in the X chromosome). If the same change had occurred in a germ cell, a white-eyed male might have been produced.

If dominant mutant genes occur in germ cells, their effects may be expressed immediately in the progeny. If they are recessive or hypostatic, their effects may be obscured by other genes. Germinal mutations, like somatic mutations, may occur at any stage in the cycle of the organism but they are more common in some stages, particularly during the gene replication process preceding gametogenesis. If the mutation arises in one of the gametes produced by an individual, a single member of the progeny may receive the mutatant gene. If, on the other hand, a mutation occurs during an earlier stage of gametogenesis, several gametes may receive the mutant gene and therefore several individuals could perpetuate it. In any case, the dominance of the gene and the stage in the cycle when the mutation occurs are major factors in determining the extent of any expression that follows.

Wright noticed a peculiar male lamb with unusually short legs in this flock of sheep. It occurred to him that it would be an advantage to have a whole flock of these short-legged sheep, which could not get over the low stonefences in his New England neighbourhood. Wright used the new short-legged ram for breeding his 15 ewes in the next season. Two of the 15 lambs produced were short-legged. Short-legged sheep were then bred together and a line was developed in which the new trait was expressed in all individuals. The mutation that gave rise to the short-legged sheep was obviously of the germinal type because the cell carrying the mutation had the capacity to reproduce the entire organism. Other examples of germinal mutations have since been described in a wide variety of animals and plants.

Mutation by Deletion or Insertion of Nucleotide Bases

Consider the sequences of bases written in the RNA code shown in Figure9-12a. Now suppose that a base is deleted or an additional base is inserted. The net effect of either a deletion or an insertions that the reading frame is shifted, because different triplets are now read after the mutation site. These are called frameshift mutations. Their net effect is that the amino acids inserted after the deletion or insertion will be different from those in the unmutated cell, and hence the

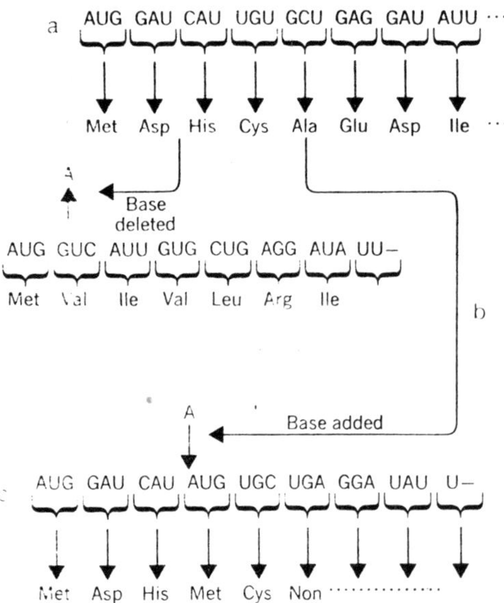

Fig. 19.1. Base deletion and base addition resulting in frameshift mutations. Translation is from left to right.

polypeptide will contain a grossly abnormal amino acid sequence and be nonfunctional. In addition, somewhere along the line it is possible for a nonsense codon to be produced and the "nonsense" chain will be terminated.

Note that the addition or deletion of two bases will have essentially the same effect as a single base addition or deletion because the reading frame will shift. A three-base change will not shift the reading frame if three immediately adjacent nucleotides are added or deleted, but simply add or delete an amino acid. This is also true for duplication or insertion of adjacent bases in multiples of three. Two or more amino acids will be affected instead of one.

Like nonsense mutations, frameshift mutations can be expected to have drastic effects on the phenotype, unless they occur in multiples of three or cause a change at a nonessential end of a protein. In the latter case, the effect may be great or hardly noticeable, depending again on the position of the deleted or added amino acid(s) in the polypeptide chain. Frameshift mutations can be suppressed by other mutations.

The Reversal of the Mutant Phenotype

So far in this chapter we have been looking at mutation in a somewhat restricted way. It has been tacitly assumed that there is a

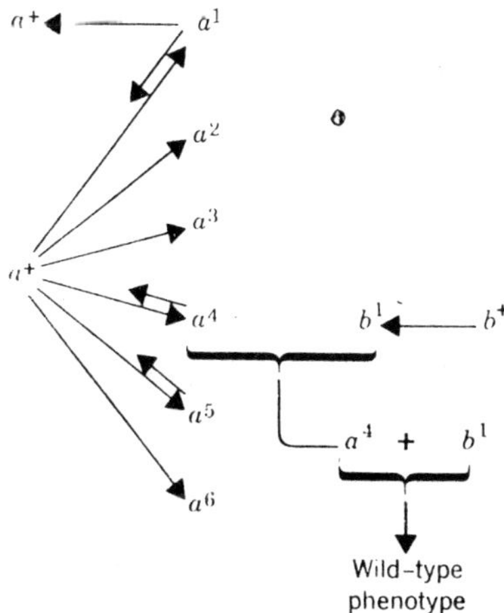

Fig. 19.2. Mutations from a wild-type allele (a^+) to mutant alleles and reversions. The reversions can either be by reversion mutations to a^+ or an isoallele ($a^{+'}$) or by mutation of another nonallelic gene (a suppressor).

wild-type allele (or group of wild-type alleles called *isoalleles*, which all produce a wild-type phenotype) for each gene that mutates to produce mutant alleles in an unidirectional way. However, mutation can occur either way. The mutant phenotype observed is a result of a forward mutation, and the change back to the original genotypic state is a reverse mutation, by definition. In addition, reversal of phenotype to the original situation, or almost so, may occur without a reversal of genotype but rather by a further mutation in the same gene or in another nonallelic gene. We consider these two situations in this section.

Reverse Mutation

If a missense mutation occurs by substitution of a nucleotide base, it should be expected that it can be reversed by a second substitution in the same codon so as to change it back to the condition in which it will code for the original amino acid, or one which will again give a wild-type phenotype. This has been demonstrated in a most convincing fashion by the use of the chemical mutagens about whose action we have some knowledge. Consider the action of 5-bromouracil, which, pointed out previously, causes transitions of the A:T–G:C type, if 5-bromouracil is able in its enol state to pair with guanine it should also be able to pair with adenine when not in the enol state, as the new chain is forming in the mutant. The net effect is another transition, G:C – A:T, the reverse of the first situation. Therefore, mutations caused by 5-bromouracil should be reversible by 5-bromouracil, and this turns out to be the case. 2-Aminopurine and nitrous acid will also cause reversion of 5-bromouracil-induced mutations.

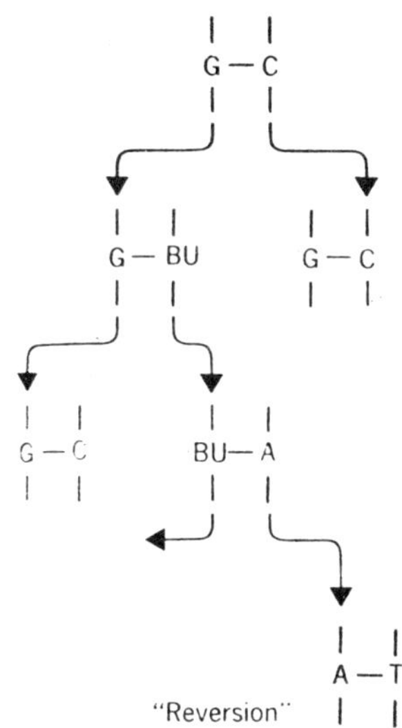

Fig. 19.3. The reversion of a mutation initially caused by 5-bromouracil.

Hydroxylamine is somewhat different from 5-bromouracil, 2-amino-purine, an nitrous, acid; mutations caused by it are not truly reverted by it. All these results agree with the theory that 5-bromouracil, 2-aminopurine, and nitrous acid should cause transitions in both direction (A:T G:C). Hydroxylamine should cause transitions in only one direction (G:C A:T).

Ethyl ethanesulfonate (EES) apparently causes transitions in the G:C–A:T direction and also transversions such as G:C T:A and G:C C:G. Some, but not all, EES-induced mutants are reverted by base analogues and nitrous acid.

The reading frame mutations caused by acridines do not revert in the presence of base analogues or compounds that later bases. And this, of course, is precisely what is to be expected, because acridines cause the deletion and duplication of bases. On the other hand, mutations caused by acridines are for the most part revertible by acridines. In this fact there resides another interesting story; revertants of reading frame mutations may occur in several ways as related in the next several paragraphs.

Intragenic Supressors

An acridine-induced mutant will usually (if not always) result in a gene deficient or duplicated for bases. This shifts the reading frame and generally results in a nonsense codon, Acridine-induced mutations can be reverted with acridines by the induction of an intragenic suppressor. If the original mutation is a deletion of a base, the suppressor will be an insertion of a base occurring nearby in the DNA. As a result, the reading frame would be brought back to the state in which the proper amino acids are now coded, except for that region between the two mutations. In an analogous way an insertion initially caused by acridine can be suppressed by a deletion. If the amino acids coded for in this region do not form a segment that is important to the protein's ability to function, a true suppression of the mutant phenotype will occur.

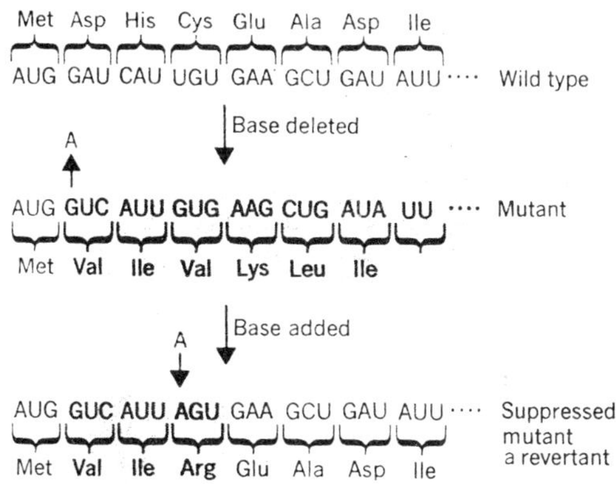

Fig. 19.4. How intragenic suppressors of frame-shift mutations act to produce a functional polypeptide.

A second type of intragenic suppressor involves missense mutations. As stated before, a single amino acid substitution caused by a missense mutation may so disrupt a polypeptide's tertiary conformation that is inactive as a protein. It has been shown that such a mutation may be rectified by a second missense mutation in the same gene. The second amino acid substitution in some way erases the "lethal" change in tertiary conformation caused by the first. This has been demonstrated in the tryptophan synthetase of E. coli and is probably a fairly common even in all organisms.

Extragenic Suppressors

When a nonsense codon is formed either by missense or frameshifts mutation, a drastic phenotypic change is expected because the polypeptide encoded by the affected gene will be incompletely synthesized during translation. Such mutants can be restored to approximately the wild-type phenotype by the occurrence of another mutation in a gene at another locus. Five such suppressors have been described in E. coli. They all suppress one or the other of the two nonsense codons: UAG and UAA or both. It has been shown that in the presence of a suppressor mutation amino acids are actually inserted during translation at points where nonsense codons exist. Hence the polypeptide chains are completed, although the completed chains may not function as well as the polypeptide chains coded in the wild type.

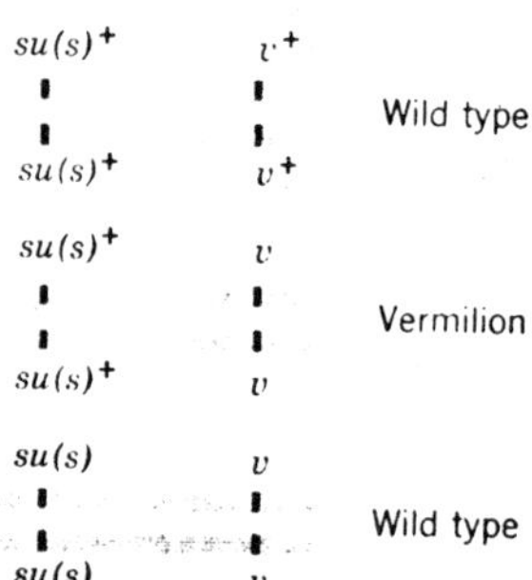

Fig. 19.5. The phenotypic effects of the suppressor of vermilion, su(s) in D. melanogaster.

As might be expected, it has been found that the biochemical basis for the action of these suppressor genes is that they code for tRNA's. The sup D gene codes for serine tRNA. In its mutant condition, it produces an altered serine tRNA that introduces serine where a UAG occurs in the messenger. Similar explanations apply for the other suppressor genes, which presumably code for other tRNA's.

These kinds of suppressors have been *supersuppressors.* They are so named because they are capable of suppressing nonsense mutations in a wide range of different genes with quite different and unrelated functions. Theoretically, any gene can mutate by the formation of a nonsense codon. Hence, a single suppressor may effectively suppress specific alleles of a large number of different genes, provided the amino acid it introduces through its altered tRNA results in functional polypeptides. Suppressors have been reported in yeast, *Neurospora,* and perhaps barley. They probably also occur in *Drosophila.* The suppressor of Hairy Wing in *D. melanogaster* suppresses at least 10 other mutant genes, all of which are at different loci and have seemingly unrelated functions.

Suppressors in diploid organisms theoretically should always be dominant, because only one dose of altered tRNA gene should be necessary to produce the tRNA necessary to recognize a nonsense codon. However, suppressors do exist that are expressed only when homozygous. The suppressor of Hairy Wing is one example. Another example in Drosophila is the suppressor of vermilion su(s). It suppresses the mutant condition, sable(s). The action of this type of suppressor, other examples of which are known in *Drosophila,* is not clear.

Missense, Nonsense, and Frameshift Mutations

Various types of mutations can conveniently be discussed in relation to the b-globin gene. It has already been described that the human b-globin gene is split into three coding regions interrupted by two intervening sequences. The nucleotide sequence of the beginning of the first coding region, the sense strand is at the top, and transcription occurs from left to right (from the 3' end of the sense strand to the 5' end). The spaces that separate adjacent triplets of bases do not exist in the actual molecule. It shows the corresponding portion of the b-globin mRNA produced after transcription and RNA processing. Although it is not shown in the figure, the next upstream codon from GUG is the AUG initiation codon, which codes for methionine. Figure shows the first eight amino acids at the amino terminal end of normal b-globin; the initial methionine with which translation begins is not represented in the finished polypeptide because it is enzymatically removed from the chain. The underlined nucleotides represent the restriction site for the restriction enzyme *Mst*II, which cleaves the DNA at this position.

As noted in the previous section, a *base substitution* mutation is one in which a base pair in a DNA duplex is replaced with a different base pair. A base substitution that leads to an amino acid substitution in the corresponding polypeptide is called a missense mutation, and there are two principal types of missense mutations. Part (*a*) again shows a portion of the first coding region of the b-globin gene, and part (B) shows the result of a base substitution in which the normal TA pair indicated is replaced with an AT pair. In this substitution, a pyrimidine base in the sense strant (T) is replaced with a purine base (A), and a purine base in the antisense sense strant (A) is replaced with a pyrimidine base (T). Any base substitution in which a purine (A or G) is replaced with a pyrimidine (T or C), or in which a pyrimidine is replaced with a purine, is called a *transversion*. Transcription and RNA processing of the mutant DNA lead to the mRNA shown in Figure; note that the sixth codon, which normally reads GAG, now reads GUG. Tranlation of this mRNA leads to a substitution of valine for glutamic acid at the sixth position in the polypeptide (black arrow). The transversion mutation is indeed the actual molecular change responsible for b^s hemoglobin. The b^s mutation also obliterates the *MSt* II restriction site, so *MSt* II will not cleave the b^s DNA at this position. This molecular difference provides a rapid and convenient method for *in vitro* diagnosis of sickle cell anemia.

There is another type of missense mutation, which is called a *transition* mutation because it involves the replacement of one purine (in this case G) with another purine (in this case A) or one pyrimidine (in this case C) with another pyrimidine (in this case T). In the transition mutation a CG base pair in the original molecule is replaced with a TA base pair. The corresponding portion of the mRNA is shown in part (C), and in this case the original GAG codon in position 6 is altered to AAG. Translation of the mutant mRNA produces a β-globin polypeptide in which the normal glutamic acid at position 6 is replaced with a lysine. This particular mutation is known as the β^c mutation. Sine the b^c mutation is relatively common in areas of the world where malaria is prevalent, it is thought that heterozygotes for β^c may have some protection against malaria, as is certainly the case with β^c

Not all base substations are missense mutation. There is also another type of mutation called a nonsense mutation. A nonsense mutation is one that creates a chain-terminating codon [i.e., uAA (orchre), UAG (amber), or UGA (opal)] in a coding region. The name nonsense mutation may seem odd, but it comes from the fact that chain-terminating codons are times referred to as nonsense codons. The transvertion to leads to a UAG terminating codon at the sixth position in the mRNA. Thus, translation of this mRNA leads to a truncated polypeptide because translation terminates at the UAG codon.

In prokaryotes such as *E.coli* and in lower eukaryotes such as yeast, certain mutations have the ability to suppress nonsense mutations. These nonsense-suppressing mutations are called nonsense suppressors, and they involve mutations in genes for transfer RNA (tRNA) that reduce the faithfulness of translation so that chain-terminating codons are sometimes misread.

Such a mutant tRNA will correctly translate serine codons, but, on occasion, will misread a chain-terminating codon (in this case UAG) as a serine codon. Consequently, in some fraction of attempts at translation of the mutant mRNA, the UAG codon will be misread as serine (stipulated arrow), and translation will be able to produce normally beyond this point. If the serine-containing polypeptide is able to function, the phenotypic effects of the original nonsense mutation will have been suppressed by the nonsense suppressor.

Of course, nonsense suppressors can occur in tRNA genes other than serine tRNA. For example, if a UAG suppressor occurs in a gene for glycine tRNA, a UGA codon will sometimes be misread as a glycine codon, and glycine will be inserted into the polypeptide at this position.

It should also be emphasized that a particular nonsense suppressor will suppress only one of the chain-terminating codons; that is, an *amber suppressor* will suppress only the UAG (amber) terminator, or *ochre suppressor* will suppress only the UAA (ochre) terminator, and an *opal suppressor* will suppress only the UGA (opal) terminator. Of course, a nonsense suppressor will occasionally misread normal termination codons, too; for example, an amber suppressor will occasionally misread a UAG codon at the end of a normal gene and the result is an abnormally long polypeptide. However, organisms that carry nonsense suppressor are able to survive in spite of the occasional misreading of normal genes.

Not all base-substitution mutations lead to missense or nonsense codons. Because the genetic code contains many synonymous codons, some base substitutions leave the amino acid sequence of the corresponding polypeptide unaltered. Such mutations are said to be silent. In this case a C/G-to-T/A transition at the third position in codon 6 of the β-globin gene leads to an mRNA that has GAA as its sixth codon. However, GAA is a synonymous codon for glutamic acid, so the amino acid sequence of the resulting polypeptide will be completely normal. As a brief inspection of the genetic code will reveal most silent substitutions in coding regions would be expected to involve transitions in the third position of a code. The situation may be very different for intervening sequence, however. Since a particular length and base sequence of an intervening sequence does not seem to be essential for proper gene function, many mutations in intervening sequences–transitions, transversions, even inversions, deletions, or insertions–may be silent.

A final example of a type of mutation that has relatively simple molecular basis is known as a *frameshift* mutation. A frameshift mutation alters the reading frame of an mRNA during translation and thereby greatly alters the amino acid sequence of the corresponding polypeptide. Frameshift mutations are caused by deletions of additions of a small number of nucleotides in a coding region; since the genetic code consists of triplets of nucleotides, any deletion or addition of a number of nucleotides other than an exact multiple of three will cause a shift in reading frame. An example of a frameshift mutation associated with a single nucleotide deletion is Part (a) is again a portion of the first coding region of the human β-globin gene, and the left side of the figure illustrates the normal course of transcription and translation of this sequence. As in previous figure involving this gene, small gaps are introduce in part (c) to show the normal triplet reading frame. The right side of the figure shows the consequence of a deletion of the indicated A/T nucleotide pair. When this deletion-bearing sequence is examined in terms of its triplet reading frame, it can be seen that all triplets beyond the second one are different from those in normal sequence. This lack of correspondence occurs, of course, because the single-nucleotide deletion causes the triplet reading frame beyond the deletion to be shifted one nucleotide to the left. The extensive lack of correspondence due to the frameshift mutation is also evident in the mRNA, and the resulting amino acid sequence in the polypeptide is hardly recognizable as a β-globin sequence. Such out-of-frame translation will continue along the mutant mRNA until a termination codon is encountered.

Mutations Resulting from Unequal Crossing-Over

It has been stated that unequal crossing-over involving duplications can lead to an increase or decrease in the number of copies of the region. When this process occurs at the molecular level, it creates new types of DNA sequences that qualify as mutations. An example of unequal crossing-over creating new mutations is found in the human b-globin gene. Part (a) shows part of the DNA sequence coding for amino acids 83 through 98, which is 18 nucleotides upstream from the second intervening sequence. The boxes indicate two nearby regions that have a perfect eight-nucleotide homology that can act as a duplication. Part (b) shows the coding sequence for amino acids 88 through 103; it has been shifted to the left relative to part (a) to show how the regions of homology can mispair (shaded boxes). As before, the gaps in the DNA sequence indicate the translational reading frame.

Sequence (c) combines the left part of (a) with the right part of (b). This sequence is deleted for the nucleotides that code for amino acids 91 through 95, and it is of some interest that this deletion, called the *Gun Hill deletion*, is found in certain rare individuals. Sequence (d) is the complementary crossover product to sequence (c); this sequence carries a duplication of the nucleotides that code for amino acids 91 through 95.

Mutagenic Agents and the Mechanisms of Mutation

Mutations occur in the somatic and germ cell of all organisms. In organisms like Drosophila, mutations are detected for specific loci in one in about 10^5 to 10^6 gametes. Therefore the spontaneous mutation rate for most genes is about 10^{-5} to 10^{-6} mutations per locus per

Table 19.1. Disorders of Human Hemoglobin

Type of hemoglobin	*Signs and cause*
S	Hemozygosity for β^S mutation (sickle cell anemia)
C	Homozygosity for β^C mutation
Lepore	δ-β chain fusion caused by unequal crossing-over in misaligned δ-β region
Anti-Lepore	Triplication consisting of normal δ gene-δ-β fusion gene-normal β gene; reciprocal crossover product of the unequal crossover leading to Lepore hemoglobin
Constant Spring	Elongated α chain due to missense mutation in terminator codon
Gun Hill	Shortened β chain due to deletion eliminating amio acids 91-95
Unstable hemoglobin	Increased blood destruction; associated with at least 70 mutations, most affecting β chain
Methemoglobulinemia	Reduced ability to transport oxygen; associated with several mutations, some affecting α, others β
Hemoglobin-induced erythrocytosis	Anemia due to increased oxygen affinity of hemoglobin causing stimulation of red blood cell production and consequent increased destruction; associated with at least 20 mutations
α-thalassemia	Reduced synthesis of α chain; principal cause is deletion of one or more α-gene copies; *hydrops fetalis* is a lethal disorder associated with absence of α chain
β-thalassemia	Reduced synthesis (β^+ form) or no synthesis (β^0 form) of β chain; some patients have β deletions; at least one β^0 form involves a mutation at the splice junction at the downstream end of the second intron.
Hereditary persistence of fetal hemoglobin	Continued synthesis of γ chains; postullated detection of DNA sequence responsible for $\gamma \rightarrow \beta$ switch

generation. First, agents that increase the mutation rate above the spontaneous level will be discussed because they help explain the spontaneous rate.

The two basic kinds of mutagenic agents are (1) physical ones including radiation (s) and (2) chemical ones, consisting of both "natural" and "unnatural" substances.

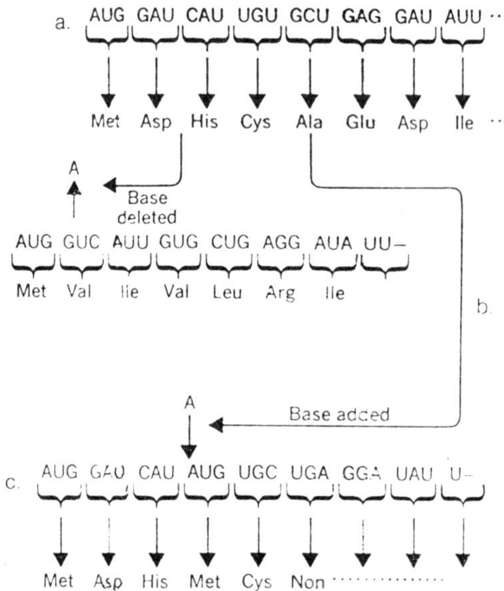

Fig. 14.6. Base deletion and base addition resulting in frameshift mutation.

The Spontaneous Mutation Rate

An organism carefully protected from known mutagenic chemicals and from known artificially made radioactive sources will still produce gametes or cells with new gene mutations, What causes these spontaneous mutations? A number of factors can be enumerated and may be divided into two categories: environmental and genetic.

Environmental Factors

Background radiation comes to mind as a factor almost immediately, because we have been sensitized since the 1950s to the hazards of ionizing radiation from atomic bombs, industrial and medical applications of nuclear energy, and so forth. However, the average background measured in roentgens per generation is insufficient to account for more than a small fraction of spontaneously occurring sex-linked recessive lethals in Drosophila. But, course, if the background radiation is raised, this could cause additional mutations that might become an important factor in the future of humans and all other organisms.

Ultraviolet light may not be of much significance in the production of mutations since the ultraviolet in sunlight that reaches the surface of the earth has wavelengths only above 300 nm. This range of wavelengths is not effective in inducing mutations to any marked extent except somatic ones in skin. In the higher animals overlying tissue protects the germ cells as already noted.

Chemical in the environment may certainly be a factor, but their effect is difficult to measure for several reasons: (1) a mutagenic chemical may be metabolized to a nonmutagenic state before it reaches the germ cells, or (2) a nonmutagenic chemical may be metabolized to a mutagenic state, and so on. Despite these difficulties, it is now apparent that close attention should be paid to the present and future effects of chemicals on the mutation rate. Our environment is becoming more and more polluted with many different kinds of natural and synthetic chemicals used as insecticides, herbicides, food preservers, medicines, beautifiers, tranquilizers, and hallucinogens. What effect do these how have on the human mutation rate? We do not know, but it is important that we find out, because an increase in the rate could have diastrous results.

For the present, we can assume that the presence of chemicals in the environment may have some effect on the mutation rate of all organisms, but it certainly cannot account for all the spontaneous mutation rate. Metabolic products, such as formaldehyde and hydrogen peroxide produced within cells, may also be involved, but here again it is doubtful that their effects are highly significant.

Genetic Factors

Organisms evolved in a changing environment to which they had to continually adapt as they evolved. Mutations were necessary for this evolution to occur, just as they are necessary for future evolution. But the mutation rate must not be so high that it results in the production of many deleterious mutation that would act as "genetic load" on the population.

It is highly probable that the main component of the spontaneous mutation rate is genetically controlled. As we have described in previous pages, enzymes are involved in the replication of DNA. Mistakes of different types may occur during replication that may lead to mutation, especially if the replicating enzymes themselves are deficient. These mistakes may be replicated or they may be repaired by the repair systems that we known to exist. Thus, two opposing systems, genetically controlled, may be postulated to exist: the replicative synthetic one and the repair system. The spontaneous rate of mutation may be determined by an equilibrium point between them.

It has long been known that different populations of *Drosophila melanogaster* collection from widely scattered areas in different part of the world show significantly different mutation rates (Table 19.2). Furthermore, genes have been identified that have an influence on the mutation rate A second chromosome gene, *hi*, and a third chromosome gene, *mu*, have been located in *melangoaster*. Both of these increase the "spontaneous" mutation rate when homozygous, although *mu* appears to act in this way only in females. Extensive work has been done in E. coli with mutant strains that show a high rate of A:T → C:G transversions. The result is that the mutation rates as specific loci are raised several thousand fold!

Table 19.2. Spontaneous mutation rates found in different populations of *D. melanogaster*, sex-linked and second chromosome recessive lethals

Stock	X chromosome			Second chromosome		
	Number tested	Number lethals	Percent lethals	Number tested	Number lethals	Percent lethals
Florida inbred	2108	23	1.09	–	–	–
Wooster	1266	8	0.63	–	–	–
Oregon R	3049	2	0.07	–	–	–
Florida No. 10	916	10	1.09	516	9	1.74
Lausanne	955	2	0.21	436	3	0.69
Leningrad	8614	14	0.16	–	–	–
Sukhami	2309	24	1.04	–	–	–

Genes that raise the mutation rate are called *mutator* genes. A number of these have been identified in yeast, *Drsosophia*, and E. coli. They have yet to be directly implicated in the DNA replicative and repair processes, but it is difficult to believe that they are not.

In addition to mutator genes, there are also numerous examples of mutable loci in a variety of organisms. Certain alleles of some genes exhibit an unstable condition, so that they mutate from the wild type to mutant condition at a high rate in either the germ cells, or somatic cells, or both.

Phenotypic Effects of Mutations

Mutations must normally cause some detectable *phenotypic change* for their presence to be recognized. The effects of mutations on phenotype range from alterations so minor that they can be detected only by special genetic or biochemical techniques to gross modifications of morphology to lethals. A gene is a specific sequence of nucleotide pairs coding for a particular polypeptide. Any mutation occurring within a given gene will thus produce a new form or *new allele* of that gene. Because of the degeneracy of the genetic code, some base pair changes do not change the protein products coded for by the genes in any way. Genes containing mutations with small effects that can be recognized only by special techniques are called "isoalleles." Other mutations result in total loss of gene-product activity. If mutations of the latter type occur in essential genes (genes required for viability), they will, of course, be lethal.

Mutations may be either recessive or dominant. In haploid (or, more accurately, monoploid) organisms like viruses and bacteria, both recessive and dominant mutations can be recognized by their effects on the phenotype of the organism in which they originated. The dominance or recessiveness of mutations in bacteria can be determined only by studying partial diploids. In diploid (or polyploid) organisms, recessive mutations will be recognized only when present in the homozygous condition. Most recessive mutations in diploids will not be recognized at the time of their occurrence, since they will be present in the heterozygous state. Sex-linked recessive mutations are an exception, since they will be expressed in the hemizygous state in the heterogametic sex (males in humans and fruit flies; females in birds). Sex-linked recessive lethal mutations will alter the sex ratio, since hemizygous individuals carrying the lethal will not survive.

The most useful mutations for the genetic analyses of many biological processes are conditional lethal mutations. These are mutations that are (1) lethal in one environment, the so-called restrictive conditions, but are (2) viable in a second environment, the permissive conditions. Such mutations allow geneticists to identify and study mutations in essential genes that result in complete loss of gene-product activity even in haploid organisms. Mutants carrying conditional lethal can be propagated under permissive conditions, and information about the functions of the gene products can be deduced by studying the consequences of their absence under the restrictive conditions. Conditional lethal mutations also provide valuable selective mutations also provide valuable selective sieve for genetic fine structure analysis.

The three major classes of mutants with conditional lethal mutations are (1) *auxotrophic mutants,* (2) *temperature-sensitive mutants,* and (3) *suppressor-sensitive mutants.* Auxotrophic mutants (as opposed to prototrophic "wild-types") are mutants that are unable to synthesize an essential metabolite (amino acid, purine, pyrimidine, vitamin, etc.) That is synthesized de novo by wild-type individuals of the species. Such auxotrophic mutants will grow and reproduce when the metabolite is supplied in the medium (the permissive condition); they will not grow when the essential metabolite is absent (the restrictive condition). Temperature-sensitive mutants will grow at one temperature but not at another temperature. Most temperature-sensitive mutants are heat-sensitive some, however, are cold-sensitive. The temperature sensitivity usually results from increases heat or cold lability of the mutant gene product, for example, an enzyme which is active at low temperature but partially or totally inactive at higher temperatures. Occasionally, only the synthesis of the gene product is sensitive to temperature, and once synthesized, the mutant gene product may be as stable as the wild-type gene product, Suppressor-sensitive mutants are viable when a second genetic factor, a suppressor, is present, but are nonviable in the absence of the suppressor. The suppressor gene may correct or compensate for the defect in phenotype that is caused by the suppressor-sensitive mutation, or it may render the gene product, altered by the mutation, nonessential.

Most of the thousands of mutations that have been identified and studied by geneticists have been found to be deleterious and recessive. This is to be expected, considering what we know about the genetic control of metabolism and the techniques available for identifying mutations. Metabolism occurs by sequences of chemical reactions, each step of which is catalyzed by a specific enzyme coded for by one or more genes. Mutations in these genes frequently produce blocks in metabolic pathway. These blocks occur because changes in the base-pair sequences of genes often (but not always) cause changes in the amino acid sequences of polypeptides, which may, in turn, result in loss of function. This, in fact has been the most commonly observed effect of easily detected mutations. Given a wild-type allele coding for an active enzyme and mutant alleles coding for less active or totally inactive enzymes, it is apparent why most of the observed mutations might be recessive, as is observed. If the enzyme is metabolically important, such mutations will also be deleterious. If the enzyme catalyzes an essential reaction, the mutations causing total loss of activity will be recessive lethals.

But why should most mutations with phenotypically recognizable effects result in decreased gene produce activity or no gene-produce activity? This result can be predicted if one accepts the effectiveness of natural selection and if one accepts the effectiveness of natural selection and if one assumes the existence of a semiconstant environment during the recent (on the evolutionary scale) evolution of life forms on earth. A "wild-type" allele of a gene coding for a "wild-type" enzyme or structural protein will have been selected for optimal activity for many generations. Mutations resulting in amino acid changes that increased the efficiencies of enzymes in carrying out particular functions will have been preserved by natural selection, and, as the most "fit," they will have become the new "wild-types". Given a sufficient period of time in semiconstant environment, most, if not all, sequences of amino acids (at least all sequences of amino acids that differ form the "wild-type" by a single mutation) will have been tried, and natural selection will have preserved the most efficient one. It will now be the wild-type. Mutations, which produce changes in these very specific sequences of amino acids, will usually result in less activity or no activity at all. As such, they will most frequently be recessive and deleterious.

An analogy can be made with any complex, carefully engineered machine. If you randomly modify any one essential component (e.g., of a watch or an automobile), it seldom performs as well as it did prior to the random change.

Radiation-Induced Mutation

That portion of the electromagnetic spectrum containing wavelengths that are shorter and of higher energy than visible light (wavelengths below about 0.1 µm) can be subdivided into ionizing radiation (X rays, gamma rays, and cosmic rays) and nonionizing radiation (ultraviolet light). Ionizing radiations such as X rays (about 0.1 to 1 nm) are of high energy and thus are useful for medical diagnosis because they can penetrate living tissues. In the process of penetrating matter, these high-energy rays collide with atoms and cause there-lease of electrons, leaving positively charged free radicals or ions. These ions, in turn, collide with other molecules, causing the release of further electrons. The net result is that a "core" of ions is formed along the track of each high-energy ray as it passes through matter of living tissues. This process of ionization (thus the name ionizing radiation) is induced by machine-produced X rays, protons, and neutrons, as well as by the alpha, beta, and gamma rays released by

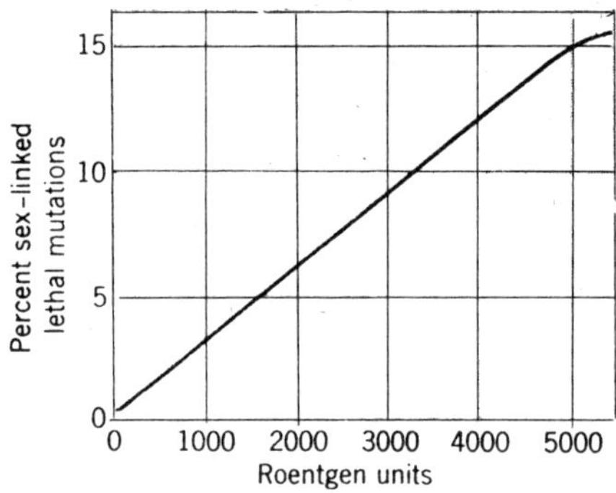

Fig. 19.7. Relationship between the frequency of sex-linked lethal mutations induced in Drosophila sperm and ionizing irradiation dosage.

radioactive isotopes of the elements (e.g., ^{32}P, ^{35}S, radium, cobalt-90 etc.). Ultraviolet rays having lower energy, penetrate only the surface layer of cells in higher plants and animals and do not induce ionization. Ultraviolet rays dissipate their energy to atoms that they encounter, raising the electrons in the outer orbitals to higher energy levels, a state referred to as excitation. Molecules containing atoms in either ionic forms or excited states are chemically more reactive than those containing atoms in their normal stable states. The increased reactivity of atoms present in DNA molecules is the basis of the mutagenic effects of ultraviolet light and ionizing radiation.

Ionizing Radiation

In 1927, H. J. Muller first demonstrated that mutation could be induced by an external factor. Muller demonstrated that X-ray treatment markedly increased the frequency of sex-linked recessive lethal mutations in D. melanogaster. Muler's unambiguous demonstration of the mutagenicity of X rays became possible by his development of a technique facilitating the simple and accurate identification of lethal mutations in the X chromosome of Drosophila. This technique, called the ClB method, involves the use of females heterozygous for a normal X chromosome and X chromosome (The ClB chromosome) specifically constructed for Muller's experiment.

The ClB chromosome has three essential components. (1) The C (for crossover "suppressor") refers to the presence of a long inversion, that prevents recombination of genetic markers on the *ClB* chromosome and alleles on the normal X chromosome. The inversion does not actually prevent crossing over, but causes gametes containing X chromosomes produced by crossing over between the *ClB* chromosome and the normal X chromosome to be inviable. Chromosomes resulting from crossing over between a chromosome containing an inversion (an inverted segment of the chromosome) and a normal chromosome will contain duplication and deficiencies (repeated and missing sets of genes). The inversion is required in Muller's experiment to assure that the markers on the *ClB* chromosome stay together through meioses. (2) The *l* refers to a recessive lethal in the *ClB* chromosome. (3) The *B* refers to the presence of the partially dominant mutation that causes the bar eye phenotype, which is a narrow, slit-shaped eye. Because it is partially dominant, it allows females heterozygous for the *ClB* chromosome to be readily identified. Both the recessive lethal (*l*) and the bar eye mutation (*B*) are located within the inverted segment of the *ClB* chromosome.

Given females heterozygous for the *ClB* chromosome. Muller's experiment was operationally quite simple. Males flies were irradiated and mated with *ClB* females. The bar-eyed irradiated and mated with ClB females. The bar-eyed daughters of this mating will carry the *ClB* chromosome of the female parent and the irradiated X chromosome of the male parent and the irradiated X chromosome of the male parent. Since the entire population of reproductive cells of the males was irradiated, each bar-eyed daughter carries a potentially mutated X chromosome. That is, each male is likely to produce some sperm that contain X chromosomes carrying new lethal mutations and some sperm without X-linked lethal mutations. The bar-eyed daughters were then mated individually (in separate bottles) with wild-type males. If the irradiated X chromosomes carried by a bar-eyed daughter contains a sex-linked lethal, all of the progeny of the mating will be female. Since males are hemizygous for the X chromosome, those receiving the *ClB* chromosome will die due to the recessive lethal (*l*) that it carries. Those receiving the irradiated X chromosome will also die if a recessive lethal has been induced in it. Matings of bar-eyed daughters carrying an irradiated X chromosome, in which no lethal mutation has bee induced, with wild-type males will produce female and male progeny in a ratio of 2:1 (only the males with the *ClB* chromosome will die). Scoring for the presence of recessive sex-linked lethals is thus unambiguous and error free using the *ClB* technique–simply scoring for the presence or absence of male progeny. By using this technique. Muller was able to demonstrate an increase in mutation rate of up to 150-fold after X-ray treatment.

Another technique that facilitates the detection of mutations in Drosophila, in this case sex-linked mutations with visible effects on phenotype (often called "visible mutations"), involves the use of **attached-X-chromosomes**. Attached-X-chromosomes undergo compulsory nondisjunction (failure of homologous chromosomes or chromatids to disjoin or separate during anaphase) since the two X chromosomes are joined to single centromere. If females with two attached-X chromosomes plus a Y chromosome (XXY) are mated to normal males, any mutation that occurs on the X chromosome of the male will be expressed in the surviving male progeny. In such attached-X matings, the male progeny receive their X chromosome from their male parent, rather that from their female parent as in a normal mating. If the male parent is treated with a mutagenic agent such as X rays, the increased frequency of recessive visible mutations can be easily assessed by screening the male progeny of the attached-X mating.

X rays and most other forms of ionizing radiation are quantitated in roentgen units (r units, pronounced "runtgen"), which are measured in terms of the number of ionizations per unit volume under a standard set of conditions. More specifically, one roentgen unit is the quantity of ionizing radiation that produces one electrostatic unit of charge in a one cm^3 volume. Note that the dosage of irradiation in roentgen units does not involve a time scale. The same dosage may be obtained by a low intensity of irradiation over a long period of time or a high intensity of irradiation for a short period of time. This is very important because in most studies the frequency of induced point mutations is directly proportional to the dosage of irradiation. In Drosophila sperm, for example, there is an increase of approximately 3 percent in the mutation rate for each 1000 r in increase in irradiation dosage. This linear relationship between mutation frequency and dosage in indicative of so called "single-hit- kinetics." That is, only one event (ionization?) or one "hit" is required to cause a mutation. Or, stated differently, every ionization has some fixed (under a specific set of conditions) probability of inducing a mutation. If the cumulative effects of many ionizations were required to induce a mutation, would plot as a curve that was concave upward.

The linear relationship between mutation rate and radiation dosage is important because it speaks directly to the frequency asked question of What is a safe level of irradiation? Even very low levels of irradiation have certain low, but very real, probabilities of inducing mutations. The question is thus meaningless. There is, no such thing as a safe level. In *Drosophila* sperm, for example, very low levels of irradiation over long periods of time (chronic irradiation) are as effective in inducing mutations as the same total dosage of irradiation administered at high intensity for short periods of time (acute irradiation). This clearly has major practical significance in evaluating the effects of the increased exposure of living organisms to radiation that results from the testing and use of nuclear weapons and nuclear reactors in generators, spaceships, and so on.

In mice, chronic irradiation has been found to induce somewhat fewer mutations than the same dosage of acute irradiation. Moreover, when mice were treated with intermittene doses of irradiation, the mutation frequency was slightly lower than when they were treated with the same total amount of irradiation in a continuous dose. It should be emphasized that all of these irrdiation treatments were mutagenic, albeit, to different degrees, to both Drosophila and mice. The different responses of fruit flies and mice to chronic irradiation may result from differences in their ability to repair damaged DNA. Repair mechanisms may exist in the spermatogonia and oocytes of mice that do not exist in Drosophila sperm.

The single-hit theory implies that one ionization can produce one mutation, but it does not imply anything about the efficiency with which this happens. Several factors have been shown to affect the efficiency with which irradiation induces mutations. The receptivity of a cell at different stages of its metabolic cycle is an important factor in determining rates for induced mutations. A. H. Sparrow has shown marked variations in numbers of chromosome fragments, assumed to be directly related to mutational changes, at different stages in meiosis and cleavage in the plant genus Trillium. Chromosome aberrations were induced about 60 times more frequently at metaphase than it interphase. Nondividing cells in Trillium showed little radiation damage, whereas rapidly dividing cells were very sensitive.

Oxygen tension and temperature change, when associated with irradiation, also may significantly alter the frequency of mutations. Low-oxygen tension decreases mutations. Oxygen can magnify the effect of radiation, but only if it is present during the irradiation. Oxygen has less effect with intense conditions than with moderate conditions of ionization. Environmental agents that protect germ cells from radiation damage often do so by lowering the oxygen concentration of tissue, and those that enhance the effectiveness of radiation and oxygen.

Ionizing radiation also induces various kinds of gross changes in chromosome structure (chromosome aberrations) such as deletions, duplications, inversions, and translocations. These changes in chromosomes structure result from breaks in chromosomes caused by ionizing radiation. Since they require two breaks, the kinetics of induction are two "hit" as expected, rather than single "hit."

Ultraviolet Radiation

Ultraviolet rays do not possess sufficient energy to induce ionizations. They are, however, readily absorbed by certain substances such as purines and pyrimidines, which then enter a more reactive or excited state. Because of their lower energy, they penetrate tissues only slightly, usually only the surface layer of cells in multicullular organisms. Nevertheless, ultraviolet light (UV) is at a wavelength of 254 nm. Maximum mutagenicity also occurs at 254 nm, suggesting that the UV-induced mutation process is mediated directly by the absorption of UV by purines and pyrimidines. In vitro studies show that the pyrimidines (especially thymine) absorb strongly at 254 nm and, as a result, become very reactive. The two major products of UV absorption by pyrimidines appear to be pyrimidine hydrate and pyrimidine dimers. Several lines of evidence indicate that thymine dimerization is probably the major mutagenic effect of UV. Thymine dimers appear to cause mutations indirectly in two ways. (1) Dimers apparently perturb the DNA double helix and interfere with accurate DNA replication. (2) Occasional errors are made during the processes that cells possess for the repair of "damaged" DNA, such as DNA containing thymine dimers.

The relationship between mutation rate and UV dosage is highly variable, depending on the type of mutation, the organism, and the conditions employed. "single-hit kinetics" are only occasionally observed, in contrast to ionizing radiation.

Mutation Frequency

Mutation is necessary to provide the genetic variability required for the evolutionary adaptation of species to environmental changes. On the other hand, most are deleterious. Thus, if mutation were to become too frequent in a species, it would create a sizable "genetic load" of deleterious effects. Clearly, if this "genetic load" became too large, the species would face extinction. Human technology has already contaminated the earth with increased levels of radiation and chemicals that are known to be mutagenic. While the consequences of the present increased level of mutagenes in the environment cannot yet be accurately assessed, most scientists agree that further increases in the levels of mutagens in the environment should be avoided. Yet significant quantities of hundreds of new chemicals are introduced into the environment each year, most of them with insufficient, if any, mutagenicity tests. Whether they will cause harmful increases in mutation rate (and/or cancer incidence) is a question of utmost concern to everyone.

Each gene probably has its own characteristic mutational behaviour. Some genes mutational behaviour. Some genes undergo mutations more frequently than others in the some organism. Those with unusually high mutation rates are called unstable or mutable, but a wide range of mutation rates exists among genes that are considered stable. The mutation rate per gene in bacteria is of the order of 1 in 100,000 to 1 in 10 million (10-5 to 10-7) per cell generation. For fruit flies, the average for mutation in a particular gene is in the order of 1 in 100,000, with a range from 1 in 200,000 to 1 in 100,000, with a range from 1 in 20,000 to 1 in 200,000 gametes. Since most of the data on mutation rates in fruit flies have been obtained from experiments with males, questions have arisen concerning a possible sex difference in overall mutation rates. B. Wallace has shown through extensive experiments that mutation rates are not significantly different in the two sexes of D. Melanogaster. However, mutation rates do differ in different strains.

Estimates of mutation rates humans indicate a somewhat greater frequency than those cited for stable gens in most other organisms. Samples collected thus far have been small, and the methods used were indirect and subject to large errors. Genes associated with such human traits as intestinal polyposis and muscular dystrophy have been estimated to mutate once in 104 to 105 people. A human generation is equal to about 50 cell generations. By expressing any mutation rate as probability of mutation per cell per generation, a mutation rate is defined independently of exact physiological conditions and stage in life cycle. This definition is based on a time unit proportional to a cell's division time. When expressed in terms of cell generations, rates for fruit and humans are generally comparable with those for bacteria.

Estimating Mutation Rates

Mutation rates are usually defined in terms of mutation events per generations or, less frequently mutations per unit time. The direct experimental determination of mutation rates is complex. Because mutation is must be sampled to get accurate estimates of mutation rates. In addition, recurrent sampling is required unless the initial population is sufficiently small that its probability of containing a mutant organism is negligible. Simply enumerating the number of mutant individuals in a population does not indicate how many mutational events have occurred since the mutants will usually undergo exponential growth like the nonmutants, although often at a slower rate than the latter.

Consider a "wild-type" allele a+ mutating to a mutant allele a with a mutation rate μ (defined as a constant a probability of a mutational event per cell duplication or per organism duplication), that is,

$$a^{+} \rightarrow a$$

clearly, the number of mutation events will depend on the number of a+ alleles in the population. For simplicity, consider a haploid organism. If one starts with population in which the number of mutant organisms equals *M* and the total population size equals *N*, then

$$dM = \left(\acute{I} + \frac{M}{M} \right) dN$$

This *assumes* that (1) M is very small relative to N, so that the difference between N and N – M is negligible; (2) the mutant and nonmutant organisms duplicate at the same rate; (3) reversion of a → a+ does not occur or is negligible. If one starts with a population in which M is zero, and makes all the same assumptions, then the relationship simplifies to

$$dM = \mu dN$$

These equations can be integerated and used to obtain estimates of mutation rates in experimental populations. More frequently, however, populations are analyzed in terms of allele frequencies, or changes in allele frequencies due to mutation, selection, or migration. Moreover, mutation is not a unidirectional process; reverse mutations occur. A more realistic picture is thus

$$a^{+} = \Leftrightarrow a$$

in which the "wild-type" allele a+ "wild-type" gene to a mutant form (usually a nonfunctional or only slightly functional form) is investigated, what is actually measured is the summation of many different mutation events at many different sites in the gene. The "average gene" is usually considered to be a DNA sequence of about 1000 nucleotide pairs coding for polypeptide that is about 333 amino acids long. Amino acid changes at many different positions in the polypeptide (resulting from base-pair substitutions at many different

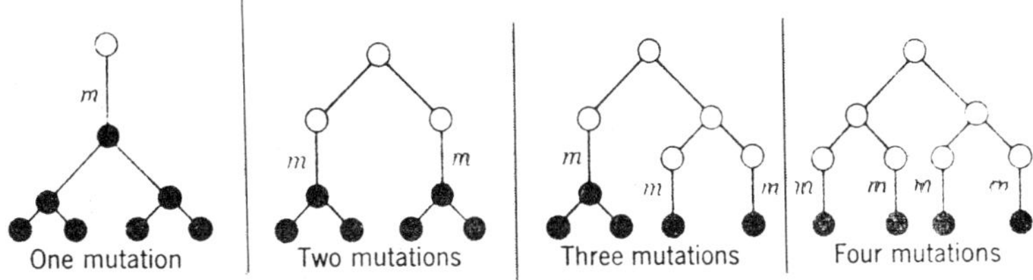

Fig. 19.8. A given number of mutant organisms in a population can result from various numbers of mutational events, depending on when the mutation occurs during the exponential growth of the population.

positions in the structural gene) may be expected to result in an inactive gene product. Thus, when one measures the mutation rate of a "wild-type" gene to the nonfunctional mutant from, one will actually be measuring the sum of many different mutations, each occurring at its own specific rate.

Mutation "Load" Versus Genome Size

In the preceding section, an average mutation rate of 1 per 100,000 (or 10^{-5}) per gene per generation was cited for Drosophila. Estimates of the total mutation rate–the summation of mutation rates of the entire genome – in Drosophila are of the order of 5 percent (or 5×10^{-2}) per generation (per haploid complement, or per gamete). These values can be used to estimate the number of genes in the Drosophila genome (haploid complement). If the mutation rate per gene in m and the mutation rate per total genome is M, then the number of genes in the genome (N), will equal M/m. Using the above estimates, N = 5 × 10–2/10–5 = 5000. This estimate agrees well with estimates of the number of genes in Drosophila obtained by other genetic criteria. On the other hand, one biochemical estimate suggests that Drosophila has about three times that many genes.

In the Drosophila genome (haploid) contains only 5000–15,000 genes, there is a major difference in the organization of the Drosophila genome and the genomes of the extensively studied bacteria and viruses. This difference, in fact, appears to be a basic difference between prokaryotes and eukaryotes. In prokaryotes, essentially all of the DNA of the genome appears to represent structural genes. Estimates of the number of genes from mutation studies (in some case, identification of all or most of the genes of the organs) agree well with estimates obtained by dividing the total DNA content in nucleotide pairs) of the genome by 1000 (103 nucleotide pairs per "average" gene). The phage lambda chromosome consists of 45×10^3 nucleotide pairs and about

40 known genes. The phage T4 chromosome contains about 200 × 10^3 nucleotide pairs. Over 100 genes have been mapped in phage T4, and a significant number remain to be identified. The E. coli genome contains about 4000 × 10^3 nucleotide pairs, while mutation analyses suggest that it contains about 3000 genes. In these prokaryotic organisms, then, the genomes are composed almost entirely of structural genes. The Drosophila melanogaster haploid chromosome complement contains about 120,000 × 10^3 nucleotide pairs. If all of the DNA represented structural genes of average size (103 nucleotide pairs), then Drosophila should have about 120,000 genes, rather than 5000 to 15,000 genes as indicated by the calculation above. It now seems clear the most of the DNA of Drosohila and other eukaryotes, including humans, does not represent structural genes. What is the function of this "excess" or noncoding DNA? Many geneticists believe that this noncoding DNA is regulatory, playing important roles in the regulaion of gene expression. Others believe that it has an important role in some aspect of chromosome structure. Still others believe that much of the noncoding DNA is just "junk" DNA with no important function, possibly a reservoir of nucleotide-pair sequences available for the evolution of new genes. In any case, this DNA does not code for proteins. This "noncoding" DNA is of two types: DNA that is transcribed, but for which the transcripts never leave the nuclei, and (2) nontranscribed DNA. These two types of DNA may have very different functions. The first type of noncoding DNA clearly includes many of the noncoding intervening sequences or introns of eukaryotic genes.

An estimate of the maximum number of essential genes per genome can also be made from considerations of average mutation rates and the maximum mutation "loads" that a species could be expected to survive. For example, consider an average mutation rate of 4 × 10^{-6} recessive lethal mutation per essential gene per generation (per gamete). J. L. King, and others, have emphasized that it is unlikely that a species could tolerate more than 0.8 new lethal mutation per zygote (or 0.4 per gamete). Dividing 0.4 new recessive lethal mutation per gamete by 4 × 10^{-6} new recessive lethal mutation per essential gene per gamete gives a maximum number of 10^5 essential genes per gamete. At 10^3 nucleotide pairs per gene, the 105 essential genes would represent a total for 10^8 nucleotide pairs. The haploid chromosome complement of mammals, including humans, contains about 3 × 10^8 nucleotide pairs. Thus, according to thee estimates, essential genes can represent a maximum of a only about 3 percent of the total DNA in the genome of mammals. The actual number of essential genes in mammals is probably closer to 1–2 × 10^4, or less than 1 percent of the total DNA.

Mutable Genes

Most gene are relatively stable and mutate infrequently, but in some organisms a few genes mutate spontaneously so often that individuals carrying them are mosaics of mutated and unmutated genes. The R locus, which is involved with anthocyanin pigment synthesis in maize, for example, was found by R.A. Emerson to undergo alteration much more frequently than other loci. The change from Rr to rr, for example, was found to occur, at the rate of 1 per 20,000 gametes. These "mutable" genes are either more unstable than others or they are influenced by other factors in the genetic environment. McClintock's conclusion from extensive studies on maize was that a mutable gene is not an autonomous entity, but is derived from an agent that is integrated at the site of the mutable gene to cause instability.

Examples of highly mutable genes have been found in both plants and animals, but they appear to be more common in plants. Mutations in highly mutable genes occur frequently in somatic tissue and occasionally in germ cells. Somatic mutations may show their effects as colour variegations (mosaics) in such plant part as endosperm, leaves, and petals. Many common plants–including the larkspur, snapdragon, sweet pea, four-o'clock, and morning glory–have colour variegations suggesting unstable or mutable genes.

The classical investigations of M. Demerec (1941) on a mutable gene called miniature-alpha in Drosophila virilis provided the first substantial data on the genetic properties of mutable gens in animals. These propertie are: (1) mutation occurs primarily before meiosis; (2) mutation occurs in both females and males, implying that meiotic crossing over is not involved (crossing over occurs only rearely in Drosophila males) and (3) mutation is strongly influenced by neighbouring genes. More recently, M.M. Green and others have described several mutable gene systems involving the white locus (eye colour) and other genes in D. melanogaster. White-crimson (w^c), for example, mutates to wild-type and phenotypes other than white-crimosn at a frequency of 10–3. This mutable system and several others at the w locus are associated with chromosome alterations, particularly deficiencies. This led to the hypothesis that a "*controlling element*" from the cytoplasm is integrated into the chromosome at the site of mutability, and that this agent is responsible for chromosome aberrations.

Many of these "controlling elements" have now been identified by G. Rubin, P. M. Bingham, and colleagues, and shown to be transposable elements with structures similar to the TN elements of bacteria. The unstable allele ww^c mentioned above has been shown to result from the insertion of an approximately 10,000 nucleotide-pair-long transposable element belonging to the class called FB (for "foldback") elements.

Another unstable allele, w^{a1} has been shown to have a transposable element called *copia* (because its sequence is present in RNA in *copious* amounts) inserted at the site of mutation. Copia has been isolated and sequenced. It is about 5,000 nucleotide pairs long with 276 nucleotide-pair perfect direct repeats at its ends. These 276 nucleotide-pair repeats, in turn, have 17 nucleotide-pair inverse repeats at each end. Note the similarity of *copia* to E. coli elements Tn 9 and Tn 10.

Mutator And Antimutator Genes

Genes in maize were shown by McClintock to influence the stability of other genes. These have been called mutator genes. A striking example of the action of a mutator gene with a specific effect on a basic colour gene in maize was described by M. M. Rhoades. The colour of maize leaves and other plant parts is dependent on a complex of three complementary genes, symbolized A, C and R. A colour other than green is produced only when the dominant alleles (A, C, and R) of all three genes are present. The colour may be purple if gene P is also present, or red (pp), or some other colour depending on what other genes are included along with A-C-R. Plant soft genotype aa, cc, or rr have green leaves regardless of alleles present at the other loci. Plants with the genotype aaC-R- would be expected to be green, but in the presence of a mutator gene called Dt, they are variegated. Light-coloured corn kernels have purple spots at locations where somatic mutations have occurred. The Dt gene produces its effect by influencing one or more of the a alleles of the aa genotype to mutate to A. Patches of cells scattered throughout the plant carry A alleles resulting from these mutate to A. Patches of cells scattered throughout the plant carry A alleles resulting from the mutations. On the leaves and kernels, these patches give a speckled appearance. The size of the spots depends on the stage of development at which the mutations occurred. Green cells contain unmutated genes; that is, they are of genotype aa. In this case, the allele (Dt or dt) present at one locus thus influences the mutation rate of an allele (a) present at another locus.

J. F. Speyer and others have found broad-spectrum mutator genes in bacteriophage T4. Temperature sensitive (ts) mutants of gene 43, for example, alter the mutation rates of point mutations in other genes. Some mutator genes exert their mutagenic effect during DNA replication by altering polymerase activity, producing mutagenic base analogs, or modifying DNA bases, thus influencing the mutation rates of other genes. One mutator in phage T4 gene 43 (ts L88) produces a DNA polymerase that utilizes incorrect nucleotides at a higher frequency than does the wild-type enzyme.

Several mutations in gene 43 exhibit powerful negative or antimutator activities, particularly during the formation of A : T → C : G substitutions. Experiments have shown that DNA polymerase (gene 43 products) that are isolated from E. coli that were infected with mutator, antimutator and wild-type strains for T4 bacteriophage discriminate between adenine and 2-aminopurine to different degrees during DNA synthesis in vitro. Significantly larger amounts of 2-aminopurine are incorporated into DNA by wild-type and mutator than by antimutator enzymes. This indicates that organisms have the potential to evolve mechanism that result in mutation rates that are lower than those presently existing in wild-type strains of somatic organisms.

When the mutator and antimutator DNA polyemerase of phage T4 were studied in vitro, they were found to have altered ratios of polymerase activity to 3'→5' exonuclease activity. Mutator have increased polymerase/exonuclease activity. Antimutators exhibit decreased polymerase/exonuclease activity. This suggests that the mutation rate is, at least in part, controlled by the relative rates of polymerization and proofreading rates decrease mutation rates (antimutator). Decreased proofreading efficiencies (or increased polymerization rates) increase mutation frequencies (mutator).

Several mutator genes have also been identified and characterized in E. coli.Some increase the frequency of only certain types of base-pair substitutions. Others cause an increase in all types of point mutations. R. C. von Borstell and others measured spontaneous mutation rates in yeast and isolated many mutator genes with different kinds of mutator and antimutator activity.

Decreases as well as increases in mutation rates, as compared with wild-type, can thus result from new mutations. With both mutators and antimutaors operating in the same system, particular spontaneous mutation rates may be optimized through natural selection. Mutators and antimutators thus become relative terms, because no standards are available for optimum rates. Results of studies on yeast have borne out the conclusions of those on viruses and bacteria; that spontaneous mutation rates are under the genetic control of the cell itself.

Insertion Sequence ("Is Elements") As Mutators

A significant proportion of the mutations that occur in bacteria and some bacteriophages are now known to be effects of small sequences of DNA called IS elements or insertion sequences. These DNA sequences, from about 800 to about 1400 nucleotide paris in length are present in E. coli chromosomes, for example, in several copies. The first five IS elements to be characterized were: ISI (768 nucleotide pairs long) and IS2, IS3, IS4, and IS5 (all between 1200 and 1400 nucleotide pairs in length). The IS elements are transposable; that is, they have the ability to move from one position in the chromosome or to other chromosomes in the same cell. They can move, for example, from the E. coli chromosome to bacteriophage chromosomes or to plasmids (small circular molecules of DNA, or "minichromosomes," often present in bacteria.

The chromosome of the strains of E. coli that have been studied contain about eight copies of ISI elements are physically and covalently inserted as linear sequences into the chromosomes when an IS element is inserted (or insert itself?) into a gene, it destroys the function of, mutates, that gene. While insertion of IS elements occur at many sites in the E. coli chromosomes, it is not completely at random. Preferred sites of insertion have been demonstrated in some case.

Detectable IS-induced mutations occur with a frequency of about 10–6 to 10–7 per cell generation in E. coli. IS-induced mutations are revertible. They can be distinguished from all mutations with similar effects in that their reversion frequency (usually in the range 10–6 to 10–7) can not be enhanced with mutagens. The fact that IS-induced mutations are revertible means that IS elements can excise themselves from the chromosome with precision to the single nucleotide pair. Reversion to the functional state requires restoration of the original nuclotide-pair sequence of the gene; no base-pairs can be added or deleted during the insertion-excision process. The failure of mutagens to hence reversion is to be recombination mechanism, not a mutation process.

The "controlling elements" and mutable loci (or unstable genes) in higher plants and animals are probably caused by sequences of DNA analogous to the IS elements in prokaryotes.

INDUCED MUTATIONS

Mutations can be induced by either physical or chemical means: such mutations are called induced mutations. Irradiation is an example of a physical mutagen, with x-rays, gamma-rays, and ultraviolet light being the most common examples used. One consequence of x- or gamma-ray irradiation is the breakage of chromosomes, which may result in chromosomal rearrangements, or the events may be lethal to the cell.

5-Bromouracil

5-Bromouracil (5-BU) is a base analog; that is, its structure closely resembles one of the bases normally found in DNA. 5-BU can exist in two states. In its usual keto state it exhibits properties similar to those of thymine and thus will pair with adenine in DNA. Rarely it switches to the enol state, and in this form it will pair specifically with guanine.

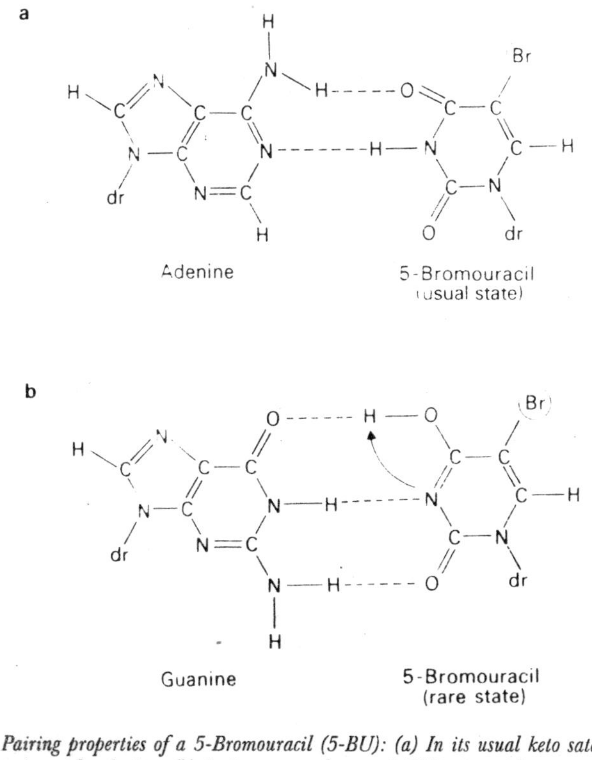

a

Adenine

5-Bromouracil (usual state)

b

Guanine

5-Bromouracil (rare state)

Fig. 19.9. Pairing properties of a 5-Bromouracil (5-BU): (a) In its usual keto sate, 5-BU pairs with adenine; (b) In its rare enol state, 5-BU pairs with guanine. (dr = deoxyribose.)

Mutations can be induced by 5-BU (and in general by base-analong muta-gens) in two ways. The first involves the incorporation of the usual form of 5-BU into DNA during replication. If 5-BU shifts to its rare enol state during the next round of replication, the result will be a transition mutation from AT to GC.

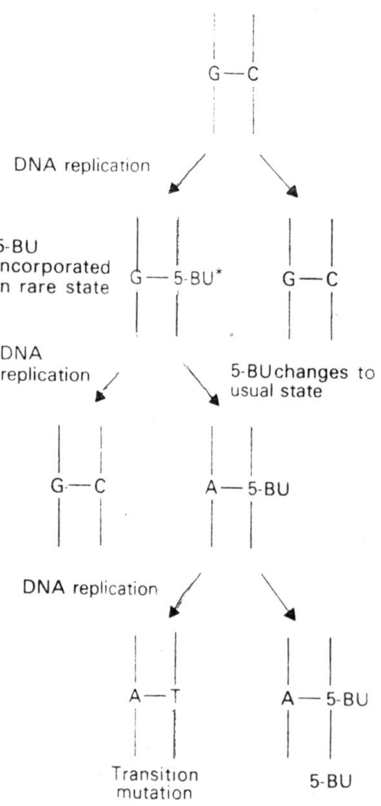

Fig. 19.10. Mutagenic action of 5-BU when it incorporates into DNA in its usual keto state and then shifts to its rare state during the next round of replication.

The second way that 5-BU can induce mutations is if the base analog g is incorporated in the DNA while it is the rare enol state. This dictates insertion opposite a G on the complementary strand, and subsequent replication with a shift of the 5-BU to the usual keto state will result in a transition mutation from GC to AT.

Thus 5-BU can induce either AT to GC or CG-to-AT transition

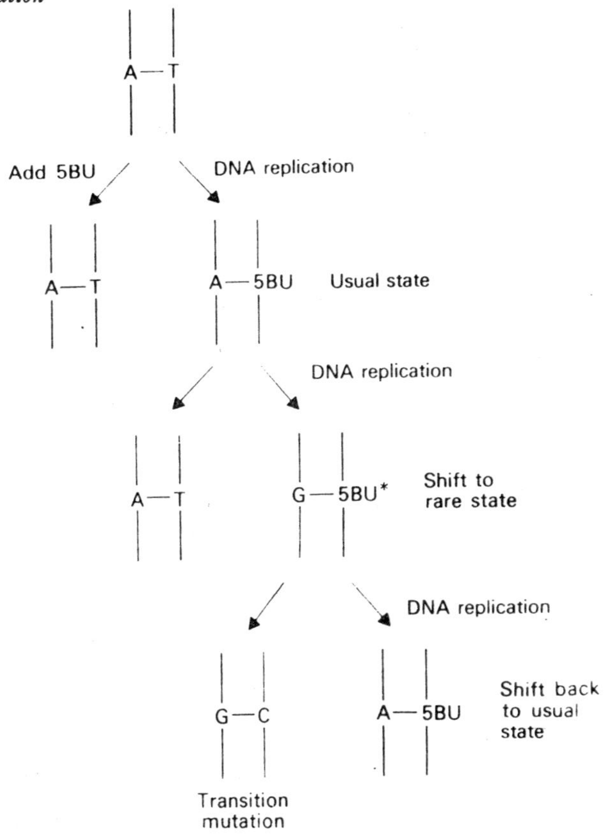

Fig. 19.11. Mutagenic action of 5-BU when it incorporates into DNA in the rate state and then shifts to the usual keto state during the next round of replication.

mutations. (In the jargon of this area, 5-BU is said to induce two-way transition mutations.) Therefore it is possible to correct a 5-BU induced transition mutation by treating with 5-BU for a second time. This is called *reversion* of the mutation.

2-Aminopurine

2-Aminopurine (2-AP) is also a base analog and, like 5-BU, it can exist in two states. In its usual state it behaves like adenine and will form two hydrogen bonds with thymine. In its rare amino state 2-AP behaves like guanine and forms two hydrogen bonds with cytosine. Thus 2-AP can induce transition mutations both from AT-to-GC and from GC-to-AT. 2-AP induced mutations can therefore be reverted by 2-AP treatment.

2-Aminopurine

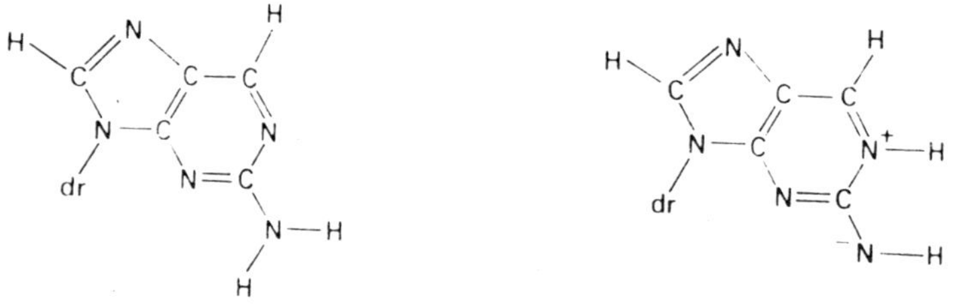

a Usual state

b Rare state

Fig. 19.12. Structure of 2-Aminopurine: (a) In its usual state (pairs with thymine) and (b) in its rare imino state (pairs with cytosine).

Nitrous Acid

Nitrous acid (NA : HNO_2) is a deminating, agent; it acts by removing amino groups (NH_2) from the bases. In some but not all

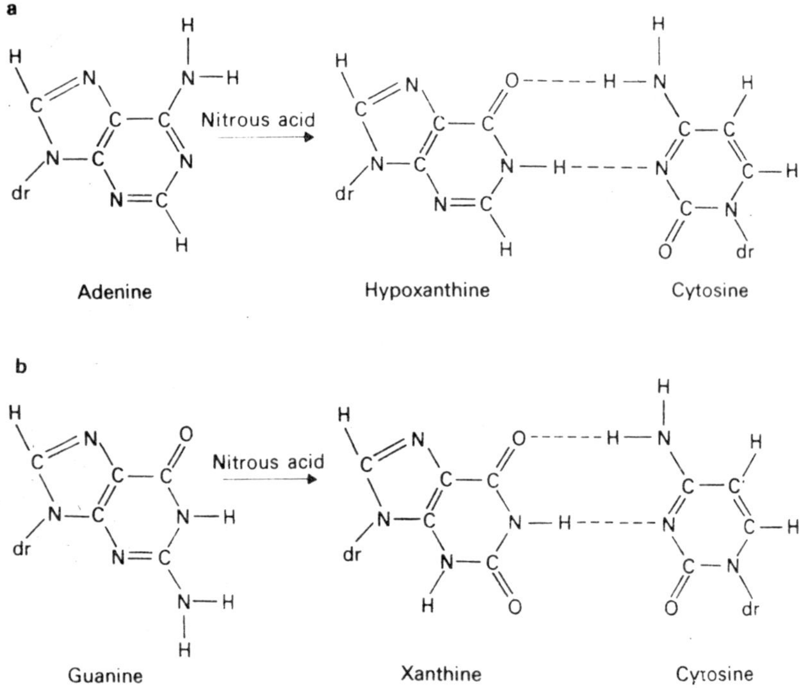

a

Nitrous acid

Adenine Hypoxanthine Cytosine

b

Nitrous acid

Guanine Xanthine Cytosine

Fig. 19.13. Mutagenic action of nitrous acid. (a) Deamination of adenine by nitrous acid treatment produces hypoxanthine, which pairs with cytosine; (b) Deamination of guanine produces xanthine, which pairs with cytosine.

instances this alters their base-pairing abilities and hence induces mutations. The three bases that have amino groups are adenine, guanine, and cytosine. When adenine is treated with NA, it is changed to hypoxanthine, which will pair with cytosine. This results in an AT-to-GC transition mutation.

Treatment of guanine with NA removes the amino group from the 2-carbon position and produces xanthine. However, since both guanine and xanthine pair with cytosine, no base-pair mutation results.

Deamination of cytosine by NA produces uracil, which, or course, pairs with adenine. This results in a GC-to-AT transition mutation–the opposite of NA's effect on adenine. Thus mutations induced by nitrous acid can be reverted by nitrous acid treatment: in other words, nitrous acid induces two-way transition mutations.

Hydroxylamine

Hydroxylamine (NH_2OH) reacts only with cytosine, hydroxylating it so that it can then only with adenine . Thus hydroxlamine induces one-way transition mutation from GC to AT. Because of this, mutations induced by hydroxylamine cannot be reverted to revert by 5-BU. 2-AP, or NA treatment since these mutagens can bring about a GC-to-AT transition.

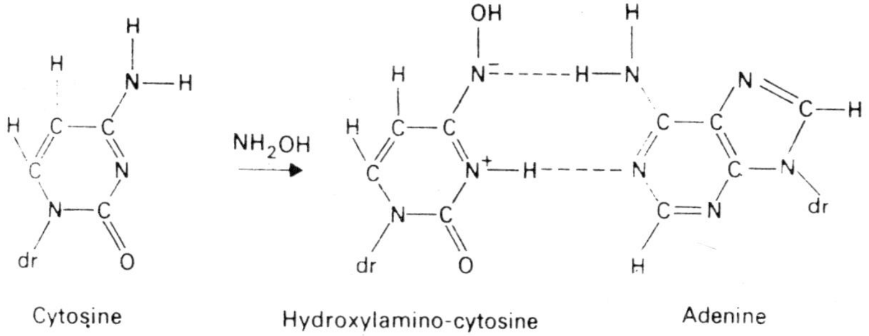

Cytosine $\xrightarrow{NH_2OH}$ Hydroxylamino-cytosine — Adenine

Fig. 19.14. Mutagenic action of hydroxylamine.

Acridines

Acridine treatment results in the addition or deletion of one base pair in the DNA. This has serious consequences, since the amino acid sequence of a protein coded for by a stretch of DNA altered in this fashion will be changed drastically. This will become more apparent in later discussions of messenger RNA translation.

When present at relatively low concentrations, acridine acts by becoming inserted between adjacent base pairs in the DNA. When this

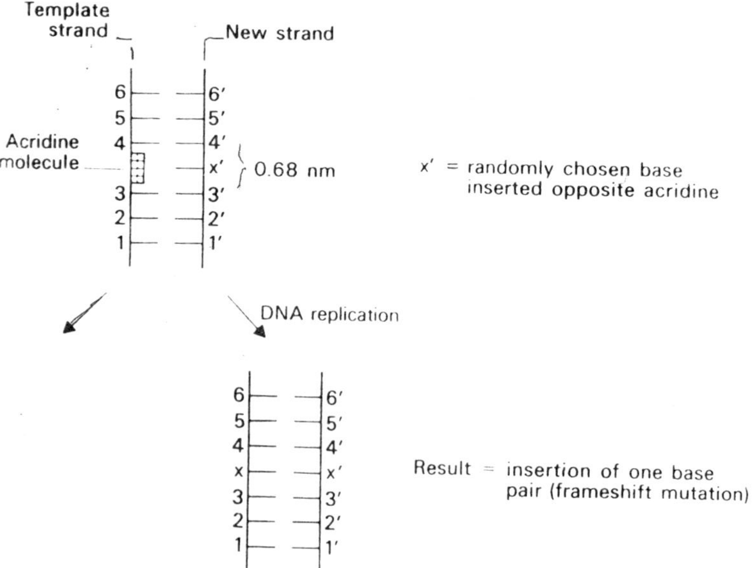

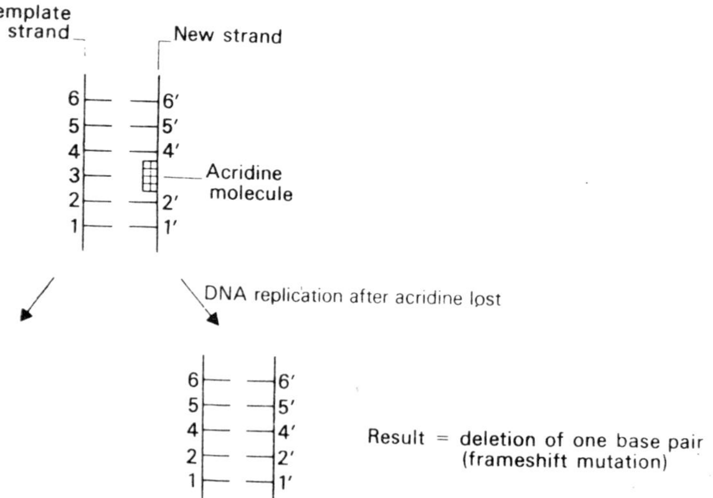

Fig. 19.15. Mutagenic action of acridines by intercalation into DNA. (a) Generation of addition mutation when acridine inserts in template strands. (b) Generation of deletion mutation when acridine inserts into newly synthesizing strand.

occurs, it "stretches" the distance between adjacent base pairs to 0.68 nm, which is precisely double the normal distance. The consequences of this depend on whether the acridine molecule is inserted into the template strand (the one being copied) or into the strand being synthesized. In the former case, a randomly chosen base is inserted opposite the acridine molecule when the replication fork passes by. At the next round of replication, the correct complementary base is paired with the inserted base, with the result that one base pair is added to the DNA in that region. This is called an insertion mutation.

Alternatively, if the acridine becomes inserted into the newly synthesized strand, it blocks one of the bases on the template strand from having a complementary base Then, if the acridine is lost before the next round of replication, the result will be a deletion of a base pair. Therefore it is possible to revert an acridine-induced mutation by a second treatment with acridine.

In summary, the mutagens described in this section have different modes of action and cause different mutation changes.

Table 19.3. Summary of the method of action of various chemical mutagens.

Mutagen	*Base-pair changes*
5-Bromouracil	AT $\leftrightarrow$ GC two-way transitions
2-Aminopurine	AT $\leftrightarrow$ GC two-way transitions
Nitrous acid	AT $\leftrightarrow$ GC two-way transitions
Hydroxylamine	GC $\rightarrow$ AT one-way transitions
Acridines	+1 or −1 insertion or deletion

Some of the chemical mutagens described in this section are commonly used in the laboratory. Many other chemicals appear to cause mutations, and public awareness of this is increasing as industrial effluents, cosmetic ingredients, food additives, etc. are examined carefully for any mutagenic activity in test organisms.

20

CHROMOSOMAL ABERRATIONS

It has already been stated that chromosomes occasionally break and that the rate of breakage can be increased by the use of various mutagens such as ionizing radiation(s), ultraviolet light, and certain chemicals. When a chromosome breaks, the two broken ends usually undergo *restitution* or healing, and everything is as before. But this does not always happens, and the result can be the formation of (1) deficiencies, (2) duplications, (3) inversions, or (4) translocations. The first three of these ordinarily involve a single chromosome, while the fourth involves two or more chromosomes.

There is very strong evidence the DNA in chromosomes that extends as a double helical thread throughout the entire length of the structure. Furthermore, we know that the genes are linearly arranged in the DNA and that at least in some instances groups of genes under the control of common regulatory units are grouped together. It is of great interest then to ask about the genetic consequences of breaks in specific regions of chromosomes. One interesting observation made by George Lefevre on Drosophila chromosomes is that perhaps as many as 70 percent of chromosome break points that have successfully reunited with new neighbouring genes have no detectable genetic effect at all. In other words, only about 30 percent of healed breaks points have mutant effects associated with them. This could mean that only a portion of the DNA of chromosomes is informational, and breaks within the regions without information are inconsequential. It could also means that these regions may be more susceptible to breakage and rejoining than others. We do not yet know, but C. S. Lee has suggested that those break points that manage to reunite are likely to be in regions of the DNA that are represented repeatedly in the genome. The reasoning behind this suggestion has to do with the healing process itself rather than with a differential fragility of chromosome regions. The union of broken ends may involve some degree of base pairing at the joint in order for the DNA strands to be properly positioned and stabilized for ligase to be effective in forming covalent bonds between the two chains. It is of course possible that unions can be formed without a perfect match between the bases of the two strands. Experiments, with a λ phage by Wu and Taylor have shown that the ends of the λ chromosome in its linear form have complementary sequences of nucleotides at the protruding cohesive ends. The λ

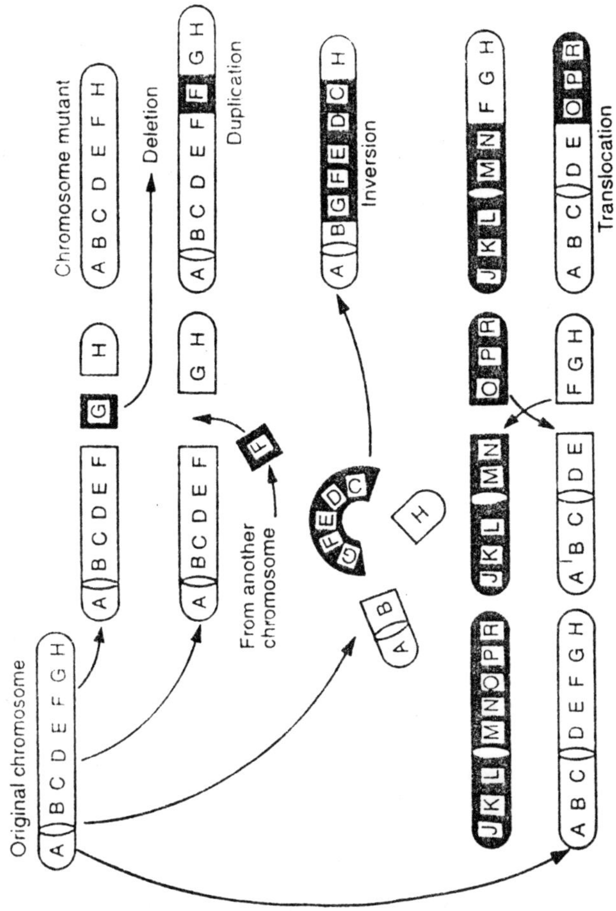

Fig. 20.1. Structural changes in chromosome.

chromosome goes from the linear to the circular form by the formation of hydrogen bonds between the complementary bases as shown in the figure. The same sort of protruding single strand ends with complementary or mostly complementary sequences may cause the joining of broken ends of eukaryotic chromosomes. Broken chromosome ends have been described for many years by cytogeneticists as being sticky. This sickness may be nothing more than the complementarily that exists between repeated DNA sequences.

It is known that in some tissues of many organisms, such as the endosperm of maize kernels, where the broken end soft chromosomes remain unhealed at the time of chromosome replication, that there is often a fusion of the broken ends of the sister chromatids. The results in a dicentric chromosome that will produce a chromosome bridge at the anaphase of the next mitotic division. In plant cells, such a bridge will often break as the cell wall is laid down between newly formed daughter cells. This produces another broken end in each of the nuclei, and the cycle may begin again. This breakage-fusion-bridge cycle may be carried on for several cell generations, and it can be used to great advantage for manipulating gene dosage in such cells. As the bridge breaks, it may not do so at the original fusion point and the result will be a duplication of some genes in one cell and their deletion in the other. McClintock used such a scheme to study chromosome organization and gene regulation in the endosperm of maize.

For most cells, however, the rule is that aberrations that persist are those that result from rejoined ends. This is because a broken chromosome, unrejoined, will generally not persist through the zygote stage. The segment without the centromere, called the acentric fragment, will be lost. The segment that has the centromere may act as a dominant lethal, depending on how much and what is lost in the acentric fragment.

Deletions

Deletions are also referred to as *deficiencies*. Some geneticists, following the historical origins of the terms, use deletion to describe an intercalary (involving two breaks in the chromosome) loss of chromatin and deficiency to describe a terminal loss. But here we use the terms deficiency and deletion as synonyms, because in either case an acentric fragment is lost. A spindle fiber cannot attach to an acentri fragment during mitosis or meiosis and guide it to one of the poles.

If the DNA in the chromatin that is lost is crucial for viability, sterile gametes or nonfunctional somatic cells may result. Some

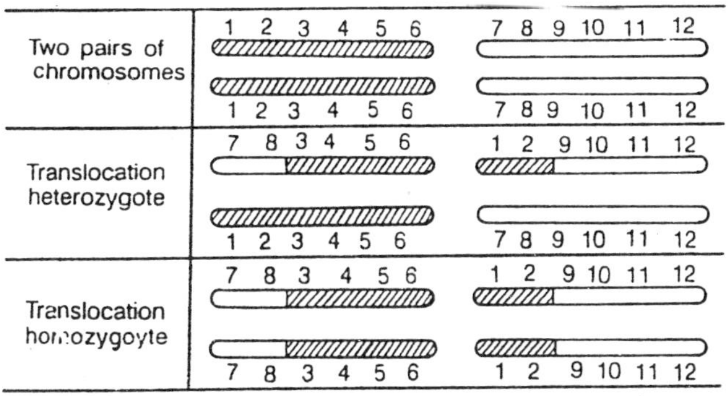

Two pairs of chromosomes
1 2 3 4 5 6
7 8 9 10 11 12
1 2 3 4 5 6
7 8 9 10 11 12
Translocation heterozygote
7 8 3 4 5 6
1 2 9 10 11 12
1 2 3 4 5 6
7 8 9 10 11 12
Translocation homozygoyte
7 8 3 4 5 6
1 2 9 10 11 12
7 8 3 4 5 6
1 2 9 10 11 12

Fig. 20.2. Chromosome constitution of a translocation heterozygote and a translocation homozygote.

deficiencies are known that are viable even when homozygous. For example, in *D. melanogaster*, a deficiency that removes the *white* locus produces, when homozygous, fully developed flies that appear normal except for their white eye colour. Most deficiencies, however, are lethal when homozygous, and these may involve no more loss of chromatin than in the example of white locus. A loss of a large segment of a chromosome may result in a dominant lethal effect even if a normal homologue is present with it in the heterozygote. If the dominant effect of a deletion is not lethal, it may result in an abnormal phenotype.

Genetic Significance of Deficiencies

The chromosomal deficiencies have following genetic effects :

(i) Lethal effect

Organisms with homozygous deficiency usually do not survive to an adult stage because a complete set of genes is lacking.

normal chromosome
a b c d e f g h
a b c d e f g h
a b c d e f g h
c d e f g h
a b e f g h
terminal deletion
interstitial deletion

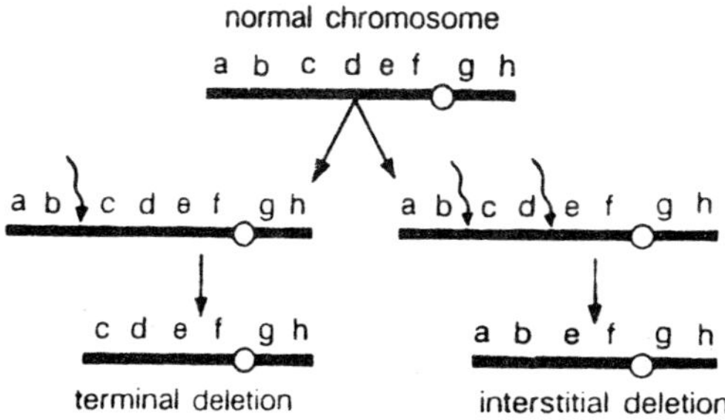

Fig. 20.3. Production of terminal and interstitial deletion. Chromosomes can be broken when struck by ionizing radiation (wavy arrows).

Examples

A mutant sex-linked condition called "notch" is lethal in *drosophila* when hemizygous in males or when homozygous in females. This "notch phenotype is a sex-linked deletion (deficiency) which acts like a dominant mutation ; a deletion at another sex-lined locus behaves as recessive mutation, producing yellow body colour when homozygous.

(ii) Pseudodominance

When an organism heterozygous for a pair of alleles, A and a, loses a small portion of the chromosome bearing the dominant allele (A), the recessive allele (a) on the other chromosome being hemizygous will become expressed phylogenetically. The phenotypic expression of recessive allele (a) due to deficiency of dominant allele (A) called *pseudodominance.*

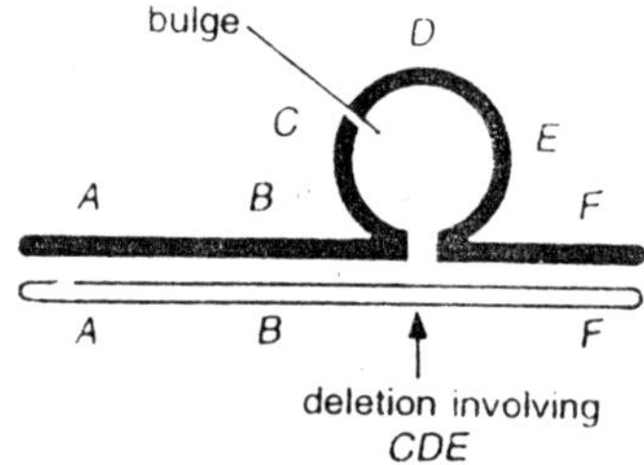

Fig. 20.4. Formation of a deletion loop during synapsis in a deletion heterozygote.

With a deficiency exchange (crossing over) becomes impossible. Disturbance of pairing causes a reduction of exchange in a segmen around the deficiency.

Occurrence of Deletions in Human Beings

Deletions are expected to occur in all organisms including humans. Some do not cause early spontaneous abortions, but unfortunately allow the fetus to come to full term and to be born as an abnormal child.

One example of this is the cri-du-chat or "cat cry" syndrome, which derives its name from the fact that the afflicted infant cries like a cat. Approximately 85 percent of children with cri-du-chat show a heterozygous deletion for the short arm of chromosome 5. The remaining 15 percent exhibit a translocation of part of chromosome 5.

DUPLICATION

Duplications are repeats of chromatin segments within a chromosome. Chromosome containing a duplication therefore possess more than the normal amount of DNA. There are several

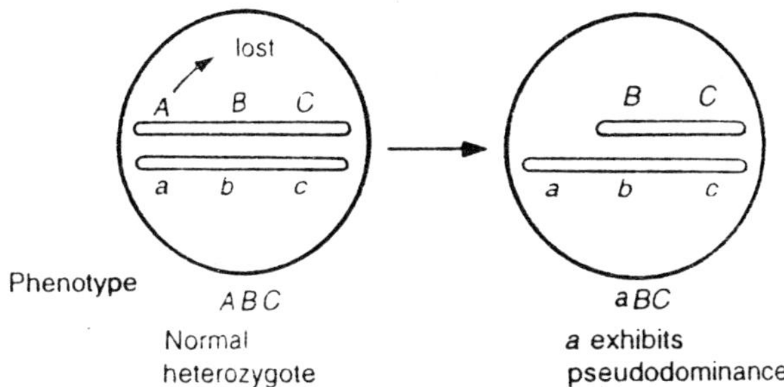

Fig. 20.5. Pseudodominance. A deficiency in the segment of chromosome bearing the dominant gene A allows the recessive allele a to become phenotypically expressed.

possible mechanisms for generating duplicate DNA segments in chromosomes.

Duplications and deletions may also be generated by the breakage-fusion-bridge cycle. A chromosome with the normal sequence of 1.23456 has a break between 4 and 5. Since the deleted 5,6 is an acentric fragment it will probably be lost. The replication of the 1.2 3 4 fragment, followed by joining of the two broken ends, results in the dicentric chromosome 1. 234-432. 1. Since homologous centromeres generally move in opposite directions, an anaphase bridge is formed and breakage will occur. The genetic constitution of the resulting chromosomes depends on where the break occurs. Figure depicts a chromosome with deletions of regions 4,5 and 6. Also it shows a chromosome with both deleted (5,6) and duplicated (3,4) regions.

The breakage-fusion-bridge cycle that has been found in maize results in a wide variety of duplications and deletions. Note that

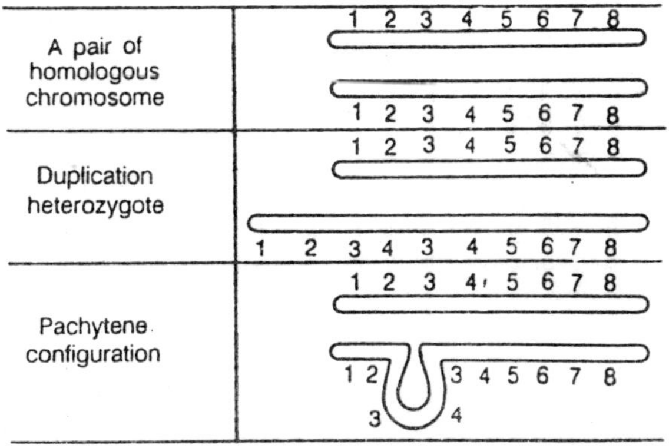

A pair of homologous chromosome	1 2 3 4 5 6 7 8 1 2 3 4 5 6 7 8
Duplication heterozygote	1 2 3 4 5 6 7 8 1 2 3 4 3 4 5 6 7 8
Pachytene configuration	1 2 3 4 5 6 7 8 1 2 3 4 3 4 5 6 7 8

Fig. 20.6. Loop formation during chromosome pairing in a duplication heterozygote.

duplications and deletions. Note that duplications created in this manner often are displaced, that is, duplicated segments are not positioned side by side on the chromosome.

Duplications play an important role in the regulation of gene activity because the multiple copies of a gene allow large quantities of RNA to be generated quickly. Examples of this type of reiteration are the cistrons coding for the ribosomal RNA's and those coding for histones. Table 20.1. gives some examples of proteins and RNA's thought to be coded by repetitive DNA sequences.

Table 20.1. Examples of repetitive DNA sequences

Group	*No repetitive DNA sequences*	*Degree of DNA or protein homology, %*
Satellite DNA	103–106	80–100
18S-28S Ribosomal RNA	100–600	97–100
5S Ribosomal RNA	100–200	97–100
tRNA	6–400	
Histones	10–1200	87–99
Hemoglobins	=10	<75–100

Satellite DNA's contain large numbers of repeating sequences which apparently are never transcribed. The nature of the mechanism(s) that generate large numbers of DNA segments in tandem linkage is unknown but it may well be by unequal crossing over. Also unknown are the factors that protect the repeated sequences from mutational changes.

Duplication occurs when a segment of the chromosome is represented two or more times in a chromosome of a homologous pair. This extra-chromosomal segment may be a free fragment with a centromere or chromosomal segment of the normal complement. During meiotic pairing the chromosome bearing the duplicated segment forms a loop to maximize the juxtraposition of homologous regions. In contrast to the deficiency loop, the duplication loop is formed by the duplicated segment, not by the normal segment. Pairing and exchange (crossing over) is inverted and displaced duplications leads to different secondary chromosome structural variants (i.e., chromosomal aberrations) such as reciprocal translocation, inversions, rings, acentric and dicentric chromatids.

Duplication of Complete Gene System

In considering gene duplications, it is important to remember that a structural gene cannot be considered an entity in itself. The expression of any structural gene is controlled or regulated by DNA not directly involved in coding for proteins. Therefore, the duplication of a structural gene without the corresponding duplication of its regulator may lead to a large array of new phenotypes including lethality.

Gene Duplication as a source of New Genetic Material

In addition to permitting a cell or organism to vary its genetic capabilities through gene dosage effects, duplications may also provide a cell with a reservoir of extra DNA. The duplicated genes could be considered as spare parts in the sense that one gene is sufficient to enable normal cell function. With one gene functioning in the normal manner, the duplicate would be free to undergo mutation, since its gene product could no longer be essential to maintain normal function.

By taking this reasoning a step further, it may be postulated that the duplicate gene may mutate and eventually evolve into a different gene coding for a protein with a new function. If so, gene duplications offer a mechanism for the evolution of new genes. There are many examples in nature that appear to fit the hypothesis that gene duplication is followed by gene changes through point mutation. Base nucleotide additions or deletions and frameshift mutations may result in genes coding for new phenotypes. The following two examples illustrate this concept.

Some of the most extensively studied human proteins that are thought to have arisen through gene duplications are the multiple hemoglobins of vertebrates. Hemoglobins are tetramers composed of two different polypeptide chains. Adult humans possess hemoglobins A and A_2. These two hemoglobins, represented by the formulas $\alpha_2\beta_2$ and $\alpha_2\delta_2$, respectively, share a common α-chain, but have distinct β and δ-chains. Thus, the hemoglobins found in the red blood cells of adults are coded by three distinct genes. Hemoglobin studies in certain families with α-chain mutants indicate that there is probably more than one α-chain gene per haploid genome. In addition, during development, embryonic hemoglobins and fetal hemoglobins occur. The Gγ, Aγ, δ and β genes have been shown to be closely linked. As we have shown the amino acid sequences of the beta and delta chains are very similar, suggesting that the genes coding for these chains arose by means of gene duplication of a common ancestral gene. Furthermore, the gene that codes for the α polypeptide is structurally related to the other hemoglobin genes, but is not linked to them. The genes coding for the δ, γ, and β chains probably arose through gene duplications from the original a gene and subsequently moved to a different linkage group by means of translocation.

Since the myoglobin chain is also structurally related to the hemoglobin polypeptides, it is thought that its genes and all the hemoglobin genes are derived from a common ancestral gene by a series of duplications over a long period of time.

Haptoglobin, one of the iron-binding proteins in vertebrate serum, is another example of a protein that probably arose by gene duplication. Haptoglobin, like hemoglobin, is a tetramer. Apparently, two genes α and β are involved. The α locus has two alleles, α_1 and α_2. Three phenotypes occur in the population Hp1, Hp2, and Hp1-2. The sera of humans exhibit either the Hp1 type (α, β) or the Hp2 type (α_2,β_n), or the Hp–2 type (α, α_2, β_n). The Hp2 haptoglobin is unique to humans, suggesting that a recent gene duplication gave rise to the Hp2. Studies of amino acid sequences clearly show that a partial gene duplication (intracistronic duplication) us have arisen resulting in a large polypeptide chain, different from the original α gene product in size and ability to combine with several beta chains to yield the polymeric haptoglobin Hp1–2.

These two examples–hemoglobin and haptoglobin–clearly indicate that gene duplication has probably played an important role in evolution.

INVERSIONS

Inversions result when two breaks occur within a chromosome and the broken fragment is rejoined to the original chromosome in a reversed order. In this example, a chromosome with a normal sequence of 12345678910 undergoes breakage where the arms are in the contact while folded over on itself, say between 3 and 4 and 8 and 9. An inversion takes place when the 3 end fuses with the 8 end and the 4 end fuses with the 9 end to give the order 12387654910. Inversions are of two types, pericentric, with breaks flanking the centromere, and paracentric, with both breaks on the same side of the centromere.

Remember that the same polarity must be maintained along a strand of DNA. Proper integration of an inverted segment could never occur in a single strand of DNA. The proper integration of a double-stranded segment of DNA occurs if the inverted segment undergoes a 180° rotation.

TYPES OF INVERSIONS

The inversions are of following types :

(i) Pericentric Inversion

When the inverted segment of chromosome includes or contains centromere, then such inversions are called heterobranchial or

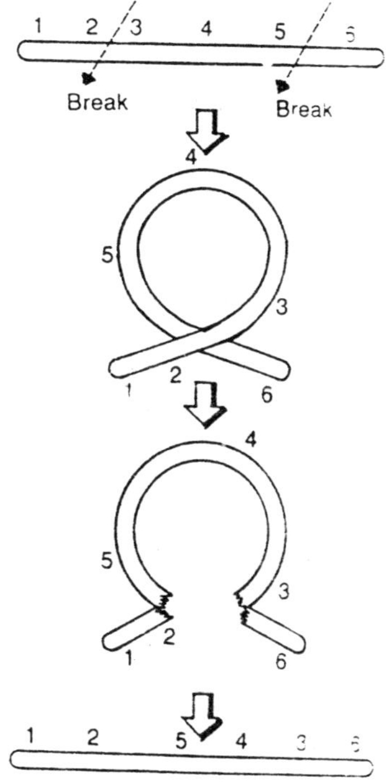

Fig. 20.7. Origin of an inversion in a chromosome.

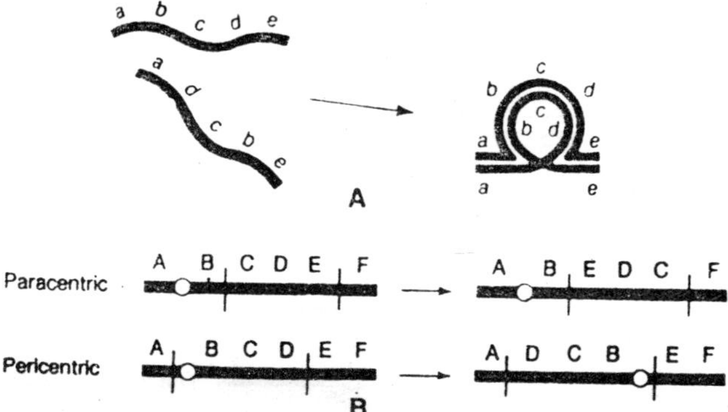

Fig. 20.8. A–An inversion loop in paired homologus of an inversion heterozygote; B–The location of the centromere relative to the inverted segment.

pericentric inversions. If crossing over occurs within the loop of a pericentric inversion, the resulted chromatids include half of the chromatids with duplications and deficiencies forming nonfunction.

The other half of the chromatids form functional gametes : 1/4 gametes have normal chromosome order, 1/4 gametes have the inverted arrangement.

(ii) Paracentric Inversions

When the inverted segment includes no centromere and the centromere remains located outside the segment, then such type of inversion is called homobranchial or paracentric inversion. Crossing over within the inverted segment of a paracentric inversion, produces a dicentric and an acentric chromosome. The dicentric chromosome contains two centromeres and forms a bridge from one pole to the other during first meiotic anaphase. When anaphase chromosomes separate towards poles, this bridge breaks somewhere along its length and the resulting fragments contain duplications and/or deficiencies. The acentric chromosome because lacks in centromere and fails to move to either pole and so, is not included in the meiotic products. Such, breakage-fusion bridge cycles of crossing over of paracentric inversions are most common in maize. The meiotic products includes

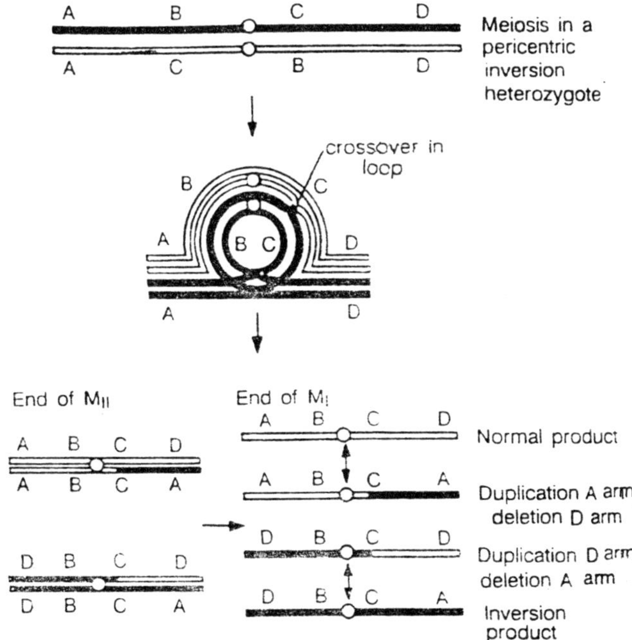

Fig. 20.9. Meiotic products resulting from a single crossover within a heterozygous pericentric inversion.

half non-function, 1/4 functional normal and 1/4 functional inverted chromosomes.

Cytological and Genetical Methods for Detection and Identification of Inversions

Since the banding patterns of polytene chromosomes are visible, is an obvious way of detecting inversions. Figure shows a polytene chromosome in Drosophila with an inversion paired with a normal chromosome. The development of techniques that differentially stain metaphase chromosomes has also made the cytological characterization of inversions possible in species other than Drosophila.

The detection of inversions also involves the study of pairing relationships between a chromosome containing inverted DNA and its noninverted homologue during the prophase stage of meiosis. Pairing would homologous chromosomes on a point-by-point basis. However, homologous pairing between an inverted arrangement and its normal arrangement is difficult because they cannot line up properly. Therefore, in order for pairing to occur, an inversion loop must be formed.

Inversions have been dubbed crossover suppressors. The results of recombination within paracentric and pericentric inversions will clearly illustrate the reason for this name, which is, in fact, not entirely appropriate. Keeping in mind that crossing over occurs at the four-strand stage, it is probable that only two of the nonsister chromatids will be involved in the crossover while the other two remain unaffected. Let us consider two chromosomes that pair, one chromosome containing the normal gene sequence, ABCD and the other chromosome containing an inverted segment, ABCD. After replication and pairing occur, and while they are still in the four-strand stage, a single crossover occurs between nonhomologous chromatids within the inversion loop (e.g,. between genes C and B). Such crossing over results in the formation of a dicentric chromosome ABCD and an acentric fragment, DCBD, both of which contain duplications and deficiencies. Dicentric chromosomes undergo anaphase bridge formation which, in some types of cells, may be followed by breakage. The result will be chromosomes with deletions and duplications. The acentric treatment probably will remain at the metaphase plate and subsequently be lost. All of the recombination products that result from crossovers that occur within a paracentric inversion will form deficient gametes. Crossing over within the inversion loop region of a pericentric inversion also produces duplication and deficiency products.

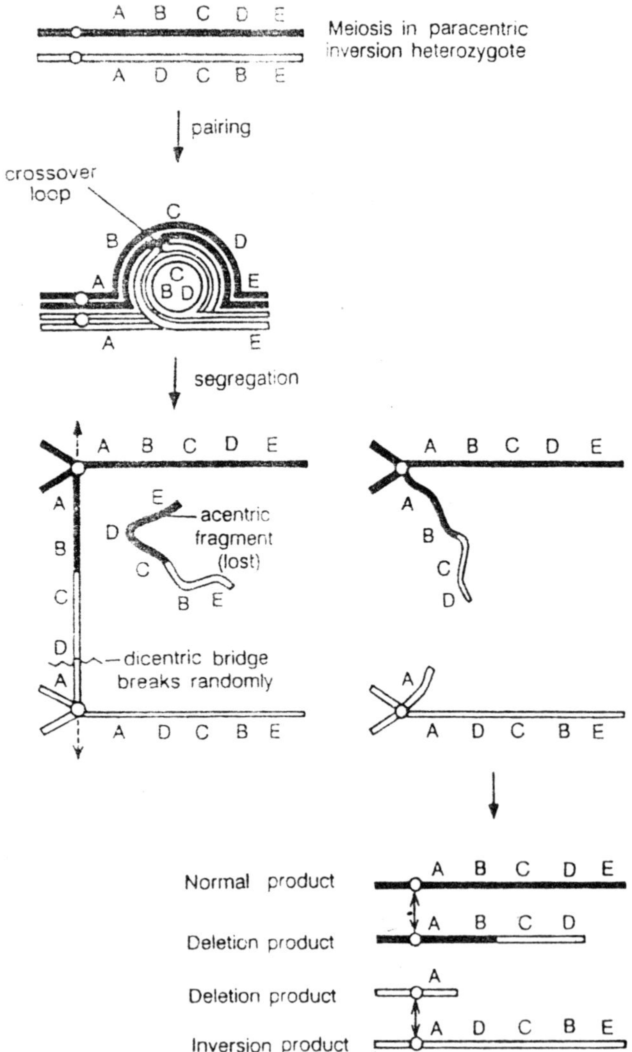

Fig. 20.10. Meiotic products resulting from a single crossing over within a heterozygous paracentric loop.

Inversions are crossover suppressors, not because crossovers fail to occur within them in heterozygotes, but because the recombinant chromosome resulting from a crossover does not permit the offspring receiving it to survive if the crossover occurs within the inversion.

As crossover suppressors, inversion. Provide useful tools foi geneticists. Inversions maintain particular seta of alleles of different loci within linkage groups by preventing recombinant offspring from surviving. Therefore, as for example in the isolation of sex-linked genes in Drosophila with the Basic stock. Inversions have also proved useful in the detection of linkage groups. A reduced number of recombinant types in a particular cross indicate that the genes are located within or near an inverted segment. And, conversely, a normal number of recombinant types indicate that the genes are located outside the inversion.

The number and type of inversions and the effects that they have in humans have not been extensively investigated until recently. Investigation has been limited by the difficulties in detecting inverted segments in human chromosomes. However, improved chromosome staining procedures now allow geneticists to detect paracentric and pericentric inversions in humans, but much work remains to be done in the field. One target area involves investigating the role of inversions in certain diseased states, such as heritable cancerous conditions.

TRANSLOCATIONS

Translocation is a broad term including all types of unilateral or bilateral transfer of chromosome segments from one chromosome to another. An important class of translocation having evolutionary significance is known as reciprocal translocations or segmental interchanges, which involve mutual exchange of chromosome segments between non-homologous chromosomes.

Cytology of a Translocation Heterozygote

If a translocation is present in one of the two sets of chromosomes, this will be a translocation heterozygote. In such a plant, the normal pairing into bivalents will not be possible among the chromosomes involved in translocation. Due to pairing between homologous segments of chromosomes, a cross-shaped (+) figure involving four chromosomes will be observed at pachytene. These four chromosomes at metaphase I can have one of the following three orientations:

1. *Alternate.* In alternate orientation, the alternate chromosomes will be oriented towards the same pole. In other words, the adjacent chromosome will orient towards opposite poles. This will be possible by the formation of a figure of eight.
2. *Adjacent I.* In adjacent I orientation, adjacent chromosomes having non-homologous centromeres will orient toward the same pole.

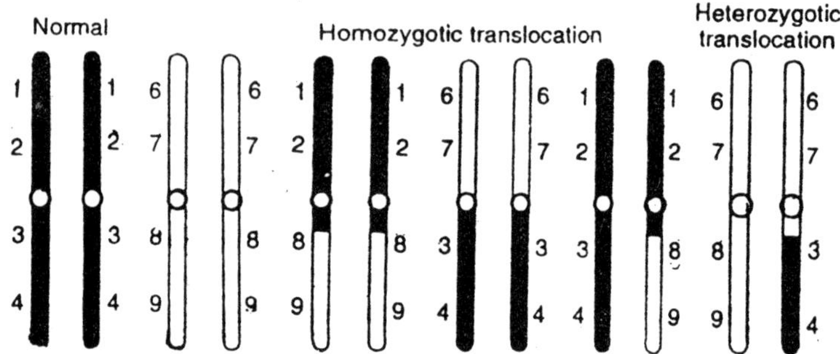

Fig. 20.11. Schematic representation of homozygotic and heterozygotic reciprocal translocations compared with the normal arrangement.

In other words the chromosomes having homologous centromeres will orient towards opposite poles. A ring of four chromosomes will be observed.

3. *Adjacent II.* In adjacent II orientation, the adjacent chromosomes having homologous centromeres will orient towards the same pole. A ring of four chromosomes will be observed.

The alternate disjunction will give functional gametes. Adjacent I and adjacent II disjunctions will form gametes, which would carry duplications or deficiencies and as a result would be non-functional or sterile. Therefore, in a plant having translocation in heterozygous condition, there will be considerable pollen sterility.

A ring of four chromosomes as described above is found under conditions when a single interchange is found. If two interchanges are involving three non-homologous chromosomes, a ring of six chromosomes is found, and the size of the ring can increase with additional interchanges. More than one rings can also be found if two or more interchanges are independently found, each involving two different non-homologous chromosomes.

The first case of translocation was found in *Oenothera*, which was originally described as mutation by de Varies while working for his Mutation Theory. Oenothera, *Tradescantia and Rhoeo* are such cases, where translocations in the heterozygous condition are frequently found in nature. In many other crop plants they have been artificially induced.

Breeding Behaviour of a Translocation Heterozygote

The presence of translocation heterozygosity can be detected by the presence of semi-sterility and low seed set. This can then be confirmed at meiosis by quadrivalent formation. As shown above only

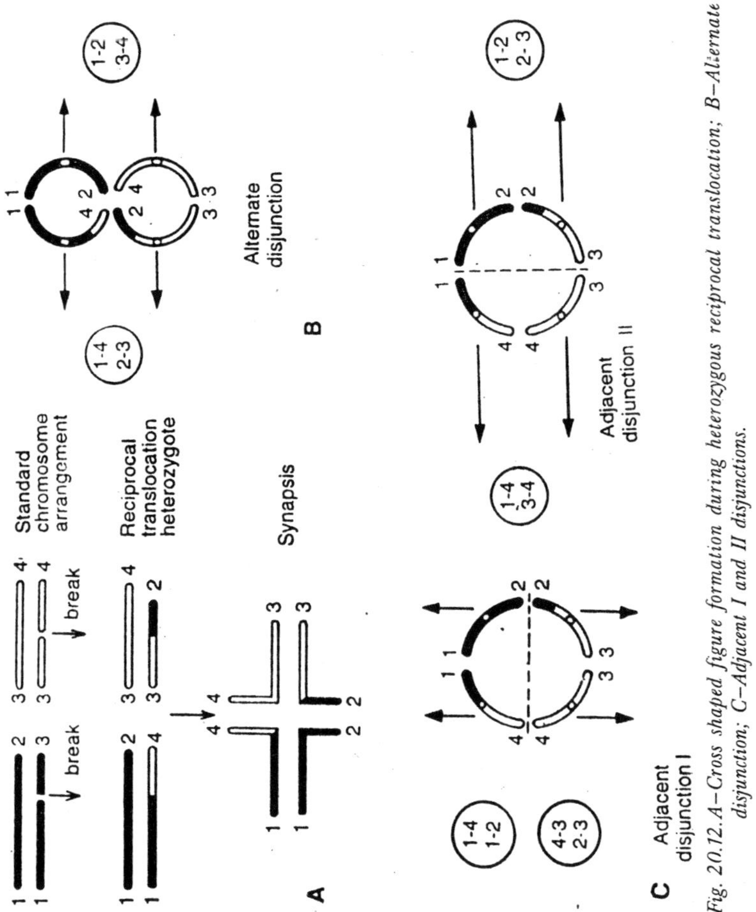

Fig. 20.12. A—Cross shaped figure formation during heterozygous reciprocal translocation; B—Alternate disjunction; C—Adjacent I and II disjunctions.

two types of functional gametes are forced which result from alternate disjunction.

The functional gametes will give rise to three kinds of progeny namely (i) normal (ii) translocation heterozygote and (iii) translocation homozygote. These three types would be obtained in 1 : 2 : 1 ratio.

Interchange Heterozygosity in Oenothera

Subgenus *Euoenothera* of the genus *Oenothera* has been studied during 1920-1930 and cytogenetic structure leading to evolution in this group was examined. This group has 2 n = 14 and all 7 chromosomes of haploid complement have median chromosomes. There are three classes. (i) First is represented with species showing bivalents or small rings at meiosis (e.g. *O. hookeri*, *O. grandiflora*, *O. argillicola*). (ii) Second class is represented by species forming rings of various sizes at meiosis indicating the presence of interchanges. These rings are not permanent but are maintained due to their superiority in adaptive value (e.g., *O. irrigua*). (ii) The third class is represented by those having permanent translocation heterozygosity involving all chromosomes, so that a ring of 14 chromosomes is regularly formed (e.g., *O. biennis*, *O. strigosa*, *O. parviflora*).

The three classes described above also differ in phenotypes like flower size etc. and can be identified. The members of third category behave like pure lines and are actually permanent heterozygotes.

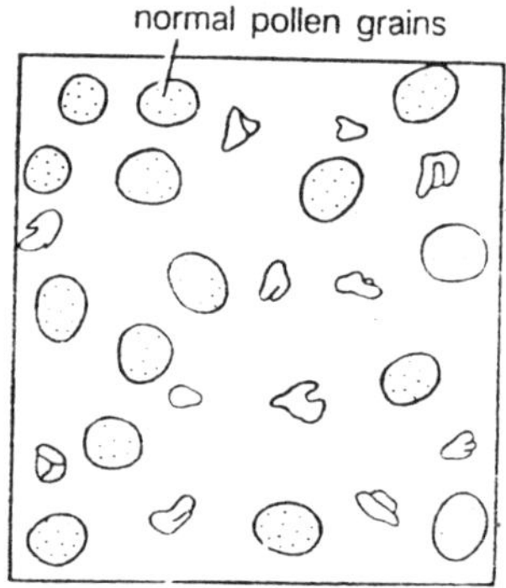

Fig. 20.13. Normal and aborted pollen in a semisterile corn plant. The small, shriveled pollen grains contain an euploid meiotic products of reciprocal translocation heterozygote.

Balanced lethals and gametic complex-permanent hybridity in Oenothera. Permanent hybridity is maintained due to operation of a balanced lethal system, which may function due to gametic lethality or zygotic lethality. Since complete rings are formed and alternate disjunction is a rule, only two types of gametes are formed showing complete linkage between chromosomes. The gametic and zygotic lethality leads to survival of only heterozygotes.

Chromosome Aberrations in Humans

Chromosome deletions are usually lethal even as heterozygotes, resulting in zygotic loss, stillbirths, or infant deaths. Sometimes infants with small chromosome deficiencies, however, survive long enough to permit observation of some of the abnormal phenotypes they express. Lejeune and his colleagues, for example, discovered a chromosome deficiency in humans that has been associated with the *cri-du-chat* (cat cry) *syndrome*. The name of this syndrome came from a plaintive catlike mewing cry from small weak infants with the disorder. Other characteristics microcephaly (small head), broad face and saddle nose, widely spaced eyes with epicanthic folds, unique facial features and physical and mental retardation IQs of cri-du-chat children studied are in the range of 20–40. The chromosome deficiency is in the short arm of chromosome 5 and is designated 5p.

Cri-du-chat patients die in infancy or early childhood and do not transfer the chromosome deletion to offspring. This chromosome deficiency, however, has been shown by Lejeune and others to become involved sometimes in a reciprocal translocation and thus to be transmitted. When the short arm of chromosome 5 became translocated to chromosome 15, the heterozygous translocation was carried in a normal healthy parent. Some gametes, however, carried only the deficient member of the translocation pair. Children inheriting the 5p chromosome expressed the cri-du-chat syndrome.

Another human disorder that is associated with a chromosome abnormality is chronic myelocytic (myelogenous) leukemia. A deletion of chromosome 22 was first described by Nowell and Hungerford and was called the Philadelphia (Ph) chromosome after the city in which the discovery was made. It was observed consistently in bone marrow preparations of patients with chronic myelocytic leukemia. Later, a translocation discovered by J. Rowley in a leukemia patient provided the correct chromosome rearrangement. A part of the long arm of chromosome 22 was translocated to another chromosome, usually chromosome 9 (46,XX.9q + 22q–), leaving a deficiency in the long arm of No.22.

More chromosomal anomalies are now being associated with malignancy. New staining methods permit more precise comparisons of (1) chromosome arms and (2) differentially stained sister chromatids. With balanced chromosomes, Burkitt lymphoma was related to a translocation of chromosome 14 [t(8q–; 14q+) (q24;q32). A deletion in chromosome 13q (band 14.1) is associated with the human embryonic tumor retinoblastoma. Williams tumor, an embryonic kidney tumor, is associated with a deletion in band 11p13. Sister chromatid exchanges, exchanges occurring between sister chromatids, occur with high frequency in somatic cells of patients with Bloom syndrome, indicating chromosome instability. Most people with this syndrome die with some form of cancer before age 30. Irregular chromosome numbers are also observed in malignant cells, particularly in later stages of cancer.

Like deficiencies, chromosome duplications are usually lethal even as heterozygotes, but sometimes they are sufficiently viable to permit observations of the abnormalities they produce. Like deficiencies, they may be associated with translocations and thus transmitted by normal healthy parents. Remember that a translocation is an exchange of parts between nonhomologous chromosomes or a transfer from one chromosome to a nonhomolog. Broken parts may be further divided and some may be lost or gained in a transfer. Thus, deficiencies and duplications may accompany a translocation carried in a parent. Such a parent may be a translocation heterozygote, with the long arm of one chromosome 21 attached at the centromere to the long arm of chromosome 14 along with normal chromosomes 14 and 21. Some gametes will contain a normal 21 and translocation 14q21q. When such a gamete is fertilized with a gamete from a normal individual, a translocated 21, along with two normal chromosomes 21, results in a viable infant with trisomy-21 and the Down syndrome.

With fluorescene microscope techniques, I.A. Uchida and C.C. Lin discovered a partial trisomy-12 that could not be identified by conventional methods. A body with some clinical features resembling the Down syndrome had been studied with conventional chromosome techniques. Because of continued lack of motor development at seven months age, further chromosomal investigations were carried out on lymphocyte cultures and slides stained with quinacrine dihydrochloride. With this technique, an additional band was identified in the short arm of chromosome 8. The mother's chromosomes were normal; but in the father, the short arm of No. 21 had become translocated to the short arm of No. 8 [46,XY, (8p+,12p–)]. The same chromosomal rearrangement was found in his daughter, an older sister of the patient. Since both father and daughter were clinically normal, the translocation was presumed to be reciprocal. The patients's No. 8 pair consisted of one normal and one translocation chromosome, similar to that of his father and sister, but both of his No. 12 chromosomes were normal. Thus, the boy had a duplication of part of the short arm of No. 12 and a deficiency of the tip of No. 8. Many translocations, addition to those cited above, have been detected in studies of human chromosomes, but most were apparently reciprocal and produced no phenotypic anomalies. At the Yale-New Haven Hospital, for example, cytological studies were conducted on 4500 infants born consecutively during one year. Lymphocytes from umbilical cord blood of each infant were grown in vitro and prepared for microscopic observation of chromosomes. Six translocations were detected. None of these was associated with a distinctive phenotypic anomaly. It is not the translocation process per se that produces abnormalities, but the imbalance of genetic material reflected in chromosome deficiencies and duplications that are produced by segregation of translocated chromosomes.

P. W. Allerdice et al. described a syndrome called the *chromo-some 3 duplication-deletion syndrome.* Phenotypically, this syndrome includes a group of morbid symptoms; stillbirths, neonatal deaths, and spontaneous abortions. Most pregnancies are lost, but two children survived and became probands for the investigation. He cannot sit up, turn over, or eat solid food. Facial malformation of the living children includes a distorted head shape; thick, low eyebrows; low hair-line; long eyelashes; persistent lanugo; distended veins on scalp; hypertelorism; oblique palpebral fissures; a very short nose with a broad, depressed bridge and anteverted nares; protruding maxilla; thin upper lip; micrognathia; low-set ears; and short, webbed neck. Port-wine stains, congenital glaucoma, cloudy corneas, cleft palate, and harelip also occur frequently. Each infant had difficulty sucking and swallowing. Internal physical abnormalities were noted in infants who died neonatally.

Giemsa- and quinacrine-banded karyotypes from a parent of an affected child revealed an inversion [inv(3)(p25q21). Fetal cells cultured for prenatal diagnosis from a subsequent pregnancy of this couple carried a recombinant chromosome 3, with the long arm described as rec (3) del p, dup q, inv(3)(p25q21). With a banded chromosome 3, the inversion was analyzed. The inverted segment included the centromere (pericentric). Breaks had occurred in the short arm (p) at band 25 and in the long arm (q) at band 21. The central part of the chromosome including the centromere had rotated 180° and become reinserted into chromosome 3. Inv 3(p25q21) had been carried in the kindred for at least four generations and 35 kindered members were presumed to carry it. Karyotype of Case 1 is 46,XYrec(3), dup q, inv(3) (p25q21). He was the second child born to a 23-year-old carrier mother with karyotype 46,XX, inv(3)(p25q21) and a 26-year-old father, karyotype 46,XY.

Crossing over within the inversion loop is presumed to produce the imbalance of genetic material associated with duplications and deficiencies and to cause the symptoms of the syndrome. Odd numbers of crossovers within the inverted are of a chromosome 3 would be expected to result in genetic imbalance. Earlier in this chapter, crossing over in *Drosophila* inverted chromosomes was shown to create irregular combinations resulting in lethals. This, in turn, resulted in "suppressing" crossing over. In humans, unbalanced zygotes from inverted chromosome crossovers sometimes survive. Some infants carrying unbalanced chromosomes are kept alive, but with varying degrees of birth defect handicaps. The degree of imbalance of genetic material may be determining factor for life or death and degree of abnormality.

21

Variations in Chromosome Number

Somatic cells of higher plants and animals usually have chromosomes in pairs (2n); that is, two of each kind of chromosome are present in each cell. Mature germ cells, having undergone reduction division, normally have one number of each pair (n). Many individual plants and animals, however, have local areas of somatic tissue characterized by a multiple of the basic chromosome number. A doubling process in cell division is the usual explanation for these deviations.

With the exception of sex differences, somatic doubling, and minor variations that occur in natural and experimental populations, all members of a species of plants or animals have the same basic chromosome number. Chromosome number can be overemphasized as a criterion for species identification, but it represents a valid characteristic to be used along with others for distinguishing species. The range or reported chromosome numbers in animals extends from 2 pair in a rhabdocoel Gyratrix hermaphroditus and some mites, midges, and scale insects, to more than 100 in some butterflies and Crustacea. The Crustacean, *Paralithodes camtschatica*, for example, has 208 chromosomes or 104 pairs. The reported range in plants is from 2 pairs in the small composite plant Haplopappus gracilis to several hundred in some ferns. A species of fern-like plants of the genus Ophioglossum is reported to have 768 chromosomes.

Where all individuals within a species, with the exceptions noted above, have the same chromosome number, different species within a genus often have different numbers. Cytological investigations of chromosomes help to unravel problems of species formation. The evolutionary path of a certain species can be followed in some cases by comparing numerical and structural relations of its chromosomes with those of other species within the genus. The essential genetic material is a major factor in determining evolutionary patterns, but chromosome number itself represents merely the number of packages into which the genetic material is divided. Chromosome number is probably more constant, however, than any other single morphological characteristic that is available for species identification.

Table 21.1. Chromosome number of some common animals and plants

Species	Number of chromosomes pair
Plants	
Garden pea, *Pisum sativum*	7
Sorghum, *Sorghum vulgare*	10
Maize, *Zea mays*	10
Johnson Grass, *Sorghum halepense*	20
Alfalfa, *Medicago sativa*	16
Barley, *Hordeum vulgare*	7
Oats, *Avena sativa*	21
Tomato, *Lycopersicon esculentum*	12
Tobacco, *Nicotiana tabacum*	24
Trillium, *Trillium erectum*	5
Animals	
Gypsy moth, *Lymantria dispar*	31
Mouse, *Mus musculus*	20
Rabbit, *Oryctolagus cuniculus*	22
Cow, *Boss tarus*	30
Horse, *Equus caballus*	32
Donkey, (ass) *Equus asinus*	31
Dog, *Canis familaris*	39
Monkey, *Macaca rhesus*	21
Chimpanzee, *Pan troglodytes*	24

Changes in the number of chromosomes may be reflected in phenotypic variations, which constitute a useful tool for identifying the influence of individual chromosomes. If, for example, phenotypically distinguishable individuals with different chromosome numbers can be identified in natural populations or produced experimentally, it is sometimes possible to determine the effect of adding or removing certain chromosomes. Some plants with increased chromosome numbers have phenotypic changes in morphological or physiological characteristics which are of practical importance to man. Grape and tomato plants with chromosome numbers above 2n are larger and produce more desirable fruit than do corresponding varieties with the usual 2n number.

Classifications of chromosome changes are arbitrary and superficial because these changes are necessarily interpreted in terms of obvious additions or eliminations of parts of chromosomes, whole chromosomes, or whole chromosome sets. The presently accepted classification system, therefore, is merely a working tool. Two main classes are euploidy and aneuploidy ("ploid," Greek for unit; "eu," true or even; and "aneu," uneven). Euploids have chromosome complements consisting of whole sets or genomes. The chromosome number of euploid organisms is basically represented by the monoploid (n). Euploids with chromosome numbers above the monoploid level may be diploid (2n), triploid (3n), tetraploid (4n), or have some other "polyploid" number.

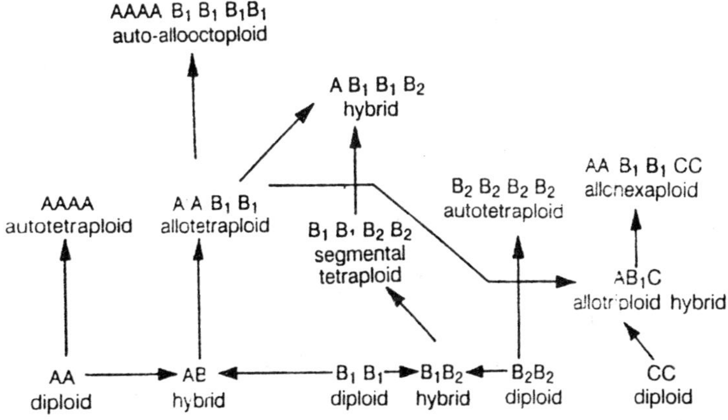

Fig. 21.1. Mode of formation of different kinds of polyploid.

ANEUPLOIDY

The first critical study of aneuploid plants was made by Blakeslee and Bellingusing the common Jimson weed Datura stramonium, which

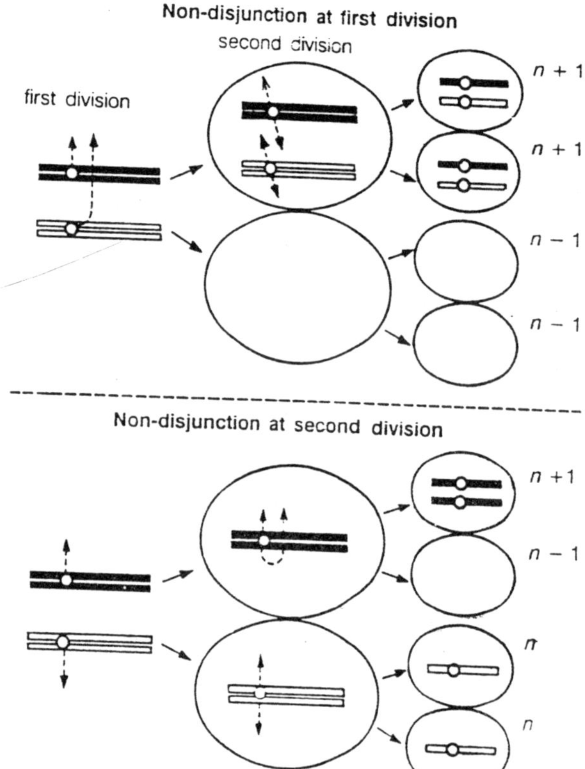

Fig. 21.2. The mode of origin of aneuploid gametes by non-disjunction at either the first or second meiotic division.

normally has 12 pairs of chromosomes in the somatic cells. In 1924, these investigators announced the discovery of a "mutant type" having 25 rather than 24 chromosomes. At the meiotic metaphase, one of the 12 pairs was found to have an extra member; that is, one trisome was present along with 11 disomes. This originally discovered trisomic plant differed from wild-type plants in several specific ways. Conspicuous deviations were observed in shape and spine characteristics of seed capsules. The chromosome complement with one extra member in addition to the regular set was illustrated by the formula 2n + 1. Theoretically, because the some have, however, been divided successfully into 7 groups identified with letters A to G. Numbers 1 to 22 are associated with the autosomes in descending order by length. All autosomes can be placed satisfactorily within a group but the numbering within the groups is more or less tentative. The X chromosome is difficult to distinguish from members of the C group and the Y chromosome shows considerable variability in different preparations.

Nullisomics (2n – 2)

Although nullisomy is a lethal condition in regular diploids, an organism like wheat (which "pretends" to be diploid but is fundamentally hexaploid) can tolerate nullisomy. In fact, all of the possible 21 wheat nullisomics have been produced. Their appearances are different from normal wheat; furthermore, most of them show less vigorous growth.

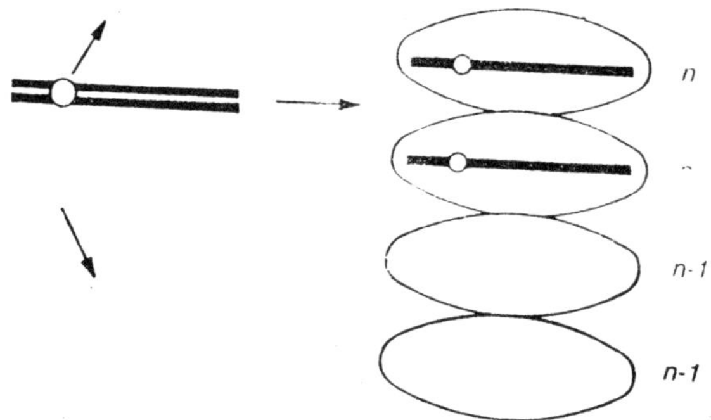

Fig. 21.3. Behaviour of a chromosome at meiosis.

Monosomics (2n – 1)

Monosomic chromosome complements are generally deleterious for two main reasons. First, the balance of chromosomes that is necessary for a finely tuned cellular homeostasis, carefully put together during evolution, is grossly disturbed. For example, if a genome consisting of two of each of chromosomes, a, b, and c becomes monosomic for c (that is 2a + 2b + 1c), the ratio of these chromosomes is changed from 1c:1 (a+b) to 1c : 2(a + b). Second, any deleterious recessive on the single remaining chromosome becomes hemizygous and may be directly expressed phenotypically. (Note that these are the same effects as those produced by deletions.)

Monosomics, trisomics (2n +1), and other chromosome aneuploids are probably produced by nondisjunction during mitosis or meiosis. In meiosis it can happen at either the first or the second division. If an n – 1 gamete is fertilized by an n gamete, a monosomic (2n – 1) zygote is produced; an n + 1 and an n gamete give a trisomic 2n +1; and an n + 1 and n + 1 give a tetrasomic if the same chromosome is involved or a double trisomic if different chromosomes are involved, and so on.

In Neurospora (a haploid), the n – 1 meiotic products abort, and do not darken like the normal ascospore; so MI and MII nondisjunctions are detected as asci with 4:4 and 6:2 ratios of normal to aborted spores. For loci on the aneuploid chromosomes, what ascus genotypes are produced?

In humans, the sex-chromosome monosomic (44 autosomes + 1X) produces a phenotype known as Turner's syndrome. Affected people have a characteristic, easily recognizable, phenotype: they are sterile females, are short in stature, and often have a web of skin extending between the neck and shoulders. Their intelligence is near-normal, although some specific cognitive functions are defective. Their frequency is about 1 in 5000 female births. Monosomics for all autosomes die in utero.

If viable, nullisomics and monosomics are useful in locating newly found recessive genes on specific chromosomes in plants. In one such method, different monosomic lines lacking a different chromosome in each line are obtained. Homozygotes for the new gene are crossed with each monosomic line, and the progeny of each cross are inspected for expression of the recessive phenotype. The cross in which the phenotype appears identifies its chromosomal location. In nullisomics and monosomics, of course, n - 1 gametes are produced. In general, these gametes tend to be more viable in a female parent than in a male. It is the union of these n – 1 gametes within n gametes, bearing the new mutation, that provides the crucial progeny types for the linkage test.

A similar approach can be used in humans. For example, two people whose vision is normal may produce a daughter who has Turner's syndrome and is also red-green colour-blind. This shows that the allele for red-green colour blindness is recessive, that it was on the X chromosome of the mother, and the nondisjunction must have occurred in the father.

Trisomics (2n + 1)

In trisomics, trivalents are regularly seen. For genes that are tightly linked to the centromere of a trisomic chromosome set, the random segregations can be represented a trisomic Aaa. All types occur equally frequently, and a gametic ratio of 1A:2Aa:2a:1aa is produced. Trisomics are sometimes recongnised by these ratios, which are also useful in locating genes on chromosomes. We have already observed a complete

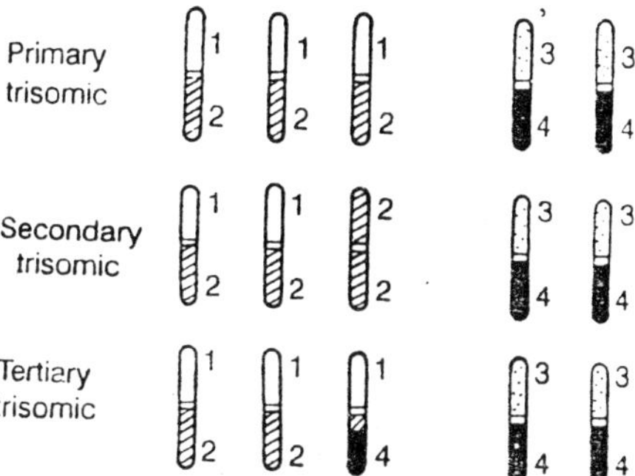

Fig. 21.4. Three kinds of trisomics.

set of trisomic lines in Datura. Once again, note that chromosome imbalance produces highly chromosome-specific deviations from the normal appearance.

In humans there are several examples of viable trisomics. The combination XXX (1/1000 male births) results in Klinefelter's syndrome, producing males that have lanky builds, are mentally retarded, and sterile. Another combination, XYY, also occurs in about

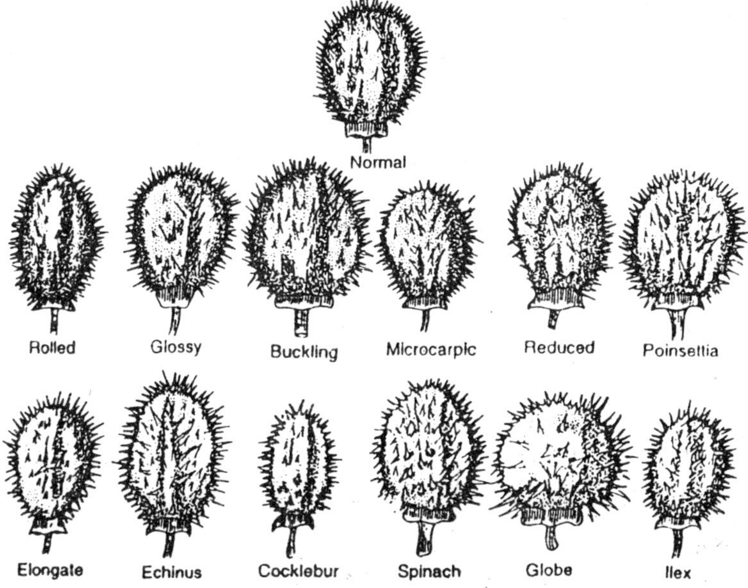

Fig. 21.5. Fruit capsules of the 12 primary trisomics of Datura stramonium, each with its particular phenotype.

in 1000 male births. A lot of excitement was aroused when an attempt was made to link the XYY condition with a predisposition toward violence. This is still hotly debated, although it is now clear that an XYY condition in no way guarantees such behaviour. Nevertheless, several enterprising lawyers have attempted to use the XYY genotype as grounds for acquittal or compassion in crimes of violence. The XYY males are usually fertile.

We have already looked at the generation of Down's syndrome through adjacent segregation in translocation heterozygotes. Down's syndrome also occurs much more commonly as a result of nondisjunction during meiosis, and it is then called trisomy 21. In this form of Down's syndrome, there is generally no family history of the phenotype; however, the frequency of this form is dramatically higher among children born to older mothers. The overall incidence of this abnormality is about 0.15 percent of births.

Down's syndrome is a severely incapacitating condition. Affected individuals are mentally retarded, and about one-third die by the age of 10 years. Recent advances in mapping the human genome will allow the identification of precisely those genes on the long arm of

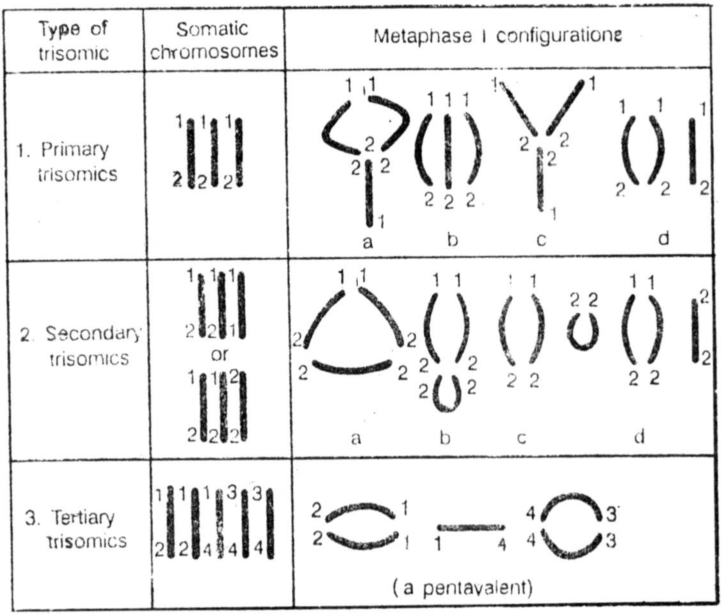

Type of trisomic	Somatic chromosomes	Metaphase I configurations
1. Primary trisomics		a b c d
2. Secondary trisomics	or	a b c d
3. Tertiary trisomics		(a pentavalent)

Fig. 21.6. Meiotic configurations formed at metaphase I in different types of trisomics.

chromosome 21 that must be trisomic to produce this syndrome; these advances offer some hope of a more precise understanding of its chemical nature and possible therapy. In humans, the only other two autosomal trisomics known to survive past birth are individuals with trisomy 13 and trisomy 18. Affected children are even more severely handicapped, both mentally and physically, and rarely survive to 1 year of age.

Chromosome mutation in general plays a prominent role in determining genetic ill health in humans. Figure summarizes the surprisingly high levels of various chromosomal abnormalities at different development stages of the human organism. In fact, the incidence of chromosome mutations ranks close to that of gene mutations in human livebirths. This is particularly surprising when we realize that virtually all chromosome mutation arise a new with each generation. This is in contrast to gene mutations, which owe their level of incidence to a complex interplay of mutation rates and environmental selection, acting over many generations of the history of the human species.

Somatic Aneuploids

Aneuploids can arise spontaneously in somatic tissue or tissue culture. In such cases, the initial result is a genetic mosaic of cell types. Good examples are provided by certain conditions in humans.

Table 21.2. Relative incidence of human ill health due to gene mutation and to chromosome mutation

Type of mutation	*Percentage of live birth*
Gene mutation	
Autosomal dominant	0.90
Autosomal recessive	0.25
X-linked	0.05
Total gene mutation	1.20
Chromosome mutation	
Autosomal trisomies (mainly Down's syndrome)	0.14
Other unbalanced autosomal aberrations	0.06
Balanced autosomal aberrations	0.19
Sex chromosomes	
XYY, XXY, and other ♂♂	0.17
XO, XXX, and other ♀♀	0.05
Total chromosome mutation	0.61

Sexual mosaics provide the first example. These are people whose bodies are a mixture of male and female tissue. One type of sexual mosaic is XO/XYY. This mosaic can be explained by postulating an XY zygote in which an early mitotic division involved a nondisjunction of the Y chromosomes, so that both went to one pole:

The phenotypic sex of such individuals depends upon where in the body the male and female sectors end up. In this case, if nondisjunction occurred at a later mitotic division, there would be a three-way XY/XO/XYY.

Aneuploidy in Man

More recent studies on individuals have related chromosome numbers below and above 46 with intersex conditions and other irregularities manifesting physical, reproductive, and mental abnormalities. Participants in the Chicago Conference agreed on a system of nomenclature for identifying numerical and structural chromosome alterations. In a description of a karyotype, the first item to be recorded is the total number of chromosomes, including the sex chromosomes, followed by the sex chromosome constitution and any autosomal aberrations. A complement of 44 autosomes and one X, for example, is symbolized 45, X. This monosomic chromosome complement has been associated with an abnormal female condition known for many years as Turner's syndrome and which occurs in one of about 5000 people in the general population.

The Turner Syndrome (45,X)

This monosomic has a chromosome complement of 44 autosomes and one X chromosome. The chromosome anomaly is associated with an abnormal female phenotype described in 1938 by H. H. Turner and associates and known as the Turner syndrome. It occurs in about 1 per 2500 live female births. More than 90 percent abort spontaneously. A rough estimate for 45,X adults in the general population is in 1 in 5000. These adults have virtually no ovaries, have limited secondary sexual characteristics, and are sterile. Microscopic sections of the ovaries show fibrous streaks of tissue representing remnants of ovaries. Affected females have short stature, low-set ears, webbed neck, and shieldlike chest. Mental deficiency is not usually associated with this syndrome. Epithelial cells of 45,X patients are X chromatin negative, as expected when only X chromosome is present.

X monosomic probably originate from exceptional eggs or sperm with no sex chromosome or from the loss of a sex chromosome in mitosis during early cleavage stages, after an XX or XY zygote has been formed. This latter probability is supported by the high frequency of mosaics that result from postzygotic events in patients with the Turner syndrome. Mosaics with X/XX sex chromosomes show symptoms of the Turner syndrome but are usually taller than X and have fewer anomalies that nonmosaic 45,X females. They show more feminization, more normal menstruation, and may be fertile. Many cases of the somatic Turner phenotype without the typical 45,X chromosome constitution are now known. Most of these have one normal X chromosome and a fragment of a second X chromosome. Both arms of the second X chromosome are apparently necessary for normal ovarian differentiation. Individuals with only the long arm of the second X are short in stature and show other somatic symptoms of the Turner syndrome, whereas those with only the short arm of the second X have normal stature and do not show as many signs of the Turner syndrome. This indicates that the Turner phenotype is mostly controlled by genes on the short arm of the X chromosome.

Patients with partial deletion of one X chromosome are X chromatin positive and therefore may be misdiagnosed if a buccal smear is the only test used. The deficient X chromosome always forms the X chromatin body. A Y chromosome also occurs in some individuals with the Turner phenotype. These patients are usually mosaic for 45,X/46,XY, with a normal Y. People with one X and a Y fragment, not including the Y short arm, have only streak ovaries but are normal in phenotype. This suggests that male-determining genes are in the short arm of the Y chromosome, and those that prevent Turner phenotype are in the Y long arm as well as the X short arm. Major features of the Turner phenotype occur in some males as well as in females. The male Turner syndrome is characterized by defective development of the testes, sterility, and limited male secondary sexual characteristics, along with somatic features of the Turner phenotype. These people have normal male karyotypes.

The Klinefelter Syndrome (47,XXY)

An extra X chromosome in addition to the usual male (XY) chromosome complement (47,XXY) has been associated with the abnormal male syndrome described (in 1942) by H.F. Klinefelter and known as the *Klinefelter syndrome.* It is estimated to occur in 1 per 500 live male births. Individuals with this syndrome are phenotypically males but with some tendency toward femaleness, particularly in secondary sex characteristics. Such features as enlarged breasts, underdeveloped body hair, small testes, and small prostate glands are a part of the syndrome. Presumably, the XXY constitution originates either by fertilization of an exceptional orginates either by fertilization of an exceptional XX egg by a Y sperm or of an X egg by an exceptional XY sperm. Studies of Klinefelter syndrome and Turner syndrome indicate that the Y chromosome in human beings, unlike that in *Drosophila*, determines male sex.

The most common karyotype (about three-quarters of the cases) for the Klinefelter syndrome is 47,XXY, but the symptoms of the syndrome will usually occur whenever more than one X chromosome is present along with a Y chromosome. More complex karyotypes associated with the Klinefelter syndrome include: XXYY, XXXY, XXXYY, XXXXY, XXXXYY, and XXXXXY. All patients with the Klinefelter syndrome have one or more X chromatin bodies in the their cells. Mental retardation is usually found when there are more than two X chromosomes. The XY/XXY mosaicism in patients with Klinefelter syndrome is associated with less severe physical and reproductive anomalies.

Aneuploidy of X Chromosomes and Mental Deficiency

Other irregular combination of X chromosomes have also been recognized among females with X chromosome aberrations. About 1 percent of all mentally defective women in institutions have been shown to have one or more extra X chromosomes. This chromosome abnormality occurs in about 1 in 700 live births in the general population. Individuals with the "triple X syndrome" are comparable in some ways with Drosophila metafemales (XXX). In Drosophila, however, such individuals are usually lethal, and those that survive are strikingly abnormal and sterile. Human XXX individuals are sometimes visibly undistinguished from normal XX females, but there is considerable range in phenotypic expression. They may be mentally abnormal.

The best-known symptoms in this syndrome are abnormalities associated with functional processes such as menstruation. One patient cited by P. A. Jacobs was a 37-year-old female who reported that the first suggestion of an abnormality was highly irregular menstruation. When the abdominal wall was opened, the ovaries appeared as if they were postmensopausal. Microscopically, they showed deficient ovarian follicle formation. Of 63 cells observed, 51 had 47 chromosomes; the extra chromosome was an X. Nondisjunction in the production of the egg from which this woman developed was postulated as the mechanism for the occurrence of extra chromosome. Buccal smears showed two sex chromatin bodies in the epithelial cells as expected when three X chromosomes are present. Individuals with tetrasomic X chromosomes (48,XXXX) are all mentally defective. The degree of mental deficiency increases with the number of X chromosomes present.

47,XYY and Behaviour

P. A. Jacobs and her associates reported in 1965 that seven XYY males were detected in a population of 197 male, mentally subnormal inmates of a penal institution in Scotland. The XYY men were unusually tall, with an average height of 73.1 inches, compared with 64 inches for XY men in the same prison. Numerous other studies, mostly in institutionalized populations, have since confirmed that a high proportion of XYY individuals are tall, subnormal in intelligence (with IQs individuals are tall, subnormal in intelligence (with IQs ranging from 80 to 95), and antisocial. The aggressive behaviour that brought them into conflict with the law was usually against property rather than people.

XYY trisomy occurs about once in 1000 live male births in the general European population. Only a few of these can be accounted for in the criminal population. Furthermore, most XYY men have been described as perfectly normal in behaviour. Hook has shown that only 3.6 percent of all XYY men are institutionalized for any reason.

Environmental factors are presumed to be involved in the development of aggressiveness. Since some XYY men are subnormal in intelligence and excessively tall in stature, particular environmental situations in childhood or adulthoood may lead to withdrawal from society or aggressive behaviour. Unfavourable social conditions such as frustration in personal accomplishment and taunting from associates may encourage physical aggression as a means of adaptation. Males with this sex trisomy have not been found to transmit the extra Y chromosome to their sons. This extra chromosome seems to be weeded out in gametogenesis. A wide range of physical and mental abnormalities has been detected in institutionalized XYY men, but most of these are irregular in occurrence and do not form a syndrome. Tallness of stature and mental dullness, however, are fairly constant characteristics among institutionalized XYY men.

Chromosomal Mosaics

Individuals who have at least two cell lines, with different karyotypes derived from one zygote, originate from nondisjunction in a cleavage division after fertilization. One daughter cell would thus receive one too many and the other would be one deficient. Each cell would give rise to cell line with its irregular chromosome number. Proportions of cells representing the different cell lines would vary in different tissues, making the extent of the mosaicism and the effect on the organism difficult to predict.

Many sex chromosome mosaics have been detected in human beings. The main phenotypic characteristic is extreme variability. Some sex chromosome mosaics have been reported –X/XX, X/XY, XX/XY, XXY/XX, XX/XXX, XXX/X XXX/XXXXY–and several other combinations reflecting two or three cell lines. Mid to severe phenotypic symptoms have been associated with these sex chromosome mosaics.

Abnormal Euploidy

Monoploids

In this section we shall consider monoploidy as an unusual condition. Monoploid individuals can arise spontaneously in natural populations as rare aberrations, but in several forms (such as bees, wasps, and ants) the males are normally monoploid, having derived from unfertilized eggs.

In the germ cells of a monoploid, meiosis cannot occur normally because the chromosomes having no pairing partners. Thus monoploids are characteristically sterile. (However, meiosis can be bypassed in some monoploid animals, such as male honeybees, which produce gametes essentially by mitotic division). If meiosis occurs and the single chromosomes segregate randomly, then the probability of their all going to one pole is $(1/2)^{x-1}$, where x is the number of chromosomes. This will determine the frequency of viable (whole-set) gametes, obviously a vanishingly small number if x is large.

Monoploids have a major role in modern approaches to plant breeding. Diploidy is an inherent nuisance in the induction and selection of new favourable plant mutations and of new selection of new favourable plant mutations and of new combinations of genes already present. Monoploids provide a way around some of these problems. In some plants, monoploids may be artificially derived from the products of meiosis in the plant's anthers. A cell destined to become a pollen grain may be induced by cold treatment to grow instead into an embryoid, a small dividing mass of cells. The embryoid may be grown on agar to form a monoploid planter, which can then be potted in soil to mature.

Monoploids may be exploited in several ways. In one method, they are first examined for favourable traits or gene combinations.

These may arise from heterozygosity already present in the parent or induced in the parent by mutagens. The monoploid can then be subjected to chromosome doubling to achieve a completely homozygous diploid with a normal meiosis, capable of providing seed. How is this achieved? Quite simply, by the application of a compound called colchicine to meristematic tissue. Colchicine, an alkaloid drug extracted from the autumn crocus, inhibits the mitotic spindle, so that cells with two chromosomes sets are produced. These may proliferate to form a sector of diploid tissue that can be identified cytologically.

Another way the monoploid may be used is to treat its cells, basically like a population of haploid organisms, in a mutagenesis-and-selection procedure. The cells are isolated, their walls are removed by enzyme treatment, and they are treated with mutagen. They are then plated on selective medium– perhaps a toxic compound normally produced by one of the plant's parasites, or an insecticide-- to select resistant cells. Resistant plantlets grow eventually into haploid plants, which can then be doubled (using colchicine) into a pure-breeding resistant type. These are potentially powerful techniques that can circumvent the normally slow process of what is basically meiotic plant breeding. The techniques have been successfully applied in several important crop plants, such as soybeans and tobacco. This is, of course, another aspect of somatic-cell genetics in higher organisms.

The another technique of or producing monoploids does not work in all organism or in all genotypes of an organism. Another useful

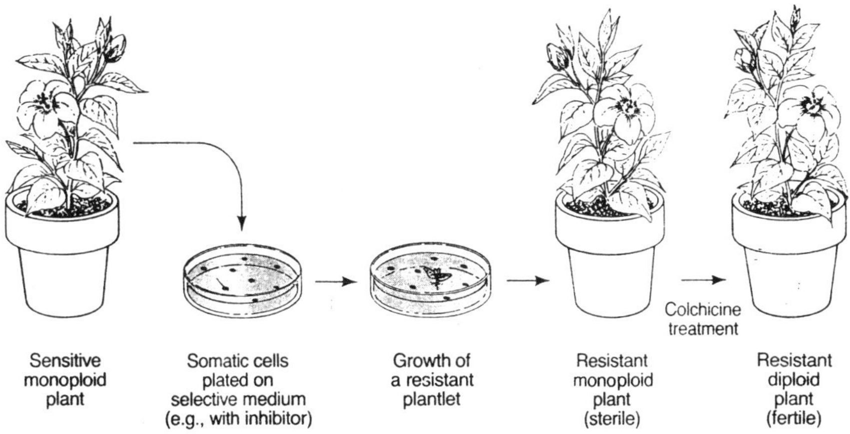

Fig. 21.7. Using microbial techniques in plant engineering. Haploid cells have their cell walls removed enzymatically.

technique has been developed in barley, an important crop plant. When diploid barley, Hordeum vulgare, is pollinated using a diploid wild relative called Hordeum bulbosum, fertilization occurs, but during the ensuing somatic cell divisions, the chromosomes of H. Bulbosum are preferentially eliminated from the zygote, resulting in a haploid embryo. (The haploidization process appears to be caused by a genetic incompatibility between the chromosomes of the different species.) The resulting haploids can be doubled with colchicine. This approach has led to the rapid production and widespread planting of several new barley varieties. It is being used successfully in other species too.

Polyploids

Once into the realm of polyploids, we must distinguish between autopolyploids and allopolyploids. *Autopolyploids* are composed of multiple sets from within one species, whereas *allopolyploids* are composed of sets from different species. Allopolyploids form only between closely related species; however, the different chromosome sets are homeologous (only partially homologous), not fully homologous as they are in autopolyploids.

Triploids

Triploids are usually autopolyploids. They are constructed from the cross of a 4x (tetraploid) and a 2x (diploid). the 2x and the x gametes unite to form a 3x triploid.

Triploids also are characteristically sterile. The problem again involves pairing at meiosis. Although pairing can take place in several ways, it usually occurs between only two chromosomes at a time. The net result is always the same, an unbalanced segregation of one of the following types:

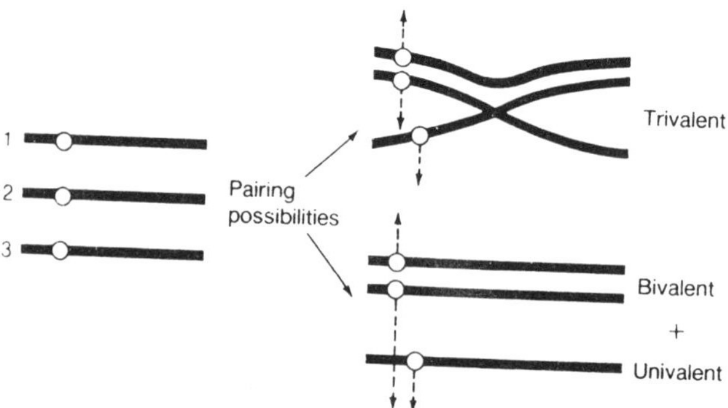

Fig. 21.8. Meiotic pairing possibilities in a triploid.

$$\frac{1+2}{3} \text{ or } \frac{1+3}{2} \text{ or } \frac{2+3}{1}$$

This happens for every chromosome threesome, and the probability of obtaining either a 2 x or x gamete is $(1/2)^{x-1}$, where x is the number of chromosomes in a set. The other will be unbalanced gametes, having two of one chromosome type, one of another, two of another, and so on, and most will be nonfunctional. Even if the gametes are functional, the resulting zygotes will be unbalanced. A practical application of the sterility associated with triploidy lies in the production of seedless varieties of watermelons and bananas.

Autotetraploids

Autotetraploids occur either naturally, by the spontaneous accidental doubling of a 2x genome to 4x, or artificially, through the use of colchicine. Autotetraploids are evident in many commercially important crop plants because, as with other polyploids, the larger number of chromosome sets is often associated with increased size of the plant. This is manifested in increased cell size, fruit size, stomata size, and so on.

Because 4 is an even number, autotetraploids can have a regular meiosis, although this is by no means always the case. The crucial factor is how the four chromosomes of one type pair and segregate. The two bivalent and the quadrivalent pairing modes tend to be most

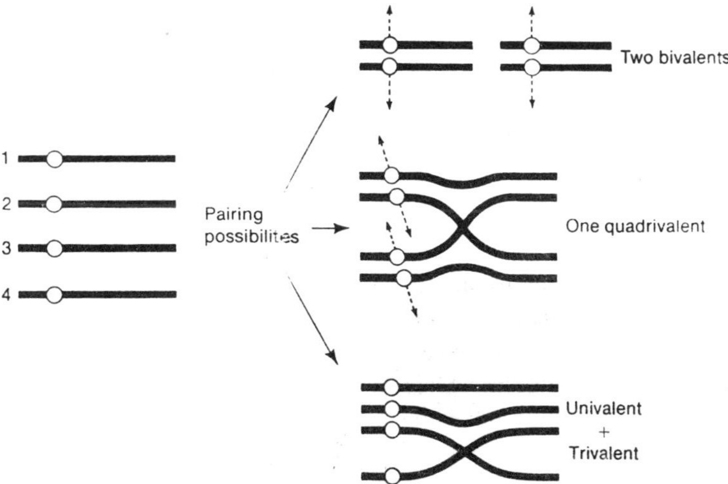

Fig. 21.9. Meiotic pairing possibilities in tetraploids.

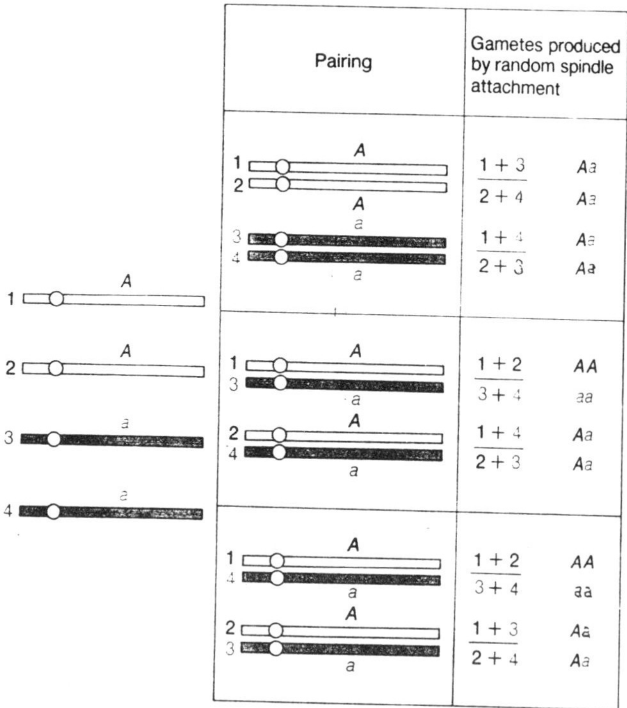

Fig. 21.10. Genetic consequences in a tetraploid showing orderly pairing by bivalents. The locus is assumed to be close to the centromere. Self-fertilization could yield a variety of genotypes, including aaaa.

regular in segregation, but even here there is no guarantee of a 2 → 2 segregation. If a regular 2 → 2 segregation is achieved at each chromosome type, as is the case in some species, than a formal genetic analysis can be developed for autotetraploids.

Let's hypothesize an experiment in which colchicine is used to double the chromosomes of an *Aa* plant into an *AAaa* autotetraploid, which we will assume shows 2 → 2 segregation. We now have a further worry because autotetraploids give different genetic results in their progeny, depending on whether or not the locus concerned is tightly linked to the centromere. First we consider a centromeric gene. The three possible pairing and segregation patterns are presented in Figures, these occur by chance and with equal frequency. As the figure shows, the 2x gametes will be *Aa, AA, or aa,* and these will be produced in a ratio of 8:2:2, or 4:1:1. If such a plant is offspring is obviously 1/6 × 1/6 = 1/36. In other words, a 35:1 phenotypic ratio will be observed if *A* is fully dominant over three *a* alleles.

If, in the same kind of plant, a genetic locus *B/b* is very far removed from the centromere, crossing-over must be considered. This forces us to think in terms of chromatids instead of chromosomes, and we have for *B* chromatids and for *b* chromatids. Because the number of crossovers in a such a long region will be large, the genes will become effectively unlinked from their original centromeres, and the packaging of genes two at a time into gametes is very much like grabbing two balls at random from a bag of eight balls, four of one kind and four of another. The probability of picking two *b* genes is then

$$4/8 \text{ (the first one)} \times 3/7 \text{ (the second one)} = 12/56 = 3/14$$

So, in a selfing, the probability of a *bbbb* phenotype will be equal $3/14 \times 3/14 = 9/196 = 1/22$. Hence there will be a 21:1 phenotype ratio of *B* – – –:*bbbb*. For genetic loci of intermediate position, intermediate ratios will, of course, result.

Allopolyploids

The "classical" allopolyploid was synthesized by G. Karpechenko in 1928. He wanted to make a fertile hybrid between the cabbage (*Brassica*) and the radish (*Raphanus*) that would have the leaves of the former and roots of the latter. Each of these species has 18 chromosomes, and they are related closely enough to allow intercrossing. A variable hybrid progeny individual was produced from seed. However, this hybrid was functionally sterile because the nine chromosomes from the cabbage parent were different enough from the radish chromosomes that homology was insufficient for normal synapsis and disjunction.

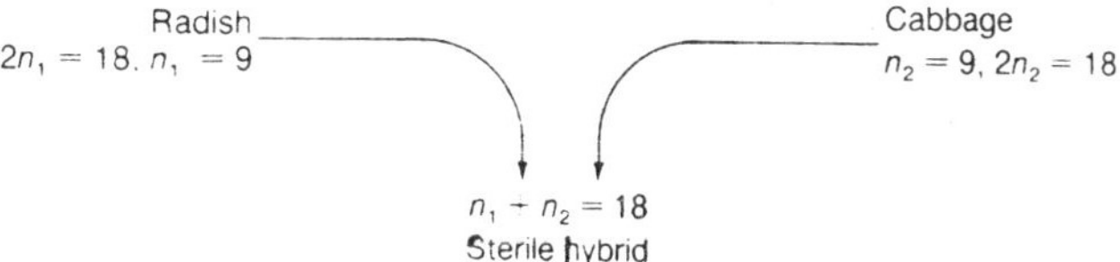

However, one day a few seeds were in fact produced by this (almost) sterile hybrid. On planting, these seeds produced fertile individuals with 36 chromosomes. These individuals were allopolyploids. They had apparently been derived from spontaneous accidental chromosome doubling in the sterile hybrid, presumably in tissue that eventually became germinal and underwent meiosis. Thus, in $2n_1 + 2n_2$ tissue, there is a pairing partner for each chromosome, and balanced gametes of the type $n_1 + n_2$ are produced. These fuse to given $2n_1 + 2n_2$ allopolyploid progeny, which are in turn fertile also. This kind of allopolyploid is sometimes called an amphidiploid. (Unfortunately for Karpechenko, his amphidiploid had the roots of a cabbage and the leaves of a radish.)

If the allopolyploid is crossed to either parent species, sterile offspring result. In the case of the cross to radish, these offspring would be $2n_1 + n_2$, constituted from an $n_1 + n_2$ gamete from the allopolyploid, and an n_1 gamete from the radish. Obviously, the n_2 chromosomes will have no pairing partners, so sterility will result. Consequently, Karpechenko had effectively created a new species, with no possibility of gene exchange with its parents. He called his new species *Raphanobrassica.*

Nowadays, allopolyploids are routinely synthesized as a major tool in plant breeding. The goal obviously is to combine some of the worthwhile features of both parental species into one type. This kind of endeavor is very uncertain, as Karpechenko found out. In fact, only one amphidiploid has ever been intentionally produced that is of potentially widespread use. This is *Triticale*, an amphidiploid between wheat (*Triticum, 2n = 6x = 42*) and rye (*Secale*, 2n = 2x = 14). *Tricale* combines the high yields of wheat with the ruggedness of rye. A massive international *Triticale* testing program is now under way, and many breeders have great hopes for the future of this artificial amphidiploid.

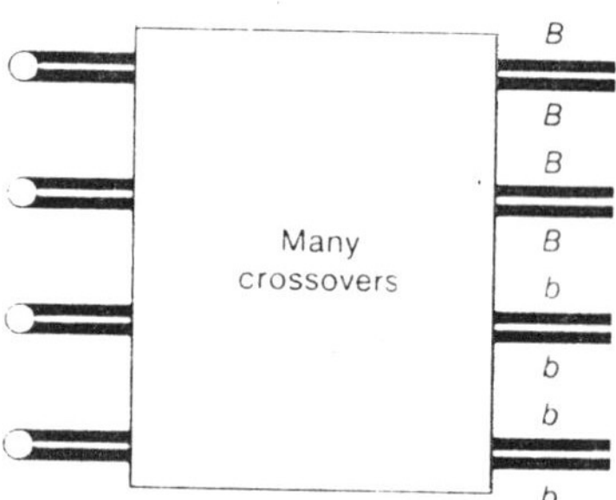

Fig. 21.11. Highly diagrammatic representation of a tetraploid meiosis involving a heterozygous locus distant from the centromere.

In mature, allopolyploidy seems to have been a major force in speciation of plants. There are many different examples. One particularly satisfying one is shown by the genus *Brassica.* Here three different parent species have been hybridized in all possible pair combinations to form new amphidiploid species. This has all taken place in nature, but *Brassica* amphidiploids also have been artificially synthesized.

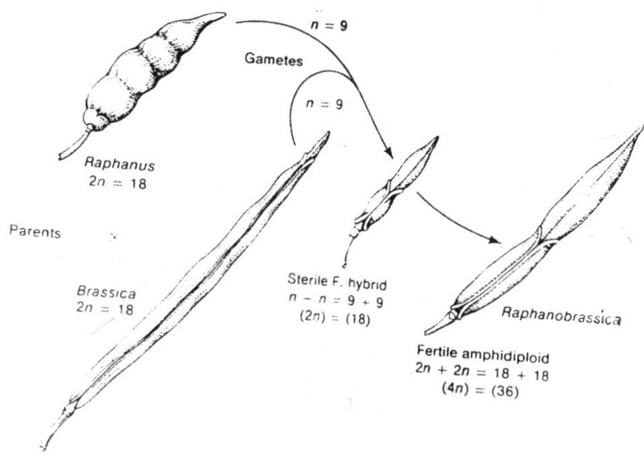

Fig. 21.12. The origin of the amphidiploid (Raphanobrassica) formed from cabbage (Brassica) and radish (Raphanus).

A particularly interesting natural allopolyploid is bread wheat, *Triticum aestivum* ($2n = 6x = 42$). By a study of various wild relatives, it has been possible to reconstruct a probable evolutionary history of breat wheat. In a wheat meiosis, there are always 21 pairs of chromosomes. Furthermore, it has been possible to establish that any given chromosome has only one specific pairing partner (homologous pairing)–not five other potential ones (homologous pairing). The suppression of such homeologous pairing (which would lead to much reduced stability of the species) is maintained by a gene *Ph* ensures a diploid-like genetics for this basically hexaploid species. Without *Ph*, bread wheat could probably never have arisen. It is interesting to speculate on whether Western civilization could have arisen or progressed without this species–in other words, without *Ph*.

Somatic Allopolyploids from Cell Hybridization

Another innovative approach to plant breeding is to try to make allopolyploid-like hybrids by asexual methods. Theoretically, such a technique would permit combination of widely differing parental species. The technique does indeed work, but so far the only allopolyploids that have been produced are those that can also be made by the sexual methods we have considered already. The procedure

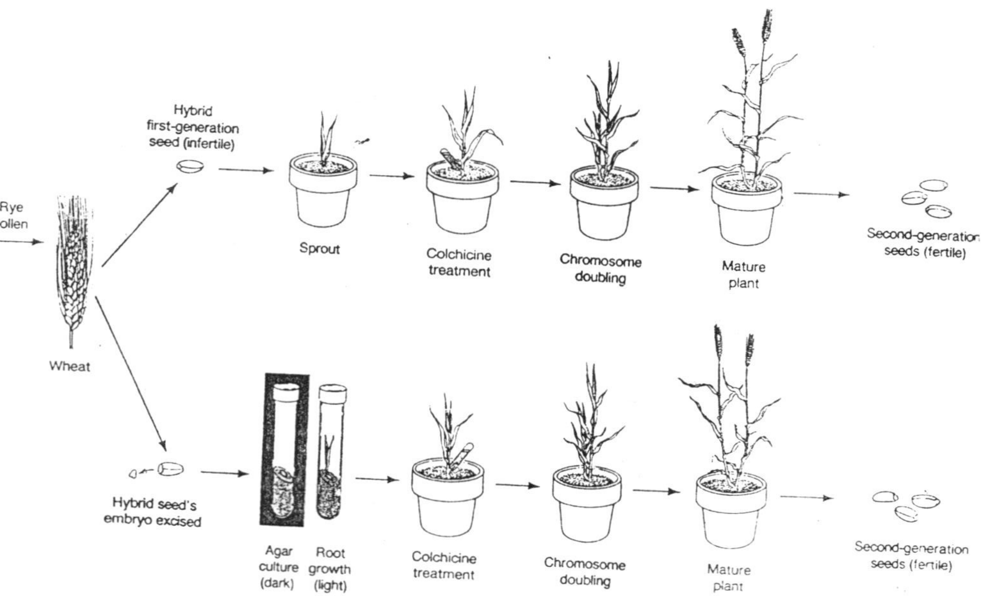

Fig. 21.13. Techniques for the production of the amphidiploid Triticale. If the hybrid seed does not germinate, tissue culture may be used to obtain a hybrid plant.

is as follows. Cell suspensions of the two parental species are prepared and stripped of their cell walls by special enzyme treatments. The stripped cells are called *protoplasts.* The two suspensions (protoplast suspensions) are combined with polyethylene glycol, which enhances protoplast fusion. The parental cells and the fused cells will proliferate to form colonies (in much the same way as microbes) on agar medium. If these colonies, or calluses, are examined, a fair percentage of then are found to be allopolyploid-like hybrids with chromosome number equal to the sum of the parental types. Thus, not only do the protoplast cell membranes fuse to form a kind of heterokaryon, but the nuclei fuse too.

A good example of an allopolyploid-like hybrid is commercial tobacco, *Nicotiana tabacum,* which has 48 chromosomes. This species of tobacco was originally found in nature as a spontaneously occurring amphidiploid. The two probable parents are N. Sylves and N. Tomentosiformis, each of which has 24 chromosomes. A sexual cross between N. Tabacum and either of the other two gives a 36-chromosome hybrid n which there are 12 chromosome pairs plus 12 unpaired chromosomes. A cross between N. Sylvestris and N. Tomentosiformis gives a 24-chromosome hybrid n which there is no pairing at all. Hence, it appears that part of the N. Tabacum genome is from N. Sylvestris and part from N. Tomentosiformis. This amphidiploid can be re-created either sexually, by processing involving colchicine as described previously, or somatically by cell fusion. When cells or the prospective parental species are fused, a 48-chromosome hybrid cell line is produced from which may be grown plants whose behaviour is identical to that of N. Tabacum. (Note that in the latter method, colchicine is not required, since the fusion product is already amphidiploid.)

The recovery of somatic hybrids may be enhanced if a selective system is available. For example, two different monoploid lines of *N. Tabacum* had light-sensitive yellowish and light-resistant, as a result of complementation between the parental genotypes. The calluses can be grown into plantlets, which then either are grafted onto a mature plant to develop or are themselves potted.

Application of Polyploidy

Among the cultivated varieties of wheat, three different chromosome numbers are represented: 14, 28, and 42 (x = 7). For example, the primitive small-grained einkorn type of Europe and Asia, Triticum monococcum, has 14 chromosomes in its vegetative cells. Its yield is low and it is of comparatively little value. An emmer wheat (durum), T. dicoccum, grown chiefly in southern Europe but also in the United States, has 28 chromosomes. It has thick heads with large hard kernels and issued mainly for macaroni, spaghetti, and stock feed. The bread wheats, T. aestivum, with 42 chromosomes, were postulated by J. Percival in England to have come from a cross between emmer wheat and goat grass (Aegilops), both of which are native to the Babylonian region where bread wheat originated.

When techniques for artificial chromosome doubling became established, investigations of the origin of bread wheat confirmed Percival's theory. Experimental evidence obtained by E. S. McFadden and E. R. Sears and separately by H. Kihara traced the pathway for the origin of one type of bread wheat, T. Spelta.

Aegilops squarrosa ($n = 7$) was found to carry a group of major characteristics that distinguish the hexaploid ($n=21$) T. Spelta from the tetraploids ($n = 1$) T. dicoccum and T. dicoccoides. Hybrids between these tetraploid species of wheat and A. squarrosa proved to have all of the major taxonomic characters of T. spelta but the hybrids were completely or nearly sterile. When the F_1 hybrids of T. dicoccoides × A. squarrosa were treated with colchicine, highly fertile allopolyploids with 42 chromosomes were obtained. These synthetic hexaploids closely resembled the cultivated. T. spelta, and they produced highly fertile hybrids with that species and with T. vulgare, known to be in the ancestry of the bread wheats. This demonstrated that the genome of the hexaploid wheats corresponded to one chromosome set of A. squarrosa. It was postulated that T. spelta is the ancestral hexaploid wheat of Europe, having arisen, possibly in fairly recent times, in southeastern Europe or southwestern Asia following chromosome doubling of natural hybrids of T. dicoccoides (or its cultivated close relative, T. dicoccum) × A. squarrosa. T. spelta is believed to have been carried over the northerly route into central and western Europe. Experiments of McFadden, Sears, and Kihara reconstructed the pathway through which a moderately useful wheat and a goat grass hybridized in nature and produced forerunners of a most valuable crop, bread wheat.

New World Cotton

Crosses can be made between distinct species of cotton, members of the benes Gossypium. The hybrids show a wide range of vigor and fertility, making the material favourable for studies of origins. Three cytological groups have been found to correspond with the major world distributional areas. Old World cotton had 13 pairs of large chromosomes. American cotton, which originated in Central or South America, has 13 pairs of small chromosomes. New world cotton (the cultivated long-staple type) has 26 pairs, 13 large and 13 small. Evidently, hybridization and chromosome duplication occurred somewhere in the ancestry of the New World Cotton.

J. O. Beasley used the colchicine technique and succeeding in doubling the chromosomes of a hybrid between the Old World and American cotton. The resulting hybrids, with four set of chromosomes (amphidiploids), crossed readily among themselves and produced fertile plants resembling New World cotton. The process by which the valuable polyploid cotton may have originated in nature was thus duplicated in the laboratory.

Primrose Hybridization

The primrose, Primula kewensis, is an allotetraploid with 36 (2n) chromosomes. It was derived from a cross between two diploids, P. floribunda (x= 9) and P. verticillata (x= 9). Plants from these two species crossed readily, producing hybrids with 18 chromosomes in their vegetative cells. 9 from P. floribunda and 9 from P. verticillata, but the hybrids were sterile. Eventually, however, a branch on a hybrid plant developed from a cell in which the chromosome number was doubled (36), so that each chromosome had a homologous partner. This branch was propagated and gave rise to a fertile primrose plant with cells containing 36 chromosomes of the two diploid parents, the sterile diploid hybrid, and the fertile allotetraploid are shown.

Tobacco Resistance

Induced polyploidy has been exploited to a great extent. Practical applications may become more common as additional data are accumulated. By artificially induced polyploidy, disease resistance and other desirable qualities have been incorporated into some commercial crop plants. Tobacco, Nicotiana tabacum, for example, is susceptible to the tobacco mosaic virus (TMV), whereas N. glutinosa appeared at first observation to be resistant. Further investigation, however, showed that in N. glutinosa the virus killed the cells that were invaded and the virus particles became isolated in the dead cell. The apparent resistance thus was attributable to hypersensitivity. When the two tobacco species were crossed, the hybrid was found to be "resistant" to the virus, but totally sterile. When the chromosomes were doubled, it was possible to secure a fertile polyploid "resistant" to the virus.

Polyploid Fruits, Flowers, and Wheat

Some varieties of plants that serve human needs more effectively than others have now been identified as polyploids. Many polyploids were selected and cultivated because of their large size, vigor, and ornamental values, before their chromosome numbers were known. Giant "sports" from twings of McIntosh apple trees that were found to be tetraploid (4n) were propagated into whole trees, which produce extra-large fruit. The texture of the giant apples is as fine as that of diploids, but the yield is inferior. Mass selection of seedlings may overcome this difficulty. Bartlett pears, several varieties of grapes, and cranberries have also produced sports with giant fruits. Some of these show promise of practical usefulness. With colchicine treatment, a number of polyploids have been developed artificially. This technique has provided a way to explore the mechanism involved in polyploid formation and to make use of the good qualities of polyploids. Tetraploid (4n) maize is more vigorous than the ordinary diploid and produces some 20 percent more vitamin A. Its fertility is somewhat reduced, but this drawback responds to selection. Polyploid watermelons have been developed from colchicine treatment by Kihara and others. The tetraploid with 44 chromosomes is large and has practical value. Triploid watermelons with 33 chromosomes are especially desirable because they are sterile and have no seeds. Among the flower garden varieties, 4n marigolds and snapdragons are widely cultivated.

Polyploid plants respond to artificial selection and hybridization, as do diploid species. The recent history of plant breeding has been characterized by a marked improvement in many polyploid plant crops. The yield of wheat, for example, has increased appreciably. This has been accomplished by developing disease-resistant strains and breeding for increased hardiness and greater efficiency so that crops may survive under the various environmental conditions found in wheat-growing areas.

A constant threat to the wheat crop is rust–a fungus that attacks the stems and leaves of the growing plants and destroys the ripening grain. Spores are borne by wind and, when conditions are right, they spread like fire through wheat fields. The disease can be combated by developing rust-resistant strains and by eradicating barberry bushes, which are hosts to the spores during the spring months. But new varieties of rust that destroy previously resistant grain keep evolving, thus perpetuating the job of plant breeders. The larger kernels at the left are from a new strain of rust-resistant spring wheat. At the right are shown kernels of wheat, similar in other respects but not resistant, that are dwarfed from infection with stem rust. The number of kernels of grain per plant as well as the size of the kernels is decreased by rust infection. Investigators in agricultural experiment stations are constantly alert for new rusts. When a new one is found, the standard wheat varieties are tested against it. If they are not resistant, breeding programs are initiated immediately to develop new strains resistant to that particular rust.

Chromosome Anomalies in Spontaneous Abortions in Humans

A wide variety of chromosome numbers and chromosome structural aberrations is found in spontaneoulsy aborted human fetuses. The types of aberrations found vary according to differences in the age of the fetus at the time of abortion. For example, 40 percent of spontaneously aborted fetuses under 90 days of gestational age (i.e., length of time since the last menstrual period of the mother) exhibit chromosome anomalies. For 91-to 120-day-old fetuses, 25 percent show chromosomal anomalies, and only 5 percent of fetuses over 120 days old exhibit chromosome anomalies. Thus, chromosomal anomalies cause most fetuses to die and be aborted in early developmental stages.

Before the percentage of spontaneous abortions resulting from chromosomal anomalies can be calculated, one must first define what constitutes an aborted fetus. Apparently a large number of conceptions occur in humans, and the rechronic myelogenous leukemia possess two cell lines; one cell line has a normal chromosome complement, whereas the other appears to be missing the long arm of chromosome 2. This condition was originally thought to be a monosomy; however, using band staining techniques to identify the long arm, it was learned that the long arm was translocated to one of the larger chromosomes, usually chromosome 9 (referred to as the Philadelphia chromosome). The role of this chromosomal aberration in the induction, development, and progression of cancer is unknown.

Inherited autosomal recessive disorders, such as Bloom's syndrome, Fanconi's anemia, ataxia-telangiectasia, and xeroderma pigmentosum, have been associated with chromosomal instability and or deficiency in mutation repair mechanisms that result in chromosomal aberrations. These individuals have a high incidence of cancer.